MALT
WHISKY
YEARBOOK
2020

www.maltwhiskyyearbook.com

First published in Great Britain in 2019 by
MagDig Media Limited

© MagDig Media Limited 2019

ISBN 978-0-9576553-6-2

MagDig Media Limited
1 Brassey Road
Old Potts Way, Shrewsbury
Shropshire SY3 7FA
ENGLAND

E-mail: info@maltwhiskyyearbook.com
www.maltwhiskyyearbook.com

Contents

Introduction

Over the fifteen years that I have been the editor of the Malt Whisky Yearbook, I never lost track (or at least I hope I didn´t) of the fact that I am a mere observer of the world of whisky. I don´t produce the golden liquid. The ones who do are the real heroes. And as much as I love to report on large companies spending millions on new distilleries, closest to my heart are the enthusiasts and entrepreneurs that dare take that leap of faith, leaving their jobs and a secure existence to pursue their dreams of building a whisky distillery. No matter how much they prepare themselves, they will face unexpected challenges and some of them will regret that bold decision they took. This year, I´ve talked to seven *Pioneers of Whisky* from all over the world, trying to figure out what triggered them into building their own distillery. They have all been succesful but it has been no easy feat.

As usual, my excellent team of whisky writers have excelled themselves this year and have contributed with some fascinating articles;

The influence of foreign ownership in the Scotch whisky industry is growing. Gavin D Smith takes a look at the pros and cons of having the headquarter overseas.

Overlooked by many of us but crucial for making a great whisky. Jonny McCormick unlocks the door to the mysteries of yeast and fermentation.

The alcohol in the whisky isn´t there just to make us feel alive. Ian Wisniewski explains the influence of C_2H_5OH on flavour and on our perception of the drink.

With more high-priced super premium malts released by the producers, have the supermarkets won the war on how to attract new consumers? Neil Ridley thinks the industry is in need of a new message.

We are all guilty of asking for a typical Islay or a classic Speyside in a bar. Charles MacLean gives us the background to whisky regions and asks the question if they are at all relevant today.

Innovate or die! A popular catch phrase in the world of business. Joel Harrison asks the question what you do if you´re hampered by rules and regulations.

Taken by surprise by the huge demand for Japanese whisky, the producers´ stock of aged spirit is low. Stefan Van Eycken reports on the industry´s inventive ways of handling the situation.

In Malt Whisky Yearbook 2020 you will also find the unique, detailed and much appreciated section on Scottish malt whisky distilleries. It has been thoroughly revised and updated, not just in text, but also including numerous, new pictures, new distilleries and tasting notes for all the core brands. The chapter on Japanese whisky is completely revised and the presentation of distilleries from the rest of the world has been expanded. You will also find a list of more than 150 of the best whisky shops in the world with their full details and suggestions where to find more information on the internet. The Whisky Year That Was provides a summary of all the signficant events during the year. Finally, the very latest statistics gives you all the answers to your questions on production and consumption.

Thank you for buying Malt Whisky Yearbook 2020. I hope that you will have many enjoyable moments reading it and I can assure you that I will be back with a new edition next year.

Malt Whisky Yearbook 2021 will be published in October 2020.
If you need any of the previous fourteen volumes of Malt Whisky Yearbook,
some of them are available for purchase (in limited numbers) from the website
www.maltwhiskyyearbook.com

Acknowledgements

First of all I wish to thank the writers who have shared their great specialist knowledge on the subject in a brilliant and entertaining way – Stefan Van Eycken, Joel Harrison, Charles MacLean, Jonny McCormick, Neil Ridley, Gavin D. Smith and Ian Wisniewski.

A special thanks goes to Gavin who put in a lot of effort nosing, tasting and writing notes for more than 100 different whiskies.

I am also deeply grateful to Philippe Jugé for his valuable input on French distilleries.

The following persons have also made important photographic or editorial contributions and I am grateful to all of them:

Scott Adamson, Alasdair Anderson, Russel Anderson, Paul Aston, Duncan Baldwin, Adam Barber, Laura Beadell, Suzi Beney, Graham Bowie, Keith Brian, Ross Bremner, Kai von Broembsen, Andrew Brown, Jennifer Brown, Alex Bruce, Gordon Bruce, Mark Brunton, Kirsten Bryceland, Simon Buley, Jessica Bullard, Neil Bulloch, Stephen Burnett, Pär Caldenby, Peter Campbell, Ian Chang, Ian Chapman, Claire Clark, David Clark, Joe Clark, Suzanne Clark, Abbie Clements, Francis Conlon, Lois Cope, Mairi Corbett, Graham Coull, Jason Craig, Andrew Crook, Nathan Currie, Paul Dempsey, Scott Dickson, Kenneth Douglas, Amber Druce, Lukasz Dynowiak, Gavin Edwards, Simon Erlanger, Graham Eunson, David Ferris, Allan Findlay, Andy Fiske, Robert Fleming, Jonathan Fletcher, Simon Ford, John Fordyce, Callum Fraser, Graeme Gardiner, Calum Gee, Archie Gillies, Colin Gordon, Paige Gordon-Stewart, Jan Groth, Pierrick Guillaume, Gary Haggart, Chloe Hall, Adam Hannett, Wendy Harries Jones, Annelise Hastings, Mickey Heads, Erik Hirschfeld, Elsa Holmberg, Fraser Hughes, Robbie Hughes, Caryn Inglis, Kevin Innes, Bart Joosten, Tara Karimian, Pramod Kashyap, David Keir, Kim King, Andrew Laing, Emily Lineham, Graeme Littlejohn, Nico Liu, Graham Logan, Alistair Longwell, Barry Macaffer, Sarah McAlaney, Iain McAlister, Tommy Macarthur, Brian MacAulay, Des McCagherty, Alan McConnochie, Alistair McDonald, Andy Macdonald, John MacDonald, Christy McFarlane, Sandy Macintyre, Doug McIvor, Connal Mackenzie, Gordon Mackenzie, Jaclyn McKie, Julia Mackillop, Paul Mclean, James MacTaggart, Ian McWilliam, Graham Manson, Ibon Mendiguren, Sophie Menzies, Andrew Millsopp, Carol More, Gareth Morgan, Jake Mountain, Neil Murphy, Ingemar Nordblom, Olof Noréus, Sietse Offringa, Anne O´Lone, Gemma Paterson, Hannah Peebles, Sean Phillips, Struan Grant Ralph, Ian Renwick, Fiona Reyner,Chris Riesbeck, David Robertson, Iain Robertson, Jackie Robertson, Stuart Robertson, Brian Robinson, Jenny Rogerson, Clara Ross, Colin Ross, Daniel Roy, Dennis Rylander, Colette Savage, Lila and Nestor Serenelli, Steven Shand, Rory Slater, Sonal Solanki, Alison Spowart, Greig Stables, Marie Stanton, Grant Stevely, Karen Stewart, Duncan Tait, Eddie Thom, Annabel Thomas, Laura Thomson, Ruth Thomson, Laura Tolmie, John Torrance, Sandrine Tyrbas de Chamberet, Perry Unger, David Vitale, Sarah Ward, Stewart Walker, Ranald Watson, Mark Watt, Iain Weir, Desiree Whitaker, Nick White, Ronald Whiteford, John Wielstra, Anthony Wills, Jamie Winfield, Kristoffer Wittström, Stephen Woodcock, Allison Young, Fiona Young, Derek Younie, David Zibell.

Finally, to my wife Pernilla and our daughter Alice, thank you for your patience and your love and to Vilda, our labrador and my faithful companion in the office during long working hours.

Ingvar Ronde
Editor
Malt Whisky Yearbook

Foreign ownership of Scotch Whisky

by Gavin D Smith

The first example of a non-Scottish company
taking control of a Scotch whisky distillery goes back to the late 1800s.
Today, the influence of foreign ownership in the Scotch whisky industry
is profound with 40% of the distilleries being
owned by companies overseas.

Knockdhu distillery stands in lush Aberdeen-shire farmland between Huntly and Portsoy. At its heart is a classic Victorian structure complete with kiln pagoda and stone-built dunnage warehouses. The brass plaque on the office door announces that the distillery is licensed to James Catto & Co, a blending firm established in Aberdeen during the 19th century. Behind the office door, one is greeted by Archie, a large Labradoodle and his more diminutive Bichon friend, Maisie, followed by distillery manager Gordon Bruce. Bruce is from Wick and has worked at three Highland distilleries for the last 31 years.

The Knockdhu kiln room features a display of agricultural implements and pieces of machinery, emphasising the intimate connection between distilling and farming. This is not the work of some on-trend agency, but a collection assembled by Bruce and his staff and laid out around the building. Bruce has also personally been responsible for designing a cyclone device to produce clearer wort and proposing the installation of a unique pre-worm tub shell and tube condenser on the wash still to ensure consistency of spirit style after a switch from five to seven day working.

So, this must surely be an independently-owned distillery where a degree of individuality and originality are allowed to hold sway? Well no, actually. Knockdhu belongs to the vast Thailand-based International Beverage Holdings (InterBev), also responsible for four other Scotch distilleries and their single malts, several Scotch blends, and a multiplicity of beer, white spirit, Thai spirit, Chinese spirit and Chinese wine brands.

According to an International Beverage Holdings spokesperson, "Our Scottish team has responsibility for developing and delivering business plans, the financial aspects of which are signed off at parent group level. The production strategy and the operational plan for the five malt distilleries and for maturation, blending and bottling is determined by a highly experienced Scottish-based management team. There are enormous operational and sales benefits, however, to being part of the International Beverage family, as we grow our brands in new and established markets around the world."

Examples of distilleries like Knockdhu, traditionally Scottish in appearance and operation, yet owned at arms' length by overseas companies are to be found the length and breadth of Scotland. In fact, it is easy to underestimate just how much of the Scotch whisky industry is actually in foreign hands.

At the last count, ownership of Scotch whisky assets was vested in 13 different overseas countries, though in some instances, there is an element of opacity regarding just who actually controls the purse string and where they are domiciled.

In 2018, 51 of the 128 malt distilleries operational in Scotland were in overseas ownership, and in terms of production capacity, accounted for some 174 mla of the 383 mla available that year. Taking into consideration the fact that UK-based Diageo alone was responsible for 121.3mla of production capacity, the overall influence of foreign ownership in the Scotch whisky industry becomes even more pronounced.

Additionally, six of the global top ten best-selling single malt brands are foreign-owned – The Glenlivet (Pernod Ricard), The Macallan (Beam Suntory), Glenmorangie (Moet Hennessy), Aberlour (Pernod Ricard), Laphroaig (Beam Suntory) and Glen Grant (Campari). Only Glenfiddich (William Grant & Sons), The Singleton (Diageo), Balvenie (William Grant & Sons) and Talisker (Diageo) are not owned by overseas interests.

Of course, Scotch whisky is far from alone in being the recipient of major overseas investments. We live in a world where business has never been more global, and ownership of key brands is concentrated into fewer corporate hands than ever before.

Consumers could be forgiven for not realising this, however, as the ultimate brand owners like to

maintain 'the illusion of locality,' as it may be termed. Clearly, they feel this is what those consumers want. Frequently, in order to make it appear that our beloved single malts are still produced by small Scottish family firms, working by hand on an artisan scale, they employ the reassuring names of old Scottish firms, hence Knockdhu's office door bearing the legend 'James Catto & Co.' It is also why you will never see the label of a bottle of Talisker declaring that 'This single malt is owned by the world's largest drinks company.'

So, what are the pros and cons of Scotch whisky's reliance on overseas ownership, and why are so many inward investors attracted to the business of Scotch whisky?

Fraser Thornton is Managing Director Distell International Ltd, and a key member of the team of senior Scotch whisky figures that acquired Burn Stewart Ltd in 1988, becoming chairman and managing director. The company was sold to Trinidad-based venture capitalist CL Financial in 2002 for £50m, and subsequently acquired by Distell of South Africa in 2016 for £160m. In Scotland, Burn Stewart Distillers owns Deanston, Tobermory and Bunnahabhain distilleries, plus the Scottish Leader and Black Bottle blends.

According to Thornton, "The fact that a sizeable portion of the Scotch whisky industry is in overseas ownership is a reflection of the success of the category. Overseas owners see the attraction and want to be part of it. It is testimony to the strengths and attractions of the industry that with uncertainty in recent years over Scottish independence and its potential implications and unknowns, and the same now with Brexit, you're still seeing increasing levels of inward investment. I don't think foreign ownership is in any way detrimental to the industry."

One senior Scotch whisky figure who did not wish to be named makes the point, however, that "An obvious concern is that in difficult economic times an overseas owner is likely to sacrifice 'peripheral' operations in a country like Scotland and focus on its heartland. You can't imagine Toyota deserting Nagoya in order to keep open its plants in Derbyshire and North Wales in the event of a major crisis in the global motor industry. The same applies with Scotch whisky."

Most criticisms that can be levelled at overseas owners may also be levelled at any organisation with a profit to turn and perhaps shareholders to please, regardless of where it is registered. It was that most British of companies, The Distillers Company Limited, that closed no fewer than 23 distilleries during 1983 and 1985, and in the past, a number of UK-based organisations have treated some Scotch

whisky businesses in their portfolios as little more than Monopoly pieces to be traded at will.

Key to a successful relationship between foreign owners and Scotch whisky assets, is the necessity of having real understanding of the Scotch whisky business and the timescales between spirit flowing from the still and a profit being returned. In the worst-case scenario, a Scottish distiller may find himself or herself having to explain to a newly-appointed corporate executive that it may be ten years or more before pay day!

Fraser Thornton of Distell makes the point that "There may well be a degree of nostalgia for the days when the industry was controlled largely by Scottish family companies – when the Walker family ran Johnnie Walker and the Dewar family ran Dewar's and so on, and that has largely gone."

England was first

The first 'foreign' incursion into Scotch whisky ownership actually came from England, with London-based gin company W&A Gilbey Ltd acquiring Glen Spey distillery in Rothes on Speyside for £11,000 in 1887, going on to purchase fellow Speysiders Strathmill (1895) and Knockando (1903).

The earliest overseas investors in Scotch whisky came from North America, with major Canadian distilling organisation Hiram Walker-Gooderham & Worts Ltd (later trading in the UK as Hiram Walker (Scotland) Ltd), buying into the industry in 1930 and subsequently expanding its interests, both pre-war and post-war, while the Distillers Corporation-Seagrams Ltd (later Seagram Co Ltd), made its first Scotch whisky acquisitions in 1935, also going on to extend its influence significantly.

It was not just the North Americans who were keen to buy up sections of the Scotch whisky industry, however. USA was the leading export market for Scotch, but another was Japan, and 1986 saw Tomatin distillery, south of Inverness, become the first in Scotland to be owned by a Japanese company, being bought out of liquidation by the conglomerate Takara Shuzo Co and Okara & Co. Three years later Nikka Whisky Distilling Company Ltd purchased Ben Nevis distillery, near Fort William, from the brewer Whitbread.

These developments were mere drops in the ocean, however, compared to subsequent Japanese investment in the Scotch whisky industry. In 1994 the Japanese distilling giant Suntory acquired Morrison Bowmore Distillers Ltd, with its Auchentoshan, Bowmore and Glen Garioch distilleries, having

Dalmunach is the latest distillery in the Chivas group owned by the French conglomerate Pernod Ricard

previously purchased 25% of what was then Macallan-Glenlivet in 1986.

2014 saw Suntory acquire US-based Beam Inc in 2014, which took Ardmore and Laphroaig distilleries, and the Teacher's blended Scotch brand into the Osaka company's fold. Today, Beam Suntory as it is known has the fifth-highest level of malt whisky capacity in Scotland, after Diageo, Pernod Ricard, William Grant & Sons Ltd and Edrington.

Along with the USA and Japan, France has long been high up in the list of countries where Scotch whisky sells best, and during the past two decades, a French company has become the second-largest operator in the Scotch whisky sector, with 13 malt distilleries and just over 17% of the industry's total capacity. This came about when Seagram's beverage assets were acquired by Diageo and Pernod Ricard in 2000, with Pernod Ricard taking on the portfolio of Scottish distilleries and whisky brands.

According to Jean-Christophe Coutures, CEO and Chairman of Chivas Brothers, "Our organisation is rooted in a guiding principle that is unique within our sector: decentralisation. This allows us to make decisions as closely as possible to consumers and brands. Chivas Brothers, as a Pernod Ricard 'Brand Company' is in charge of production and overall strategy for the group's Scotch. With a single-minded focus on Scotch whisky, Chivas Brothers has been instrumental in the continued growth of the Scotch category around the world and most recently its Strategic International Scotch brands – Chivas,

Ballantine's, Royal Salute and The Glenlivet – reported strong organic growth in net sales at plus nine per cent."

Distell of South Africa follows a similar policy, with Fraser Thornton declaring that "Distell is on a programme of decentralisation. There is a reporting structure common to the whole company to go through, but the thinking and marketing for Scottish brands remains in the UK."

Distell purchased Burn Stewart Distillers in 2013 from Trinidad-based CL Financial, and Thornton explains that "Distell had a joint venture in Africa with our Scottish Leader blend and they had previously been a third-party distributor, so we had enjoyed a 15-year trading partnership. I think that's why they were interested in buying Burn Stewart in particular. They knew the company. Also, our brands were relatively small, but with room for growth. Burn Stewart was on the path to premiumising its spirits offering and the growth of single malts generally also made it attractive to Distell.

"They had Amarula cream liqueur and a fledgling portfolio of South African whiskies, and it seemed logical to them to have a Scottish portfolio as part of that. Culturally, we speak the same language and operate in pretty much the same time zones, so integration of the two operations was relatively straightforward. Distell had no previous production assets outside Africa, so there were no overlaps to deal with."

When it comes to the benefits accruing from

Alan McConnochie, distilleries manager for BenRiach Distilleries Company, sees only benefits since Brown-Forman took over

ownership of a multi-national operator like Distell, Thornton says that "A great advantage is the quality of marketing and the amount of money invested in the infrastructure at distilleries and other sites. We've invested a lot in Tobermory distillery, and are currently spending on upgrading Bunnahabhain and creating a new brand home there."

Improved distribution

Another group of distilleries benefitting from overseas investment is BenRiach, GlenDronach and Glenglassaugh, trading as the BenRiach Distillery Company, and previously owned by whisky industry veteran Billy Walker and his South African business partners.

In 2016, Kentucky-headquartered drinks company the Brown-Forman Corporation bought the firm for £285 million, making its first foray into the world of Scotch whisky in the process. Distilleries Manager Alan McConnochie notes that "Scotch whisky, and particularly premium single malts, has cachet. People want to be part of it, and for Brown-Forman it completes their portfolio of high-quality brands."

Musing on the changes brought about by Brown-Forman's ownership, he says that "Brown-Forman is a foreign owner but also gigantic. The sheer size of it has more impact really than the fact it's an overseas company. Policy decisions are made by Brown-Forman in Louisville, and we meet targets

they set. But when it comes to spending money, the capital is there if you can make a sound case for why the expense is justified."

McConnochie says that "The fact that the owners are 'foreign' has no impact on the day to day running of the Scottish operation. We're left to our own devices, and they're happy for us to carry on almost as before. We call each other bad names at times, like everybody does, but we really work together well."

He sees one of the key advantages as being "Access to their importing chain, which has a huge impact for us. We're still working on that, but it will have major advantages in the future. Another advantage is a fantastic supply of the best casks available from their own cooperage in Kentucky."

"Brown-Forman is a family-owned company, and the best thing about that is that there is no short-termism. With Bourbon they are looking at four to five-year timescales anyway. They're in it for the long haul and they have stability. They don't set targets that aren't achievable. I've had owners in the past – Scottish owners at that – who've basically asked 'Can't you turn on the 12-year-old tap?' Not Brown-Forman."

He considers that if there is a downside it is that "Small companies can be nimbler. With a big organisation based overseas it can take longer to get decisions made, you can't just get approval for something straight away."

Like Brown-Forman, Remy Cointreau is a family-

Since Remy Cointreau took over Bruichladdich, the have invested close to £23 million in the distillery

controlled company, but when the French cognac and liqueur specialist acquired the Islay distillery of Bruichladdich in 2012 there were fears that the unique nature of Bruichladdich, its passion for provenance and pride in proving local employment might be sacrificed, or at least diminished.

The opposite, however, has proved to be the case. £22.8million has been invested in the distillery over the last five years, and plans for future projects could see further investment of the same magnitude by 2024.

Adhering to the existing ethos of keeping as much activity as possible on Islay, Remy Cointreau has built two new warehouses, with another four to follow, so that all maturation can take place on the island. Plans are also in place to create an on-site maltings, while Bruichladdich has recently acquired Shore House Croft, a 30-acre farm adjacent to the distillery, where trials of barley varieties are being undertaken to determine which might be most viable for cultivation on Islay.

As Communications Manager Christy McFarlane notes, "The projects under current management clearly evidence how Remy Cointreau is investing behind the values of Bruichladdich. Perhaps in contrast to other examples, where a new parent company may be inclined to strip a distillery of supposed inefficiencies, Remy Cointreau have very much allowed Bruichladdich to retain their entrepreneurial attitude and agile mindset."

"Their fundamental principles of provenance, transparency and now sustainability are clearly being supported. What is particularly positive is the long-term view that is being taken by both parties; their

commitment to Islay has resulted in the permanent employment of 85 people on the island – ensuring Bruichladdich continues as the largest private employer on Islay - despite being the third-smallest producer."

Bruichladdich Distillery CEO, Douglas Taylor adds that "It very much feels like Bruichladdich continues to be an independent business, but one with benefits. Remy Cointreau has extended a massive degree of trust and support, for us to continue to drive the business forward in line with our values. In truth, the change from 80-plus independent shareholders to one new owner has allowed us to move further and faster with our ambition to be an Islay-centric, community minded, sustainable and transparent business."

And the growth of overseas ownership within the Scotch whisky industry continues. The first half of 2019 saw two more foreign acquisitions, the first coming when Glenturret distillery in Perthshire was bought from Edrington by Glenturret Holding – a company formed by luxury glass specialist Lalique and its second-largest shareholder, Hansjörg Wyss.

Glenturret was always destined to play fourth fiddle in the Edrington single malts orchestra behind The Macallan, Highland Park and Glenrothes, with Edrington's main focus being on promoting Glenturret as the brand home of its Famous Grouse blend, rather than a single malt worthy of serious attention. The investment already planned for the distillery and brand will surely allow Glenturret to flourish and fulfil its true potential.

The news about Glenturret's change of ownership was followed by an announcement in June that the

Loch Lomond Group was being acquired by Asia-based investment firm Hillhouse Capital Management. Loch Lomond had been in the ownership of UK-based private equity management firm Exponent since 2014, when it was acquired from the Bulloch family.

Assets include Loch Lomond malt and grain distillery in Alexandria, Glen Scotia distillery in Campbeltown, a bottling plant in Catrine, Ayrshire, plus the remaining stocks of whisky from the closed Lowland distillery of Littlemill.

During the last five years, Loch Lomond's business has grown internationally, with overseas markets now representing 70% of the business, as opposed to less than ten per cent in 2014. According to Loch Lomond Group CEO Colin Matthews, "We believe now is the right time to move forward into the next stage of our growth strategy as we look to innovate further, extend our portfolio of brands and continue to expand our international presence, particularly in Asia, where Hillhouse has significant experience."

While whisky – and in particular high-end single malt Scotch whisky – remains such a buoyant commodity, and one with the same cache and kudos as a premier cru vineyard, individuals and companies with money to invest, whatever their nationality, will always desire a piece of the action. Given that they are not just playing Monopoly with their precious Scottish assets, but treat them and the people who work for them with respect and integrity, then their attentions should surely be welcomed.

Gavin D Smith is one of Scotland's leading whisky writers and Contributing Editor Scotland for Whisky Magazine. He regularly undertakes writing commissions for leading drinks companies and produces articles for a wide range of publications, including Whisky Magazine, Whisky Magazine & Fine Spirits – France, Whisky Etc, Whisky Advocate, Whiskeria, Whisky Quarterly and The Cask.

He is the author and co-author of some 30 books, and recent publications include The Microdistillers' Handbook, Ardbeg: Heavenly Peated, and World of Whisky (with David Wishart and Neil Ridley)..

OVERSEAS OWNERSHIP OF SCOTCH WHISKY DISTILLERIES

Australia
David Prior
Bladnoch

Bermuda
Bacardi (John Dewar & Sons)
Aberfeldy, Aultmore, Craigellachie, Macduff, Royal Brackla

Cayman Islands
Hillhouse Capital Group (Loch Lomond Group)
Glen Scotia, Loch Lomond

Denmark
Copenhagen Fortuna (John Fergus & Co)
Inchdairnie

France
Moet Hennessy
Ardbeg, Glenmorangie

Pernod Ricard
Aberlour, Allt-a-Bhainne, Braeval, Dalmunach, Glen Keith, Glenburgie, Glentauchers, The Glenlivet, Longmorn, Miltonduff, Scapa, Strathisla, Tormore, Strathclyde

Picard Vins & Spiritueux
Tullibardine

Remy Cointreau
Bruichladdich

La Martiniquaise
Glen Moray, Starlaw

Italy
Campari
Glen Grant

Japan
Beam Suntory
Ardmore, Auchentoshan, Bowmore, Glen Garioch, Laphroaig

Nikka
Ben Nevis

Takara Shuzo, Kokubu, Marubeni Corp
Tomatin

South Africa
Distell
Bunnahabhain, Deanston, Tobermory

Sweden
Haydn Holdings (Mossburn Distillers)
Torabhaig

Switzerland
Art & Terroir
Glenturret

Thailand
Thai Beverages (Inver House Distillers)
Balblair, Balmenach, Knockdhu, Pulteney, Speyburn

The Philippines
Emperador (Whyte & Mackay Distillers)
Dalmore, Fettercairn, Isle of Jura, Tamanvulin, Invergordon

USA
Brown Forman (BenRiach Distillery Co)
BenRiach, GlenDronach, Glenglassaugh

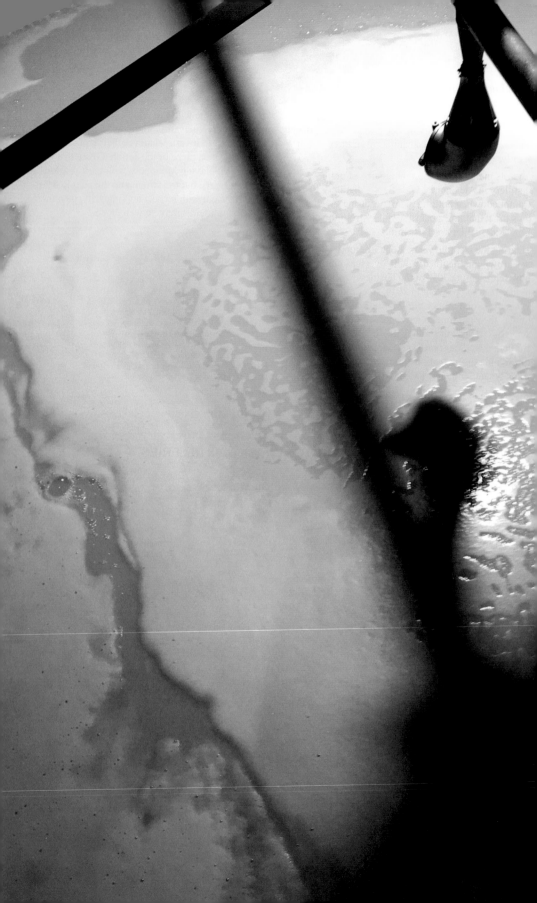

The Feast from the Yeast

by Jonny McCormick

Few of us give yeast a second thought,
but it's worth putting under the microscope for once because
the latest scientific developments offer the potential to radically alter
the direction of malt whisky distilling.

Before we begin, let me tell you the proper scientific name for distilling yeast: *Saccharomyces cerevisiae*. That looks a bit of a mouthful when you only came here to read about whisky, doesn't it? Don't be put off: just switch it to something more familiar with the same initials and say that instead. Replace it with Sean Connery, Sofia Coppola, or Stewart Copeland and suddenly, it doesn't look quite so intimidating. Now, if we're all sitting comfortably, then I'll begin…

We owe yeast a debt of thanks. This plucky, self-sacrificing single cell is the powerhouse that produces not only all the alcohol in whisky, but also a feast of delicious aromatic and flavoursome congeners that whisky connoisseurs find pleasurable. Trust me, fermentation is the most critical stage in whisky production. Malted barley merely provides fuel for the yeast, while distillation and maturation simply concentrate and refine the fruits of their labours. Without yeast, there's no alcohol, no flavour, and no whisky. It's time someone stood up for the little guy.

Fermentation appreciation

"Yeast is fantastic," agrees Marie Stanton, the distillery manager at Highland Park. "But distillers aren't nice to their yeast; they're cruel to it. Having a brewing background, I pity my poor yeast, but you have to get over that. I think yeast is brilliant, amazing, interesting stuff and what you can do with it is fantastic."

Yeast gives us the building blocks of the flavour profile, and distillers can choose to adjust those building blocks by playing with their fermentation. In brewing, they control their fermentation in different ways to make a different end product.

"Distillers can do the same thing; it's just that nine times out of ten, they choose not to," shrugs Stanton. "It's history that's given them what they've got, not manipulation."

Stanton was particularly drawn to the job at Highland Park because of the maltings and fermentation. Although batch distillation evens out any variations, fermentation ranges from just less than 60 hours during the week to 90 hours at weekends.

"The team on the shop floor are interested because I've got a science background, because I get malting, I get yeast, and I ask questions; 'What's it doing this

time? What temperature was that?' and they look at me and say, 'We're not used to being asked these questions!' But I just need to know! Yeast is my thing, I love yeast," she says grinning. "And there's still a lot of science to be digging around in."

Fermantation explanation

Professor Graeme Walker has been known to ask his students, "Where do you think the flavour of whisky comes from?" Walker is professor of zymology at Abertay University, Dundee, and a world-renowned expert on the study of fermentation. "Some students say the raw materials, some say the water, some say maturation in oak wood," Walker replies, "but none of them say yeast and fermentation."

Commercial distilling yeasts are strains of *Saccharomyces cerevisiae*, which are supplied in three forms: creamed, pressed into cakes, and dried. These have a shelf life as short as two weeks for creamed yeast and up to four years for dried yeast: the latter being a useful attribute for remote distilleries where deliveries can be disrupted by raging storms and blizzards. Different strains of *Saccharomyces cerevisiae* form subtly different concentrations of ethanol and congeners. "Although distillers look upon their

Marie Stanton - Highland Park's Distillery Manager

Photograph by Søren

Professor Graeme Walker and Martina Daute exploring different kinds of yeast at Abertay University, Dundee

yeasts as just alcohol factories, they are flavour factories as well," reminds Walker.

To release the genie from the bottle, yeast is pitched into the worts, the liquid which contains the extractable sugars from the mash of malted barley. Finding surroundings with an agreeable temperature and an abundance of sugars, it begins to multiply. Let fermentation begin! Over the next couple of days, the vigorous, proliferating multitude of yeast generates alcohol, glycerol, and carbon dioxide. The ability of yeast to ferment and produce the desired flavours in an environment where the temperature is increasing, sugar levels are dropping, the yeast population is multiplying, and the alcohol levels are steadily rising around them is called yeast tolerance. They finish their main sugar uptake and fermentation within 48 hours, and secondary flavour compounds begin to appear such as higher alcohols, esters, and aldehydes: some created by the yeasts' own metabolism, and others modified by further reactions in the wash.

"The endogenous heat of fermentation, together with the alcohol reaching 8–10%, creates a double stress on the yeast that results in their viability declining," explains Walker, sugar-coating the reality of when conditions become unsustainable for the yeasts' endurance. "Basically, the yeast die with the double stress. Dead yeast won't make alcohol, that's for certain."

Fermentation adaptation

In Scotland, washbacks are not temperature-controlled, but temperature is critical to yeast performance, so the inescapable question is how distillers manage in warmer countries? In Taiwan, Kavalan use commercially available distiller's yeast, specially adapted to the subtropical climate.

"Firstly, we use a strain to produce a maximum yield, then half-way through fermentation, another one is added to provide Kavalan's desired aromatic flavours," says Ian Chang, the master blender. Inside Kavalan's impressive second distillery, there is a gigantic room full of gleaming washbacks that look like futuristic spacecraft hunkered down inside the hangar bay of the mother ship.

"To adapt to Taiwan's heat, we use custom-made stainless steel washbacks, which help us to properly maintain and control the delicate fermentation process," explains Chang, pointing out the cooling system inside the double-jacketed washbacks. The purpose of this adaptation is to reduce the stress on the yeast, and improve its performance over the 60 hour fermentation. "The yeast feed off the potion of minerals contained within our water to create incredibly rich congeners, which intensify the flavour of our whisky. Kavalan's signature fruitiness – the mango, apple, cherry, and pineapple – can all be

In the hot and humid climate of Taiwan, the owners of Kavalan need to use temperature-contolled washbacks for the fermentation

traced back to fermentation."

Admittedly, yeast isn't a sexy way to sell whisky. If you quizzed the two million people who visited a Scotch whisky distillery in 2018 and asked them what they came hoping to see, probably none of them came looking for yeast. Most MWYB readers are well versed in the improvements in wood quality, cooperage technology, and malting barley varieties over the last two or three decades. Contemporary whisky labels display prominent age statements and vintages, highlight the casks used for maturation or finishing, and a few even mention the origins of the barley. Unsurprisingly, yeast never receives any credit, but this may be slowly changing. When the 2017 Diageo Special Releases were announced, collectors fawned over the latest Port Ellen, Brora, and Convalmore expressions and almost overlooked a seemingly inconspicuous detail on the Glen Elgin 18 year old: part of the recipe used whisky made with *Schizosaccharomyces pombe*, an alternative yeast to *Saccharomyces cerevisiae* [Author's note Uh-oh, another tricky name: try substituting it with Sean Penn, Simon Pegg, or Sarah Palin, whoever takes your fancy].

Fermentation experimentation

Dr. Bill Lumsden frequently met Michael Jackson (1942 – 2007) during his tenure as Glenmorangie distillery manager in the 1990s, and tackled him about a passage in The World Guide to Whisky

where Jackson mentions Glenmorangie propagating its own unique strains of yeast.

[While many distillers feel that the use of both beer and whisky yeasts ensure the best fermentation, Glenmorangie opts only for the latter, in two pure cultures. Their house yeasts impart an estery character that may have something to do with the whisky's distinctive bouquet. – *Michael Jackson, The World Guide to Whisky, Dorling Kindersley, 1987*]

"Michael was very vague about where that [information] had come from," he recalls.

Lumsden researched the issue in depth, even enlisting the expertise of the company archivist, but unfortunately, no evidence was found to substantiate Jackson's claim. The practice of adding brewer's yeast was once commonplace at many distilleries, but died out during the 1990s and early 2000s.

"I am not able to back this up with hard scientific evidence, but it is my feeling that new make spirit in Scotland lost a little bit of character when we stopped using brewer's yeast," states Lumsden. Despite being unable to corroborate Jackson's statement, an ambitious quest for Glenmorangie had begun to form in Lumsden's enterprising mind. "I vowed there and then that we would have our own unique yeast strain."

Lumsden had long been fascinated by yeast and fermentation, since it was the subject of his PhD thesis, but by his own admission, the first experiment was an utter disaster. Around 15 years ago, he attempted Belgian lambic-style fermentation, but much of the flavour came from the action of lactobacillus and other bacteria.

"To no-one's surprise, certainly not mine, we ended up with a rather curious flavour profile," he admits euphemistically, wrinkling his nose in distaste.

Although yeasts and fungal species are ubiquitous in the environment, Lumsden endeavoured to conduct a properly designed trial isolating natural microflora in the vicinity of Glenmorangie distillery.

"Early on, I ruled out the idea of culturing up a yeast from my own body," delivers Lumsden with deadpan humour, "I didn't think that was necessarily going to be a good selling point."

Instead, he jumped the fence into the adjacent field where the local Cadboll estate barley grows, and pressed the green ears of the cereal on to agar plates. These were sealed and sent to Lallemand, a commercial yeast manufacturer.

"All sorts of monsters grew up on the agar plates as you can imagine, including many types of yeast." Lallemand isolated a number of *Saccahromyces* species and ran trial fermentations to narrow it down to a single species. Glenmorangie chose to name this *Saccharomyces diaemath*, which is a Gaelic approximation for 'God is Good' [Author's note Again, don't be put off, just exchange it for Stormy Daniels, Snoop Dogg, or Sophie Dahl]. Lumsden explains, "The ancient Egyptians, who are largely credited with discovering liquor, knew that something was transforming their cereals into alcohol but they didn't know what, so they referred to this magical phenomenon as 'God is Good.'"

This was wild yeast propagation, but the term 'wild yeast' normally refers to the unintentional actions of yeast in the fermentation, one that was not introduced, nor under the control of the distiller. *Saccharomyces diaemath* was deliberately used as the distilling yeast in this fermentation, but it was isolated from the distillery surroundings and not otherwise commercially available: domesticated wild yeast, if you like. Lallemand cultured sufficient supply for a week's production at Glenmorangie and delivered it to Tain, where a happy coincidence led to a reunification that Lumsden describes as 'the most beautiful and serendipitous part of it all'. "The Cadboll barley that the yeast had come from had been harvested, dried, malted, and delivered to the distillery," he

Bill Lumsden started his trials with wild yeast at Glenmorangie ten years ago

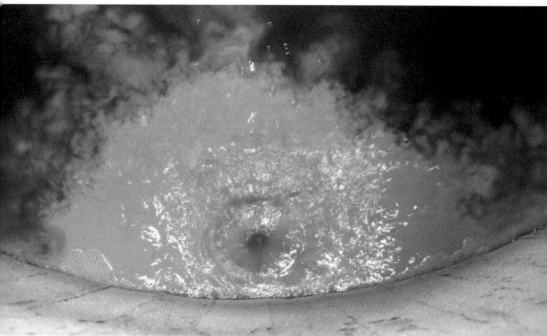

Yeast is added to a washback at Oban distillery

says. "We used the new culture of *Saccharomyces diaemath* to ferment the malt from the very barley it came from."

Commercial yeasts are bred for their desirable characteristics. Just as elite athletes set new world records by working on their nutrition, training regimes, and sports psychology, so yeasts are selected to produce ever higher ethanol yields and more desirable flavours while bolstering their tolerance to ethanol, temperature, and higher substrate levels. Predictably, switching to *Saccharomyces diaemath* would be unlikely to match the elite athletic abilities of the main commercial strains of *Saccharomyces cerevisiae*. Ethanol yield was 350-360 litres per ton compared with the 410 litres per ton that Glenmorangie normally achieves with their regular Lallemand yeast or AB Mauri Pinnacle yeast. *Saccharomyces diaemath* showed a lower stress tolerance in the later stages of fermentation as anticipated.

"We believe it stopped fermenting a lot sooner than a classic yeast; the increased temperature and alcohol probably popped it," admits Lumsden breezily. "The fingerprint of the key volatiles were different from *Saccharomyces cerevisiae*, and a number of the acetate esters were slightly down."

He reports quite a marked difference in the new make spirit character between the *Saccharomyces diaemath* and classic Glenmorangie, though the primary purpose was not the pursuit of new aromas and flavours.

"The priority was to undertake a kind of terroir experiment," says Lumsden. "Of course, I sincerely hoped that I would get different flavours, and I did, but not dramatically different."

Glenmorangie Allta, the tenth in the trailblazing Glenmorangie Private Edition series, was the first major Scotch whisky to be launched where yeast took centre stage. To showcase the influence of *Saccharomyces diaemath* in the mature whisky, it was important not to overpower the flavours with oak and allow the distillate character to flourish. This meant filling less active wood by choosing third fill bourbon barrels and bottling earlier at around nine years old.

"The other reason was I was getting bored reading about all these new so-called craft distilleries and the 'ground-breaking innovation' that they were going to do. When I read through their plans, I thought, 'Boring, did that ten years ago, did that fifteen years ago,' and I thought 'I'm not going to be usurped by these people.'"

Bill Lumsden has confirmed to the MWYB that further Glenmorangie yeast trials are planned, but is he prepared for a slew of yeast-driven whiskies to follow in the wake of Glenmorangie Allta in the Twenties?

"I most definitely am," he affirms. "Now, whether or not they're all good to drink…" he adds, trailing off mischievously. Lumsden is wary about how much flavour will shine through in these whiskies after maturation in cask for ten or twelve years.

"The whole nature of the Private Edition series was to be the first to do something a bit different and take it to market. You see people talking about doing all sorts of wacky things, but is that actually going to give you a nice taste at the end of the day? Or are you just doing it for sensationalism?"

The major Scotch whisky companies have already explored wine and brewing yeasts extensively, but they seldom talk about it publically because they assume it's not of interest to the average whisky drinker (that's not you, you're reading this book). Indubitably, you will be hearing more about yeast from certain producers. Professor Walker predicts, "If any whisky producer in Scotland was going to go back to the old days of using surplus yeast from brewing, it will be the craft guys that do it."

Fermentation speculation

Starch-degrading enzymes are permitted in fermentation in the U.S., Canada, and Ireland, whereas only enzymes derived from malted barley are allowable in Scotland: a major divergence in practice.

"You may get a completely different fermentation profile depending on whether or not you add commercial enzymes," says Professor Walker. When the mash bill contains unmalted cereals, supplementary enzymes are particularly prevalent, and that changes the flavour as yeast preferentially take up glucose over maltose. "A maltose fermentation is quite different to a glucose fermentation in terms of its progress and flavour," reports Walker.

What are these supplementary enzymes doing? Alpha-amylase is a viscosity breaker used during high temperature cereal processing. Glucoamylase is a slow, starch-degrading enzyme that debranches starch, liberating free glucose while fermentation is progressing: a process called simultaneous saccharification and fermentation. This is common in the huge U.S. corn distilleries working for the biofuel industry.

Bioethanol is driving innovation in improving yeast tolerance to harsher conditions through genetic modification. Commercialised genetically modified (GM) yeasts have been engineered to excrete starch-degrading enzymes. GM yeasts have been used to create hoppy flavoured, hop-free beers, leading Walker to enquire, "Would you need malted barley in the future, if you used a yeast that was a starch degrader? It may sound a silly question, but maybe not. Why do you carry out malting? To create the enzymes that will degrade the starch. If you used starch-degrading GM yeast, then perhaps you wouldn't need to go through malting? The main barrier to new technologies such as these yeasts will be public perception. I don't think the regulations will change, but it's a public perception barrier, not a technological barrier."

Aside from genetic modification, yeast is one part of the Scotch whisky regulations where there is still plenty of scope for innovation. Martina Daute studied molecular biotechnology and is working on her PhD with Professor Walker at Abertay University to investigate non-Saccharomyces strains for flavour diversity. She has a useful canine analogy to help people understand her project. Imagine all *Saccharomyces* are a group of Border Collies: they look and behave in a similar fashion to one another, but they're not identical clones. If all yeasts are dogs, Daute is investigating the equivalent of how a German Shepherd or a Poodle performs compared to a Border Collie.

"It's unlikely that any of the non-*Saccharomyces* yeasts will produce higher levels of alcohol, but their metabolism is different so they can produce different flavours." Daute uses the nano-brewery at Abertay University to make batches of worts, run fermentations, and then distils them using a miniaturised still in the laboratory. To see if flavours are preserved over the wash, low wines, and new make spirit stage, the liquid is sent to sensory panellists and analytically profiled. At present, her focus is on individual fermentations with alternative yeasts.

"The brewing yeasts are performing quite well, because they already produce enough alcohol, and the different flavours come over to the low wines stage where you can smell the differences."

If successful, she hopes her project will enable distillers to blend alternative yeast species with *Saccahromyces cerevisiae* to run either co-fermentations or sequential fermentations for greater flavour diversity. One day, distilleries may be able to match raw materials to a specialised blend of yeasts and mature it in specifically designed casks to best enhance these flavours through maturation. Then, thanks to yeast, we'll have produced the perfect dram.

Whisky writer and photographer Jonny McCormick is Contributing Editor of Whisky Advocate magazine and one of their leading whisky reviewers. He is known as a specialist in the field of rare and collectable whiskies, and his prolific writing on the topic has made him an authority on the secondary market. He is a Keeper of the Quaich and has presented Scotch whisky tastings in Europe, North America, and Asia.

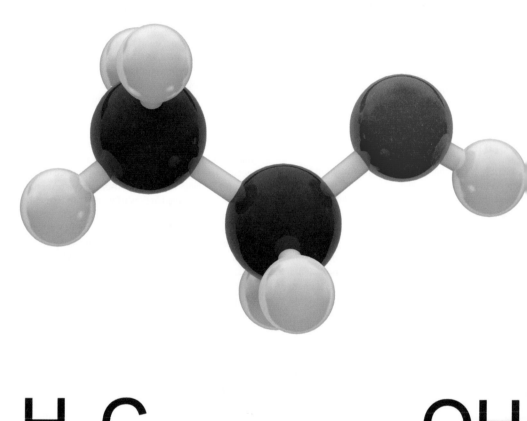

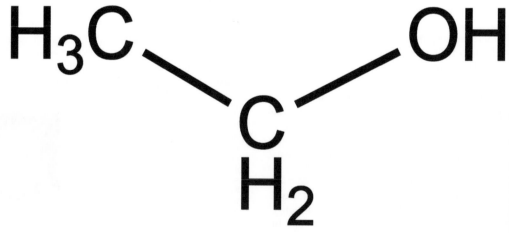

H₃C — C — OH

H₂

The role of Alcohol

by Ian Wisniewski

From production to tasting
- alcohol has a huge impact on whisky. It brings out more nuances during fermentation and extracts more flavour compounds from the cask during maturation. And on the palate, alcohol is a delivery system for flavour, mouthfeel and roundness.

A malt whisky can enchant and enthral the palate, but the experience exceeds flavour. Able to create intense, transcendent emotions, a sip of malt whisky can provide a momentary sense of eternity in a sublime world. Integral to this transformative ability is alcohol. Having an appreciable impact on the mind and body, alcohol also has a comprehensive influence on malt whisky throughout its development. Sometimes the influence is direct, sometimes indirect and difficult to define, which makes alcohol an essential and elusive element of malt whisky.

"From a sensory perspective alcohol makes malt whisky what it is, binding everything together and anchoring it. Alcohol adds texture, body, roundness, complexity and completeness. But from a scientific perspective, I can't really quantify what alcohol contributes to malt whisky," says Dr Bill Lumsden, Glenmorangie's Director of Distilling, Whisky Creation & Whisky Stocks.

If an alchemist could produce two versions of the same malt whisky, one with alcohol and one without, we could compare and contrast. In the absence of this possibility, there are two areas to consider: the role of alcohol during production and ageing, and the subsequent role of alcohol when drinking malt whisky.

Alcohol owes its existence to yeast, although yeast's sole objective is to grow and reproduce. The energy required for this is gained by metabolising sugars drawn from the wort, and then emitting the residue of digestion through the cell wall back into the wort. Each yeast cell comprises separate 'digestive systems,' one emitting residue in the form of flavour compounds such as esters, though the volume is very minor. Another digestive system emits residue on a far greater scale in the form of alcohol (expelling alcohol is a survival mechanism as it's toxic to yeast).

Alcohol emissions span a range that includes 'higher alcohols,' which contribute some fruit notes for example. The type of alcohol produced in the greatest quantity is ethanol, but this makes a very modest contribution to flavour, merely some sweetness and pepper notes.

Alcohol is, however, able to create an additional range of flavours during fermentation, through a liaison with acids (a range of which are naturally present in the wort, derived from malted barley). It's a case of mutual attraction, with one alcohol molecule physically integrating with one acid molecule. Alcohol and acid molecules span various dimensions, and it seems logical that molecules of the same size would seek each other out for integration. But does logic ever apply in cases of attraction ? Certainly not among the molecular community. Differently-sized ethanol and acid molecules readily rendezvous and integrate, which creates a range of esters.

The type of acid that integrates with an alcohol molecule determines whether the result is a short-chain ester (comprising a few linked molecules); a medium-chain ester (several linked molecules), or long-chain fatty acid ester (around 10 or more linked molecules).

Esters are renowned for providing fruit notes, but their repertoire is more extensive. Short-chain esters also include 'nail varnish' and 'solvent' notes, while long-chain fatty acid esters can be 'soapy.' As long-chain fatty acid esters are physically the largest, this suggests a natural dominance. However, they are less flavour active than short-chain esters, which consequently have a greater impact on the resulting wash.

"Initially ethanol and fatty acid interaction creates grassy and lighter fruit notes, whereas later in the fermentation process this is complemented by more intense fruit flavour creation, such as tropical fruit," says Emma Walker, Whisky Specialist and blender, Diageo.

Meanwhile, esters are also being produced by the yeast. How the volume of esters created by yeast compares to that of alcohol-acid integration is uncertain (counting during fermentation is hardly feasible, though it would be nice to know).

As yeast continues feasting on sugar and emitting alcohol, the alcoholic strength of the wort keeps rising. This imposes ever greater levels of stress on the yeast, which consequently behaves differently (don't we all, when stressed out).

"As the alcoholic strength of the liquid increases, the yeast produces a different result. I think the proportions of flavour compounds vary at different alcoholic strengths of wash. It's the old adage, you are what you eat! If you give the yeast a lot of food in the form of sugar and, consequently following on from that, high alcohol strength, it will produce other flavour congeners compared to if it was under less pressure from a weaker liquid. This is not necessarily a detrimental effect however. You get some complex aromas and character from the yeast having to work hard," says Peter Holroyd, Kingsbarns' Distillery Manager.

The majority of alcohol is produced within about 48 hours, and this marks the end of fermentation at some distilleries, with the resulting wash showing distinct cereal and biscuit notes. Allowing fermentation to continue, up to 100 hours or longer, sees some additional alcohol produced. But the greatest difference is a broader range of flavour compounds, notably esters (including fruit notes). Needless to say, longer fermentation also provides more time for alcohol-acid integration.

The end of fermentation results in expiring yeast cells beginning to autolyse (rupture), which liberates their contents into the wash.

"There might be some alcohol within the yeast, but it's fatty acid esters that are locked inside the yeast, and released by autolysis. Yeast is the catalyst during fermentation, while copper is the catalyst during distillation. When ethanol and acids are in contact with copper this can catalyse a small level of esterification. However, this is a minor contribution to the overall level of esters," says Rachel Barrie, Master Blender, The BenRiach Company.

Alcohol-acid integration during distillation is on a reduced scale compared to fermentation, and at an even lower level during maturation. Moreover, some master blenders consider the level of integration to be so low during maturation that its influence is insignificant.

...ter Holroyd, Kingsbarn's distillery manager, finds that the yeast produces different flavour profiles when the alcohol levels are changed

Creator of flavour

Meanwhile, alcohol has a highly significant influence in terms of extracting flavour compounds from the cask. Being soluble flavour compounds 'dissolve' when in contact with the spirit, then proceed along this liquid pathway to join the bulk of the spirit. But it's a process that should be considered in conjunction with temperature.

"Solubility is instantaneous, and the movement of flavour compounds from the staves is continuous, with the rate rising in warmer temperatures. Heat increases the rate of reactions, and an increase of 4-5 degrees centigrade in the temperature of the maturing spirit, for example, results in significantly greater movement of flavour compounds. An increase of 10 degrees centigrade generally doubles the rate of movement, with this usually the maximum temperature change in the spirit. However, while the air temperature can change quite dramatically and rapidly in an ageing warehouse the temperature of the spirit

within casks changes more gradually," says Brian Kinsman, Master Blender, William Grant & Sons.

The specifics of extraction are also influenced by the filling strength (alcoholic strength of new make spirit when filled into casks for ageing). This is typically 63.5% abv, reached by adding water to new make spirit which is usually around 70% abv. Consequently, spirit at a standard filling strength comprises 63.5% alcohol. The balance is essentially water, together with a tiny amount of flavour compounds. However, filling strengths vary and can even reach distillation strength.

"We see variations in maturation using different filling strengths. The highest has been around 70-72% abv which is distillation strength, but I think there's too much alcohol and not enough water at this strength. We've filled between 50-60% abv to see what happens, and there's not enough alcohol. Above 60% abv is a sweet spot, but it's very difficult to nail down exactly which strength is optimum, as it depends on the flavour profile you would like to

The alcohol extracts the flavour compunds from the wood during maturation

achieve," says Brian Kinsman.

One way of determining filling strengths is on the basis of cask type.

"I wouldn't fill a cask which is a first fill at more than 63.5% abv, this ensures a gentle extraction of flavour compounds and avoids the spirit ripping everything out of the cask in that fill. The first fill has to be gentle, if the filling strength is too high the spirit does too much, it's not a measured release of flavour. With a second or third fill I would nudge the strength up to 67% abv, enabling the spirit to extract slightly more. It's about attaining a balance between extracting the required flavours, and providing longevity for the cask. I've inherited this approach, we've done a lot of trials and revisited the results, and everything backs up this protocol," says Sandy Hyslop, Director of Blending and Inventory, Chivas Brothers.

The role alcohol and water each play in the extraction process depends on their affinity with individual flavour compounds. Opinions on this vary. One belief is that all flavour compounds are, ultimately,

soluble in alcohol and water, though quantifying where 'ultimately' lies on the solubility scale is hardly straightforward. Another belief is that water can extract all flavour compounds, with some being exclusively water-soluble. And then there's the case for alcohol.

"Ethanol is much more active in terms of extraction than water. Vanillin, for example, is more alcohol soluble than water soluble. Many tannins are water-soluble, but there is a range of tannins and more complex tannins come out in alcohol," says Stuart Harvey, Master Blender, Inver House Distillers.

A related factor is the chemical structure of a flavour compound, whether short-chain, which includes fruity esters, while medium-chain and long-chain flavour compounds provide progressively richer characteristics, such as certain cereal notes.

"Short-chain flavour compounds may be extracted more easily in water or lower strength alcohol, while longer-chain compounds are more soluble in a higher strength water-alcohol combination. The mix

BenRiach's master blender, Rachel Barrie, uses different bottling strengths depending on the cask type and the age

of water and alcohol ensures a range and balance of different flavours are extracted, and as the alcoholic strength changes in maturation, the range of flavours extracted from the cask may also vary," says Emma Walker.

Alcoholic strength typically decreases (in Scotland) by about 0.5% abv per annum in the first 15 years of ageing, then the rate slows down. Evaporation from the cask takes the credit for this decrease, though other factors are involved.

"When ethanol molecules react with acids there is a consequent loss of ethanol molecules, which contributes to a reduction in the alcoholic strength. However, the level is too tiny to measure. Ethanol and acid reactions also create a tiny amount of water, which also reduces the alcoholic strength, but again the amount is too small to measure," says Brian Kinsman.

Evaporation is a vital aspect of maturation, which refines and concentrates the flavour profile. Quantifying evaporation beyond this generalisation is tricky, as some aspects are understood, while others remain elusive.

"A higher abv means higher evaporation losses, but this doesn't increase the rate of development compared to a lower evaporation rate, it's a quicker route to a different result," says Sandy Hyslop.

Whether to bottle the resulting malt whisky at cask strength, or a pre-determined strength such as 40%, 43% or 46% abv is a significant decision.

"When I look at different bottling strengths for a new expression, which strength is the best balance varies depending on the cask type and the age. Alcoholic strength is another tool to be able to fine tune the flavour profile. It's a balancing act, the weight of flavour compounds in balance with alcohol, which amplifies the whole experience," says Rachel Barrie.

Malt whisky bottled at 40% abv, for example, comprises 40% alcohol and 60% water (together with a tiny level of flavour compounds). But even when alcohol is the minority partner it still displays 'alpha behaviour.' Alcohol molecules are larger than water molecules and consequently assume priority, arranging themselves in spherical configurations.

A key role for alcohol is to add texture and roundness to the bottled whisky

Water molecules subsequently slot into the remaining spaces.

"Water and alcohol are miscible and result in a uniform solution. However, combining 50 ml water and 50 ml alcohol doesn't produce 100 ml of liquid, it's slightly less as the water molecules fit tightly amidst the alcohol molecules, which reduces the volume," says Brian Kinsman.

Moreover, molecular arrangements have a particular significance.

"Alcohol adds weight and body to malt whisky, and I think it's the configuration of ethanol and water molecules that creates this impression," says Rachel Barrie.

While master blenders decide on the bottling strength, they can't stipulate 'drinking strength,' which is equally vital as the abv helps determine the characteristics a malt whisky shows.

"Alcohol is a delivery system for flavour, mouthfeel and roundness. At a higher strength the experience is determined more by the alcohol than the flavour, which is concentrated and intense. Reducing alcoholic strength takes heat away from the tongue and taste buds, making various flavour compounds more available to taste. For each individual there's a particular alcoholic strength at which a malt whisky works best," says Brian Kinsman.

Water or neat

Adding water is said to 'open up' malt whisky. I prefer to say it changes a malt. Flavour compounds come 'into' and 'out of' solution at various alcoholic strengths. A flavour compound 'in solution' is effectively 'dissolved,' rendering the flavour inactive. When a flavour compound is 'out of solution' it becomes 'solid,' and therefore discernible. Consequently, malts taste different at varying strengths, depending on which flavour compounds are 'in' or 'out' of solution.

"Reducing the alcoholic strength decreases richer flavours such as dried fruit which show more clearly at a higher strength, and shows more of the lighter compounds, such as floral aromas and citrus," says Stuart Harvey.

Inevitably, drinking strength is a trade off. I don't add water as I want to experience ultimate intensity, and I find undiluted malts tend to show an individual sequence of flavours. Mouthfeel can also be a revelation, and amazingly delicate even higher up the scale.

"Alcohol is akin to the backbone of a whisky, it can hold the whisky together. The higher the strength the firmer the grip," says Stephen Rankin, Director of Prestige, Gordon & MacPhail.

When I dilute it's out of curiosity to track changes. Diluted malts (in my experience) tend to have more 'integrated,' evenly-balanced flavours appearing simultaneously rather than sequentially. Mouthfeel also becomes progressively mellower.

Whether or not to dilute is of course subjective, but there are objective considerations.

"My opinion of American oak matured malt is that the alcohol dominates at 46% abv, so I recommend adding water until you reach the right balance for your palate. Spanish oak matured malts at the same strength need less water as the impact of the alcohol tends to be mellower, being concealed by a higher level of oak-derived characteristics, including tannins, chocolate and spice notes. Adding more water to a Sherry matured malt also entails the risk of it ending up unbalanced, or even collapsing," says Stuart Harvey.

Tasting malt whisky can be on a casual, purely pleasurable basis, or elevated to an academic exercise comprising techniques, an analytical mind and strict discipline (which is also a pleasure). Identifying and evaluating flavours is integral to the brain's work load, but the brain goes beyond what's in the glass.

"Alcohol can put you in a different frame of mind, and evoke memories, taking you to a place where you've previously enjoyed a particular malt whisky. In this way the brain intensifies the experience, making it much more than the flavour alone," says Sandy Hyslop.

Dr Bill Lumsden adds, "A physiological pleasure in the taste profile triggers something in the brain, and the overall experience is much greater than the sum of its individual parts."

Stuart Harvey, the Inver House master blender

The role of alcohol within a malt whisky is certainly complex, and trying to quantify it is ambitious. But this only makes it all the more intriguing and compelling.

Ian Wisniewski is a freelance drinks writer focusing on spirits, particularly Scotch whisky. He contributes to various publications including Whisky Magazine and is the author of eleven books, the latest being The Whisky Dictionary published in September, 2019. He regularly visits distilleries in Scotland, in order to learn more about the production process, which is of particular interest to him.

Picture of Ian Wisniewski courtesy of Finlandia vodka

Supermarket Sweep

by Neil Ridley

The budget 'Multiples' have certainly had a bumper year
when it comes to cleaning up at the plethora of whisky awards out there.
However, is this success justified and is the media hysteria surrounding
the wins in fact helping to potentially reinforce a growing
consumer disconnect in Scotch whisky?
Neil Ridley grabs a trolley and goes on a shopping spree.

I can't say I was surprised when I read the headline whilst sipping my morning coffee. It even raised a wry smile from my lips, as I pored over the article in more detail. It shouted, rather importantly, in a big, bold typeface:

£13.49 WHISKY NAMED
BEST IN THE WORLD

'Here we go again,' I thought, digging in, in earnest, reminiscing about practically the same article at a similar time last year.

The piece, in the Independent, a UK-based newspaper, was effectively a complete wet-dream-of-a-headline for both the author and the retailer of the whisky: on this occasion, it was the giant German discount store, Lidl, triumphing at the 2019 World Whiskies Awards. This astounding news was reinforced – neigh – galvanised with the cast iron fact that the supermarket's whisky, Queen Margot, an eight-year old blended Scotch had beaten the likes of the globally renowned Johnnie Walker Black Label, before going on to give a brief tasting note: "It contains dried apricot and plum, which the supermarket says

Lidl´s Quen Margot 8 year old blended Scotch scooped gold in the 2019 World Whiskies Awards

imbue it with 'a rich sweetness and depth of flavour' and is mellowed for eight years in oak casks."

Attention grabbing stuff indeed. However the only slight issue was that several of the key facts were particularly accurate.

In the ensuing days, many more articles followed, each one seemingly upping the ante in terms of spurious claims about the lofty accolade the whisky had scooped, before the collective whisky writing community stopped rolling its eyes, sat down at the computer keyboard and started to craft a response – by far the most compelling coming from drinks writer Billy Abbot on his excellent 'Spirited Matters' blog.

What really occurred is that Queen Margot was indeed triumphant at the aforementioned awards, but only in one particular category: the Scotch Blended Whisky – 12 Years and Under.

To put the win into some context: if this were the Crufts annual international dog show held here in the UK, it would be like a youthful Jack Russell entrant taking the 'Best Puppy' prize in the Terrier category: still a significant achievement, but some way off

winning the 'Best Of Breed' against Terriers of all ages and no where near taking the overall 'Supreme Champion' prize, where it would have had to beat off some pretty stiff competition from a range of pedigree Corgis, Akitas, Tibetan Matiffs, Chow Chows and Schnauzers etc.

Today, the article still exists online, but it has been dramatically post-edited; altered to reflect the true nature of the achievement.

Fast forward several months and I'm still thinking about what this all means for the whisky business in general. The story, as I mentioned, is strikingly similar to 2018 when Lidl again scooped gold, this time at the highly prestigious IWSC awards with its GlenAlba 23 year-old blend, which, according to the IWSC's awards website entry for the product, is produced by Whyte & Mackay. It also mirrors the achievements of the other German giant, Aldi, whose Glen Marnoch single malts from the Highland, Islay and Speyside regions all won silver medals at The IWSC awards back in 2017. Heady days indeed for the big boys then.

Taking nothing away from the whiskies themselves (I actually reviewed Queen Margot at the time of the win for my column in the Daily Telegraph and for a sub-£20 whisky it was surprisingly well-balanced,) I began to wonder if this success is translating within a new breed of consumer and quite what legacy the supermarkets hope to lay down within the category, alongside the combined thoughts of the brand owners and distillers – the actual producers of the whiskies themselves.

The response I got was not what I expected at all.

In fact, in 12 years of writing professionally about spirits, I have never received such a muted and guarded attitude than that exhibited by the main supermarkets mentioned above.

After contacting Lidl I was told categorically that the chain would be 'unable to respond' to any of my questions. Aldi also declined to answer them too, but did manage a token quote from Julie Ashfield, Managing Director of Buying, who explained that 'our own-label bottles often scoop medals at internationally renowned competitions and offer the same great taste as premium brands.'

A certain popular phrase, ending in the word '…Sherlock', springs to mind, but there you go.

For Anita Ujszaszi, Awards Director of the World Whiskies Awards and the person with the seemingly impossible task of organising, grouping, categorising and distributing the 700+ different whisky entries for the judging panel each year, there's no surprise as to why the larger multiples are finding fame at last.

'Supermarkets have experts at hand that source the very best within a certain price range. They work closely with industry experts and distilleries and due to the quantity they buy, can afford a large volume discount,' she explains.

Ujszaszi also predicts that this is really just the start of things to come for the supermarkets where surely we'll start to see a greater competition between more and more of the larger industry players each year.

'We see a diversity in blends as well as single malts coming, she points out. The judges are looking for well-produced spirits that not only fit the flavour profiles perfectly but have this little bit extra that makes them special.'

Supermarket whisky benefits all

One person who I was very keen to speak to about the new found success in this area was Richard Paterson, a man who has spent his entire illustrious career constructing some of the most globally well-received and awarded blends and single malts for the Whyte & Mackay group. He started by telling me just how far back the supply relationship between the supermarkets and distillery groups actually spans.

'Back in the early 80s, the 'supermarkets' didn't really exist and whisky company directors were all being tasked with 10% sales increases year on year,' he explains. 'Everything in the garden was rosy. Then in 1989, what happened… the industry overproduced and you ended up with the Loch of Whisky scenario, at the same time as Wine Lakes and Butter Mountains. So with all that extra Scotch sloshing around, the bubble burst. The supermarkets happened to be looking for liquid products, and whisky filled that category.'

'However, it started a price war with the brands,' he continues. 'Because of industry production cutbacks, the supply became limited and eventually started to dry up. That's when the supermarkets started to look for more long term supply agreements. Today, the whole thing is about genuine quality: from the packaging to the liquid itself, but its been a very slow, almost painful process.'

Loch Lomond Group Chief Marketing Officer, John Grieveson is a man who knows more than a thing or two about how to balance the stock supply from an existing proprietary brand portfolio – in Campbeltown's Glen Scotia and Loch Lomond – and in continuing to honour the group's highly successful supply agreements. I asked him for his thoughts on how he viewed the recent spate of trophy wins, especially if the supermarket products were beating his own products to the major prizes.

'The bottom line is that we welcome the success of the supermarket brands,' he explains. 'Anything that brings new consumers into the category is good news. We find that as consumers' knowledge grows they become increasingly interested in discovering other high-end whiskies such as our own award-winning malts. As in so many categories, a strong own label sector will push brands to raise their game and this can only be good news for the consumer.'

At the back of my mind, I wonder if there's any pangs of jealousy from either two men, when they see the potential PR impact of the supermarket own-label competition successes, but Paterson is in no doubt that it all adds to the greater good of Scotch.

'There's no frustration, as it highlights that 'Scotch' is the Ultimate Gold Medal Winning Spirit. Yes, there's increased competition but it means we all have to try that little bit harder. There's always wins and losses, but generally speaking, it's 'Scotch Whisky' which is winning and that's a key factor for all of us in the industry.'

Grieveson is also quick to point out how important the quality angle is for the long term success of all players. 'Our focus is on producing the best whisky possible for all of our customers. At Loch Lomond Group we have a unique capability to produce varying styles of new make spirit thanks to our combination of straight neck and swan neck stills. This flexibility allows us to meet the different needs of all our customers. Add to that that our own brands have been particularly successful in recent competitions, for example our Loch Lomond 12 and 18 year old expressions both received a Double Gold at the 2019 San Francisco World Spirits Competition.'

'For us it's not a problem at all,' adds Gordon Doctor, Operations Director for Ian Macleod Distillers: home of Glengoyne and the recently resurrected and much-loved Tamdhu, as well as a host of proprietary blends such as Smokehead. 'In my view, if it draws attention to the category and brings more consumers in to try single malt, we all benefit from this in the long term. In addition, if we were to supply a supermarket with an award-winning product then it tends to benefit us in terms of our contract and relationship with that supermarket,' he points out. 'Also please keep in mind that most prizes/trophies are categorised by price so we do not tend to see a £20 bottle of supermarket malt competing with a £40 branded product.'

Given that we all know just how successful the whisky category is doing globally right now (recent S.W.A figures again look very rosy, with the contribution of the Scotch Whisky industry to the UK economy growing by 10% since 2016 to £5.5bn and global exports reaching a staggering £4.7bn in 2018,) stock must surely become more scarce for the brand owner/supplier groups?

'Not really, as these are two very different things, continues Doctor. 'We do not use our brands for supermarket or customer bottlings. This stock has always been sourced on the open market. The success of our own proprietary brands presents a different set of problems. For example our single malts are all naturally coloured, so by its nature, this limits the availability of young stock.'

Has whisky lost its soul?

For me, there is potentially a worrying element emerging in the aftermath of all these awards wins, which feels like it could be the makings of a broader industry discussion. Yes, arguably, the profile they raise for the supermarkets may drive curious new consumers through the doors to try the products – if they can indeed find them – and then on into the world of other more well-known whisky brands, meaning a bigger win for the whole category.

However, the issue at the heart here is one of a growing divide between the bottom, 'budget end' and that of the very top, 'ultra-premium' element in whisky, compounded by the reaction of the mainstream media towards the entire whisky category.

It feels like there's a growing, lazy mass-sensationalism towards whisky at the moment, unlike, let's say gin, tequila and rum, where stories about the more positive lifestyle elements of the spirits comfortably and regularly appear in non-specialist newspapers and media outlets. In many ways, this positivity simply and clearly promotes the 'soul' behind the spirits and the open-handed nature of inviting new consumers in without any pretence or prior knowledge: be it the transparency of the ingredients or the botanical recipe particularly in gin, the person actually making the spirit, or the wondrous place it comes from.

However, when it comes to mainstream stories about whisky – mostly Scotch and Japanese – they almost always seem to focus on being 'the world's most expensive…', 'the world's rarest/exclusive/impossible-to-find' etc, or, as we've seen right at the start of this article, similarly sensationalist, non-substantial, inaccurate, attention-grabbing guff.

As someone who has regularly pitched more lifestyle-orientated whisky pieces into traditional media outlets, there seems to be the bizarre feeling from a number of commissioning editors that the public simply won't have an appetite for them, which I would say is a deeply questionable mentality. This in turn surely creates a growing potential 'elitism' in the spirit alongside a growing resentment from those readers and would-be consumers who feel they'll never truly be accepted into such an exclusive club.

Are we therefore entering an age where the divide between approaching whisky for the first time as a consumer and the 'perception' of whisky from a luxurious POV are poles apart? If so, how does the newbie find their way in these days?

'For me, the newbie is faced with a number of choices,' points out Whyte & Mackay's Head Of Whisky Experience, Daryl Haldane, a man coming at the category with an extensive background in brand advocacy and from the ultimate 'coal face' of presenting whiskies and new serves directly to consumers, bartenders and retailers.

'Budget is one that I recommend they establish quickly and the other is the purpose; are they looking to like whisky? Are they looking to feel extravagant? Are they looking to learn? Answering these questions will define where they should go to and what they should buy.'

Like the others, I'm keen to find out Haldane's

Richard Paterson of Whyte & Mackay has seen a change in the super markets´ buying behaviour from acquiring any surplus Scotch in the 1990s to the search for genuine quality today

A good way to make whisky more accessible for new consumers is to "have a whisky cocktail on every drinks menu in the world", says Daryl Haldane, Whyte & Mackay´s Head of Whisky Experience

thoughts on the growing presence and positioning of the supermarkets and their award wins.

'The big grocers have a great offering of single malts for a new drinker. It's clear though that high value bottles are sold in quite sophisticated environments. Niche products sold in specialist whisky retailers by great sales consultants, to those looking for something different offer a very different experience to a high-end luxury retailer where a 'boutique-esque' retail space is adopted.'

Is Scotch becoming elitist?

So how do we stop Scotch whisky from becoming too elitist and try to move the mass media opinion towards one of excitement about the spirit; one which focuses on its truly wonderful cultural elements and exquisitely unique flavour, as opposed to simply anchoring it to eye-wateringly high price points and extreme exclusivity – ultimately bringing more value and brand satisfaction back into the category?

'The growth of super premium malt whiskies is certainly raising the profile of the category and this is a good thing,' thinks John Grieveson, 'however,

it is important for brands to remember that there are multiple segments of consumers with varying tastes and differing needs. Keeping things accessible through simple, understandable language which focuses on flavour and character will draw people into the category. Not being prescriptive on how to serve a malt whisky but instead talking about individual preference is key,' he continues. 'Ultimately, make the category aspirational, but talk about craftsmanship – and avoid style-over-substance.'

'I don't think the industry is being elitist' feels Neil Boyd UK Managing Director at Iain Macleod, 'though we have to remember the issue of supply and demand particularly when dealing with a category that uses ageing as a differentiator. The industry has reacted to the demand for less expensive products by, in some instances, removing age statements and selling products at a younger age. In addition the industry has begun to focus more on the style of whisky rather than purely age, for example more smoky styles.'

'It's great that we're seeing more reports published about investment in distilleries and whisky tourism,' thinks Daryl Haldane. 'What I will say is that whisky has apples in many parts of the tree. Some nice and low, others right at the top. It's a complex

Photo: W Grant & Son

Neil Ridley wants the producers to focus more on the passionate people, the sense of place, the unique flavour and the wonderful cultural elements rather than highlighting exclusive bottlings at "eye-wateringly high price points"

category with many facets and the secondary market and soaring prices are just one of those in my eyes.

So if he could change the whisky business to make it more accessible for new consumers, what is the first thing he would you do? 'I would have a whisky cocktail on every drinks menu in the world,' he smiles. 'The way I see whisky is how I found coffee. Espresso is bitter and intense. At 18 I would have added milk, cream, syrups etc to soften that flavour. Whisky is quite similar as it is high ABV and the flavours are generally intense, especially for a new drinker. How can we ensure people have a great first experience with Scotch whisky? A starting point is to make it easier for people to drink it. Whisky is, after all, a learned flavour and it can take time; mixing is the simplest route to appreciating all the unique flavours whisky offers.'

Award wins or not, the current supermarket offerings in whisky have certainly given the rest of the industry something bigger to think about. Whilst its not at all likely that their intention is to challenge the specialist retail sector, it has certainly opened up the debate about the origins of whiskies which are gaining such high regard. One thing almost certain in today's global marketplace is the modern consumer's desire for transparency, alongside genuine value for money and it's potentially an area where we could see a greater level of effort and understanding from the major multiples when it comes to openly discussing their sourcing with the consumer.

For me though, the fundamental area where the industry needs to do its homework is how best to articulate the finer points of whisky to the wider media, because clearly, something is going wrong when its best attributes: namely quality ingredients, truly passionate people and a genuine sense of place are being sidelined by excessive pound signs and perceived rarity by many influential headline writers.

Neil writes about whisky and other fine spirits for a number of publications globally, including The Daily and Sunday Telegraph. He is the former chairman of the World Whiskies Awards and regularly presents a drinks feature on the popular TV show, Channel 4 Sunday Brunch. His first book, (written with Gavin D. Smith) 'Let Me Tell You About Whisky' was published in 2013 and since then, he has co-authored 'Distilled', with Joel Harrison, which is now printed in thirteen languages, winning the Fortnum & Mason Drink Book Of The Year in 2015. His latest book, 'Straight Up...' celebrates the finest drinking experiences on every continent.

HIGHLANDS
SPEYSIDE
HIGHLANDS
ISLAY
LOWLANDS
CAMPBELTOWN

Whisky Regions
~ are they still relevant?

by Charles MacLean

In whisky bars, Scotch are generally presented
according to the region where they were produced. We also like
to talk about "a typical Islay" or a "classic Speyside". But where does
all this come from and is it at all relevant today
to group Scotch geographically?

The first regional division was defined by the Wash Act 1784, which drew a notional diagonal line across Scotland between Dumbarton on the Clyde and Dundee on the Tay, with different duties and other legal requirements applying to distilling above and below the 'Highland Line'. The Highland region was adjusted twice, then abolished in 1816, and although the provisions of the Act satisfied neither Highland nor Lowland distillers, they reflected the differences between the way whisky was made in the two regions, and the impact these practices had on flavour.

In the Highlands, malted barley was used almost exclusively; the more fertile Lowlands made it possible to use mixed grains, and thus avoid the onerous tax on malted barley. As Adam Smith noted in The Wealth of Nations (1776) in relation to Lowland distilling: "In what one called malt spirits, it [the malt] makes but a third of the raw materials; the other two thirds being raw barley or one third raw barley and one third wheat".

An Ammending Act in 1786 assessed the duty payable according to still capacity, based on estimating the quantity of spirits a still might be expected to make in a year. The House of Commons Committee which set the tax, informed by declarations made by the leading London distillers, "settled upon the supposition that stills could be dischanged about seven times a week".

It was clearly in the distillers' interest to speed up the rate of distillation. Shallow stills with broad, saucer-shaped bases and tall heads, used in conjunction with stronger and thicker washes, were developed by the larger Lowland distillers. The London distillers complained to the Lords Commission of the Treasury that "the Scottish distillers had, by the ingenuity of their contrivances, found means to discharge their stills upwards of forty times a week".

Distilling at such a rate makes for little or no reflux and permits no interaction between the alcohol vapour and the copper sides of the still, which is essential in removing unpleasant, even harmful, empyrheumatic flavours in the spirit. In short, Lowland whisky was rough and fiery and, at best, fit only for punch or rectification into gin (which much of it was).

Among the provisions of Wash Act and its successors was a ban on licensed Highland distillers selling their product in the Lowlands, but given the poor quality of Lowland whisky, it was much sought after, and fetched a higher price. Made from weak mashes of malted barley, in small stills which were operated slowly, it was cleaner and more flavourful.

Blending

Until the invention of continuous stills in the late-1820s all whisky was made in pot stills by a batch process. Now very pure, high strength, spirit could be produced, cheaper and at a furious rate, but because the process stripped out flavour-bearing congeners, it was not appropriate for malt distillation, although ideal for grain whisky, and because the product was blander, licencensed grocers and wine & spirits merchants began to explore the blending of malt and grain whiskies.

Until Gladstone's Spirits Act 1860, duty had to be paid prior to blending; now both malt and grain whisky could be held and blended under bond.

"Blending was initially intended to reduce the cost of a well-flavoured whisky, but the ability of a skilful blender to produce a consistent spirit, bringing out the best characteristics of the component malts and grains, proved a decisive factor in the industry's development" [Professor Michael Moss, The Making of Scotch Whisky (1981)].

Blended whiskies can be made to be less highly flavoured than single malts, and could thus be designed to have broader appeal beyond Scotland. Furthermore, and crucially, the flavour-profile can be maintai-

ned from batch the batch: without consistency you cannot brand and promote individual products.

By the mid-1860s spirits merchants and blenders were sending their own casks to be filled at distilleries of their choice – a practice which soon became standard – in order to have some control of the quality and consistency of the whiskies they required for their blends.

Indeed, it might be argued that it was the blenders who determined the character and quality of malt whisky. When the owners of Cardhu Distillery (then Cardow) sold to their principal customer, John Walker & Sons, in 1893, Alec Walker (then aged 26; later Sir Alexander Walker) camped in the distillery and sent daily reports on the quality of the spirit to his brother, George, reporting on spirit character and adjusting the process until he was happy with it.

Henceforward the fortunes of both malt and grain distillers were increasingly tied to the blenders' needs.

Quality and consistency

At the height of the 'Whisky Boom' of the 1890s, many blending houses built distilleries to meet their specified needs: in all thirty-three new distilleries were commissioned during the decade.

By this time the leading blenders were categorising their 'filling' whiskies by region which varied from blender to blender, but only slightly. The four acknowledged regions were: Highland, Lowland, Campbeltown and Islay. To which, of course, was added Grain whisky.

Although, as Tom Bruce-Gardyne reminds us in his article on www.scotchwhisky.com: "These whisky regions meant nothing to your average Scotch drinker, and existed purely in the minds of the blenders".

They also graded the malts for their usefulness as blending whiskies into 'Top Class' (or 'First Grade'), 'First, Second and Third Class/Grade'. Scottish Malt Distillers (now Diageo) listed the following as First Grade in 1954, and other blenders' lists I have seen are similar:

Aultmore, Benrinnes, Cragganmore, Glenfiddich, Glen Grant, Glenlossie, Glenrothes, Linkwood, Macallan, Mortlach.

I would emphasise that this ranking was for blending purposes. It was not a measure of single malt quality. Interestingly, the term 'single malt' was not used until the late 1970s: single whiskies were described as 'Pure Malt', 'Straight Malt', 'Unblended' or simply 'Highland/Lowland/Islay Malt'.

As well as regional differences/usefulness for

Single malt from Linkwood (picture taken before the latest upgrade) was, and still is, considered Top Class by blenders

blending, 'quality' evaluation was reflected by the slight differences in the cost of new make spirit and its insurance value.

I have a Harpers Whisky Calender 1924 on my wall, which lists trade prices for new make spirit (per gallon):

Highland Malts 6/6d to 4/6d
[30.5p (Royal Lochnagar) to 20.5p]

Lowland Malts 4/3d to 4/-
[20.2p (Rosebank) to 20p]

Campbeltown Malts 4/6d to 4/-
[20.5p (Hazelburn) to 20p]

Islay Malts 6/3d to 5/6d
(30.1p (Laphroaig) to 25.5p]

Grain whiskies 2/9d to 2/6d
(10.75p (Caledonian and Cameronbridge) to 10.5p]

Happy days!

Speyside region

You may have noticed that the malts listed as First Grade above are all from the region we would now term 'Speyside'. In fact this was only legally re-cognised as a region separate from 'Highland' by the Scotch Whisky Regulations 2008. Indeed, the word itself was not widely used. In Whisky (1930) Aeneas MacDonald remarks:

"…it is impossible to discover any reasonable geographical basis for the district-names assumed by different distilleries. No amount of research can determine why some of these are called 'Strathspey' whiskies and others 'Speyside'. Both are equally beside the Spey and in Strathspey. And it is not as if the names are interchangeable; they are jealously guarded".

The malt whisky distilled illegally in the Glenlivet district had long had a high reputation for quality, but it was not until the late-1800s that the region came into its own: of the thirty-three distilleries commissioned during the 1890s, twenty-six were on Speyside. Over twenty-five distilleries adopted the name as a suffix to their own (Macallan-Glenlivet, Aberlour-Glenlivet, Cragganmore-Glenlivet, etc.), some of them a very long way from the Glen itself – Glenforres-Glenlivet (now Edradour) was near Pitlochry, ninety miles from Glenlivet! In other words, the appelation was broader than the place: it indicated a style of spirit character – sweet, estery, only lightly peated – and this was just the style of malt required by the blenders.

There were several reasons for favouring the Glenlivet/Speyside region as well as its historic reputation. Once remote, by the late 1800s the region was well connected by rail, north to Elgin and

The Speyside region grew in importance when the railway network expanded

Inverness and south to the Central Belt. There was a deep reserve of distilling expertise in the district. High quality barley was available from the Laigh o'Moray; pure water and peat were abundant.

Before it was eclipsed by Speyside, Campbeltown was described as 'the whisky capital' – in the 1890's journalists named it 'Whiskyopolis'. There were around thirty distilleries here, and the sea route to Glasgow facilitated transportation. But the Campbeltown malts were highly variable and blenders increasingly turned against them; there was even a rumour that ex-herring barrels were being filled, giving rise to them being termed 'stinking fish', but this calumny might have been generated by Speyside distillers!

Terroir

The very fact that blenders divided their malts by region surely implies differences in flavour from region to region. Until the 1980s writers and distillers believed that spirit character was determined by four factors.

The first and most important was the water source – pure, copious, preferably soft; 'flowing through peat and over granite'. In the 1950s Bill Smith Grant, owner of The Glenlivet Distillery, famously told Time Magazine, when asked to account for the high reputation of his whisky, that it was "99 per cent the water and a certain fiddle-faddle in the manufacture"!

Next was air quality. The vast majority of malt distilleries are rural, "far from smoke and dust… Uplands cleanse the winds which sweep across them and impart some of their own purity to the whiskies of hallowed name," writes Aeneas MacDonald, "Nor is it without significance that about fifty per cent of the Scottish distilleries stand within a few miles of the sea. There seems to be some obscure association between the qualities of sea air and the whisky produced in it".

Third was peat. All Highland and Island distilleries dried their malt over peat fires (including those on Speyside, apart from Glen Grant and Strathisla, which used a mix of peat and 'silent coal' [i.e. coke]) which imparted varying degrees of smokiness, by design. Lowland malts were unpeated. Peats from different parts of the country have different vegetable compositions and thus impart different flavours. Around 2000, Highland Park, one of the few distilleries to retain its own maltings, experimented with buying in malt from the mainland, ready-peated to precisely specified levels. However the spirit character was discernably different, owing to the different composition of Orkney and mainland peat.

Until the 1980s, the water was considered an important factor giving the whisky its flavour

Finally: barley. But Aeneas MacDonald admits in 1930 that "It is a little too late in the day to pretend that all whisky malts are made from barley grown in the district where the distillery is situated". He goes on to mention 'the proprietor of one well-known Speyside distillery' who had "examined whisky made from Californian, Danish, Australian, English and Scotch barley and had not been able to detect any very marked difference between them. Still, he confessed that…he preferred to use a home grown barley from the Scots seaboard lying between the rivers Deveron and Ness".

Regional differences today

It will be noted that no mention is made in the classic sources of any influence that barley varieties or yeast might have, let alone production process and maturation. Since the rediscovery of the virtues of single malts, which gathered momentum in the 1980s, much work has been done in exploring the scientific basis of flavour, especially in regard to the influence of copper and oak-wood.

Copper is a purifier and strips out undesirable compounds, especially sulphur, in alcohol vapour. So the way the stills are operated – charged high or low, run hot or cool, cut points, rested between charges, fitted with worm tub or shell and tube condensers,

etc. – will determine to a high degree the character of the spirit. It goes without saying that these factors have nothing to do with the regional location of the distillery.

In his magesterial 600 page application of Victorian science to distilling, The Manufacture of Whisky and Plain Spirit (1893), J.A. Nettleton makes no mention whatsoever of maturation and the influence of wood in developing flavour. Today, as readers will be aware, the type of casks used – American or European oak, sherry, bourbon or wine seasoned, first fill/refill/rejuvenated, etc. – is of paramount importance. Some distillers claim that 85% of the flavour in mature whisky comes from the casks.

Again, the beneficial effects of cask maturation is independent of the casks' location. Some claim that the style of warehouse (dunnage or racked) and its geographical location (especially if this is close to the sea) will influence the mature character. However, most chemists maintain that even if this is the case, the influence is tiny compared with other factors (spirit character and wood type).

In short, with greater understanding of what determines flavour in malt whisky, distillers are able to create the flavours they want or need for their blends irrespective of the distilleries' location.

So where does that leave terroir and regional differences?

One aspect of the popular term 'terroir' is often neglected - the people who craft the whisky

In his excellent World Atlas of Whisky (2010) Dave Broom looks at the topic from an unusual angle. He writes:

"The more you talk to distillers and ask them how their character is created, the more they talk about fermentation times and reflux and oxidation and still shapes and copper conversation… and then they shrug and say 'to be honest, we don't know'.

"In other words, there is something that happens at each distillery which helps to set it apart… while you can set up a still to make grassy or fruity or peaty whisky, you can never dictate the specific flavours the the distillery will produce. That's akin to the type of terroir you encounter in Burgundy…

"The one element of terroir which is always overlooked is the human element. There is a cultural terroir at work here – the founding distiller's approach being dictated by his own personal preferences, by the smells which he inhales every day, by his use (consciously or unconsciously) of elements within his landscape. These whiskies taste this way because of the people who first crafted them."

Is this too romantic? I think not, although I would qualify the idea of 'cultural terroir' by saying that the spirit character at each distillery, and indeed within each region, was originally (and continues to be) driven by the requirements of the key customers – i.e. the blenders. And the last thing blenders want is a change in spirit character.

The continuing success of single malt whiskies may well be shifting 'key customers' to consumers. This is particulary the case for recently commissioned distilleries, few of which are supplying spirit or mature whisky to blenders, and which are free to create the style (and styles) of whiskies they choose – within the constraints mentioned by Mr. Broom above – irrespective of their location.

Ironically, the use of regions to distinguish differences between malt whiskies was also consumer driven. Not only did it provide a useful way to divide up a book's contents – Wallace Milroy's Malt Whisky Almanac and Michael Jackson's World Guide to Whisky (both 1987) are early examples – it helped consumers realise that single malts were not all the same, and pegging them to regions made this clear, especially to those familiar with the wine regions of France.

4

United Distillers' Classic Malts Collection (1988) must be credited for opening up the whole sector with a range of six single malts which stressed regional differences – Lagavulin (smoky Islay), Talisker (spicy Island), Oban (maritime West Coast), Cragganmore (complex Speyside), Dalwhinnie (typical Highland) and Glenkinchie (lighter Lowland).

U.D.'s predecessor, the Distillers Company Limited, had made a half-hearted attempt to launch a collection of malts in 1982 which they named 'The Ascot Malt Cellar' (Ascot was the company's home trade base). It comprised Rosebank (Lowlands), Linkwood (Speyside), Talisker (Island), Lagavulin (Islay), Strathconnan (vatted malt) and Glenleven (vatted malt), but was done reluctantly and not widely promoted.

Other companies followed U.D.'s lead. Allied Distillers launched its 'Caledonian Malts' collection in 1991 – Tormore (Speyside, later replaced by Scapa (Orkney)), Miltonduff (Speyside), Glendronach (Highland) and Laphroaig (Islay) – but pulled it in 1994. That year Seagram launched their 'Heritage Selection' – Longmorn, Glen Keith, Strathisla and Benriach (all Speysides) – but this did not last.

Useful though they may be in shelf stacking and the arrangement of books, tastings and competitions, the Scotch whisky industry is generally distancing itself from regional differences and encouraging consumers to focus on 'flavour'.

Mark Thomson, Glenfiddich's U.K. Ambassador, tells me:

"People, thankfully, are moving away from listing whiskies by region. Whisky may have at one point been simple to classify, but it's all becoming too complex with variants popping up from this distillery or that (which I applaud) - but we really do need to stop pigeon holing a distillery style just because of its geographical location. No customers walk into a bar and ask for a whisky that only uses waters from granite hillsides in the production…."

In conclusion, there are differences in the flavour of malt whiskies from one part of Scotland and another, for historical reasons and driven by blenders' requirements. These differences are by no means rigid and may be varied: regional styles may be replicated to a degree, beyond their original region – many distilleries are now producing peated variants of their traditionally non-peated spirit.

Notwithstanding this, the regional concept is still useful way of demonstrating that malt whiskies vary in flavour, especially to consumers unfamiliar with the category.

Charles MacLean has been writing about Scotch whisky for 38 years, and has published seventeen books on the subject. He was the founding editor of Whisky Magazine, writes regularly for several international magazines and is a frequent commentator on TV and radio. He was chief presenter of the TV channel www.singlemalt.tv and starred in Ken Loach's film The Angels' Share, which won the Jury Prize at the Cannes Film Festival in 2012. In 2009 he was elected Master of the Quaich, in 2012 won the I.W.S.C.'s 'Outstanding Achievement Award', in 2016 was inducted into the Whisky Hall of Fame and in 2019 was named 'International Ambassador of the Year' at the Spirit of Speyside Festival.

The Classic Malts Collection, based on regions, opened up the entire sector of malt whisky to consumers

Innovation within boundaries

by Joel Harrison

From the liquid inside the bottle, to the bottle itself,
the world of whisky is awash with innovation despite heavy
regulation in some quarters. Joel Harrison looks at how companies
innovate within ranges, and with entirely new products,
to keep the industry both engaging and exciting
for existing and new customers alike.

Sitting in one of those 'this-could-be-a-budget-hotel' airport lounges that litter the airports of Europe and America, a mid-range whisky and soda in hand, I'm flicking through one of the many high-end magazines on offer. You know the sort, the ones which showcase a range of lifestyle products, from the latest watches (some are genuine lifesaving devices for divers and pilots; others the horological equivalent of Patric Bateman's business card, but for middle management), through to cars, suitcases, and, a lot of the time, bottles of whisky.

As I take in the descriptions of some of the non-liquid items, between the obviously copy-written aspirational platitudes cribbed from press releases by a journalist short on both time and imagination, there is, on the whole, a genuine sense of innovation.

For example: luxury luggage. Aside from using strong and lightweight aluminium casing, you'd think that so long as luggage has the basic ideal of locking, and keeping your underwear safe from spilling out across the baggage carousel at your destination airport (something that happened when the zip was invented over 100 years ago) innovation in luggage was either unnecessary or impossible. Or both.

The collaboration between Macallan and Lalique Crystal is a good example of packaging innovation and, not least, brand building

Yet take the latest German-made luxury suitcases, the sort being wheeled along by creatives looking as if they'd just left a photoshoot for Monocle Magazine. These now feature USB charging slots to make sure your phone/tablet/headphones/ereader/ecigarette is never out of charge. Some come with GPS trackers, meaning you can see that your luggage is still at Heathrow, when you've arrived in Aberdeen. And a few even have 'electronic paper' screens to which you can upload your bag tags, designed so your bags aren't lost and you don't need to check your phone to see that your bag is still inside the M25, when you're not. Something which you can't do anyway, as you can't charge your phone because you've lost your bag with the built-in to it USB charger…

What this does prove is that innovation is alive and well in the world of luggage design. So too, when it comes to watches and cars, but not really in whisky. Of the portfolio laid out in these lifestyle journals, whisky seems to be the one left behind to focus on history and heritage, packaging and bottle design, or left to lean on a celebrities from the world of sport, fashion or film.

A new side of Scotland

Part of this is down to the regulations that ties Scotch, in particular, and whisky in general, to a certain set of standards, setting it apart from rum ('lies, damn lies and rum labels', as the saying goes), while giving it some flexibility, meaning it is thankfully not as rigid as cognac.

One brand that has pushed the boundaries within the rules of Scotch whisky, is The Macallan. Eschewing a standard distillery expansion which defies the traditional ideal of the white-washed walled, oft-used cookie-cutter image of a distillery shrouded in the morning mist as a stag takes full advantage of his native 'right to roam', the sort of thing that might be the perfect fit for the front of a chocolate box (or, more aptly, a shortbread tin). They have instead opted for something reflective of a modern, 'Twenty First Century' Scotland: a Scotland that has a thriving digital economy; one that nurtures artists and designers and Michelin-starred chefs; one with forward-thinking architecture such as the new V&A museum in Dundee.

It was this side of Scotland that inspired Ken Grier, recently retired from his role as Creative Director at

The unusual setup of equipment at Loch Lomond Distillery make way for versatile distillation and a variety of different styles

The Macallan and now working as a consultant, to think outside the box when it came to both the new distillery itself and wider innovation in the Macallan portfolio; an ideal that saw the brand grow to become the biggest Single Malt by value in the world.

"Great innovation that I have been involved in really changed the face of the market to the benefit of the brand and its cachet," Grier notes. "The Macallan in Lalique did this for the secondary market and The New Macallan Distillery did it for whisky visitor experiences".

But good innovation isn't easy, so what was the key for Grier while at The Macallan?

"Good innovation is a combination of a number of important components. It needs a rich understanding of your core consumer's life", he says.

"It must have a feel for the macro-cultural zeitgeist; a deep understanding of your brands world and how it emotionally connects with drinkers; a distinctive and heart pumping idea; strong economics with powerful content; presence and liquid to lips execution; and a peerless introduction to the market."

The ideal that innovation is not just about having a

great idea, it must be supported through great liquid, too is also key.

"The relationship that I have had with Bob Dalgarno [the former whisky-maker] on The Macallan, for instance, has been tremendous. Bob is a real genius, with a huge knowledge of the whisky-making process, and wood management", Grier comments.

"We always had a really challenging and fun dialogue about how to produce outstanding liquids that made distinctive propositions magical and were newsworthy in their own right".

Now both working as consultants with The Glenturret distillery, Grier says the pair "aim to continue that, with great innovation with a superb and differentiated liquid at its core".

Doing things differently has always been in the DNA of the Loch Lomond distillery. Equipped with a range of different still shapes, enabling them to produce a kaleidoscope of spirit styles, Chief Marketing Officer John Grieveson makes sure any innovation they have is rooted in their unusual spirit styles.

"At Loch Lomond innovation has always been

at the heart of our distillery. Different spirit stills, including our straight neck stills, and varying collection strengths provides different flavours of new make spirit. This makes innovation easier and provides a broader pallet for our Master Blender to work his magic", he notes.

For Grieveson, innovation is an important element of the whisky business.

"Consumers are increasingly buying across categories and therefore Scotch Whisky drinkers are exposed to innovations in multiple categories", he continues.

"This puts pressure on Scotch Whisky to remain relevant in the face of increasing competition. In simple terms, innovation keeps the category fresh. It is a way of educating consumers and acts as a magnet to keep people close to your brand".

Recently, the Scotch Whisky Association (SWA) has moved to relax some of the rules and regulations regarding the maturation of Scotch whisky in more exotic and unusual casks, opening up the category to greater liquid innovation. But would a further relaxing of the rules help or hinder Scotch?

"This is a double edged sword," says Grieveson. "On the one hand the rules of single malt help to protect the provenance and credibility of the category. On the other hand, Scotch Whisky is competing with other categories where the rules are less tight, enabling greater flexibility for innovation. It is important for producers to continually debate and challenge the scope of the regulations to ensure the category remains relevant".

Grier agrees, saying that, "it is a highly prestigious category that should be carefully curated, ensuring only the highest quality liquids are purveyed to keep us at the apex of world spirits. However, recent steps to take a more relaxed view on the use of wood from former tequila casks and other spirits I think are positive".

Flexible regulations in Ireland

One area which boasts a more relaxed attitude towards the maturation of whiskey is Ireland. While still having to measure their products against the wider European legislations that allow them to use

The Method and Madness range was created to show the breadth and diversity of Irish whiskey

Brendan Buckley, marketing director at Irish Distillers, was one of the driving forces behind Method and Madness

the term 'whiskey', greater creativity is allowed.

"Granted, the regulations governing Irish whiskey are slightly more flexible than Scotch," says Brendan Buckley, Marketing Director at Irish Distillers, whose background is in innovation, both reviving classic Irish whiskey brands such as Red Breast and Green Spot, and developing new-to-world ranges like Method and Madness.

"Particularly as it relates to the kinds of barrels in which we can mature our whiskeys so we have been able to innovate with wood types in order to deliver new, interesting taste profiles. For example, as part of our Method and Madness range we released a whiskey which we finished in barrels made from Chestnut, imparting a unique and appealing flavour".

Despite these relaxed ideals, Buckley admits that there is a "distinction between 'innovation' and 'successful innovation'. Lots of brands can do innovative stuff, irrespective of the rules of a particular category, but the real challenge is in bringing something new to the world of whiskey which brings genuine value - to the consumer, to our customers and to the brand".

The Irish whiskey category has been through huge changes over the last decade, with one flagship brand, Jameson, dominating the market, just a handful of distilleries producing liquid, while indecent bottlers have been reliant on small parcels of mature stock. This has thrust innovation to the fore across the whole of the Irish whiskey market, making it vital to the growth, health and wealth of the category.

"Up until quite recently, the Irish whiskey category was dominated by a handful of brands, with one brand very much to the fore; Jameson. As category leader, Irish Distillers recognised that there was an opportunity and indeed an obligation to show that there was much more to Irish whiskey than these leading brands and started investing behind other brands such as Redbreast, the Spot Range, Midleton Very Rare and latterly, Method and Madness, in order to demonstrate the breadth and diversity that Irish whiskey can offer," Buckley states.

"In particular, we made a concerted decision to raise awareness for single pot still Irish whiskey which is widely accepted as the definitive style of Irish whiskey and hence, we have innovated widely in this area," he continues.

Shinji Fukuyo, Suntory's chief blender and Fred Noe, master distiller at Beam created the new Legent bourbon

"The arrival of new entrants to the category is to be welcomed as they too can add to the canon of choice."

This remarkably open-armed opinion, one which sees Irish Distillers also run a Whiskey Academy for anyone wanting to learn about the production and history of whiskey in the Emerald Isle, underscores an understanding that within Irish whiskey there is huge opportunity for growth; this is evidenced by the raft of new distilleries opening up, as well as those creating independently bottled blends with existing mature stocks. As such the rising tide which keeps many good boats afloat, also brings a lot of driftwood to the surface too. Their aim is to educate and inform, in order to ensure that the port is filled with good quality seafaring vessels, not a load of old junk boats.

Widening the net away from Scotch, and Irish whiskey to a more global view of innovation, there has recently been some groundbreaking new products which meld together the whisky-making expertise from a number of established whisky-producing nations.

5

Richard Bates, Senior Director of Consumer Strategy & Innovation at Beam Suntory, has experience with two key releases in the last year, Ao and Legent, which look to break new ground when it comes to innovative whisky products.

Ao is a world whisky blend, bringing together mature liquid from Irish, Scottish, American and Japanese whisky distilleries, with the aim of creating a mixable, global whisky.

Legent focuses on just two of these components: American bourbon distillation with the art of Japanese blending.

"Legent combines our Japanese blending skill and finishes, and applies that in bourbon. It blurs the edges and plays with your perceptions a bit," Bates says.

"It was a pure partnership between Fred Noe and Shinji Fukuyo. They're both passionate about what they do and it is our business coming together; East and West. It is the first whisky articulation of that."

"If you think about it, it is a really simple idea," Bates says with a smile.

"There is depth to it, but it is just really simple. It's that Japanese focus and eight generations of bourbon making coming together. It is a lovely reminder that whisky is a very human thing."

And this is the key to whisky; it has a human connection, be it through time, through place, through flavour, or simply through the ideal that it has been made by people in a time-honoured way. As much as it has become an iconic status symbol, like a good watch, it is still a product with a huge emotional attachment. More than just a piece of luggage to keep your belongings safe while you travel.

Bates is also adamant about this side of whisky. "That's why whisky innovates; it keeps people interested, it keeps people engaged", he says passionately.

But this emotional contract with the consumer is also a delicate one, and one that was under threat with the heavy focus on 'no age statement' products over the past five years or so, not just in the world of Scotch, but the wider industry too.

No one could have anticipated the global growth and appeal of Japanese whisky so it is not surprising that many in the category had to move away overt age stated products.

However as Bates' comments, "for Suntory, quality is everything. We don't take shortcuts. So the business had to think carefully about how it addressed this demand."

"Scotch has had the same challenge but seems to be coming out the other side and coming back to age now," Bates points out.

"The interesting thing is that no age statement products are always a challenge for whisky, because the Scotch industry has done a wonderful job educating consumers about age. Age and price have become an easy guide to quality. Brand, age and price: you put the three together and that's what many consumers are using as a guide. So to make a non-aged innovation appealing you have to think differently "

In Scotch, it opened the door for wider innovation in both packaging and story telling. Grier recalls that no age statement "gave access to tranches of great Single Malt stocks for manufacturers and consumers that previously were not used in products because of their age, not their quality".

He notes that no age statements "gives whisky makers a greater colour palette to play with using whiskies that are aged for depth, and younger for liveliness, to make really interesting vattings. This has opened up more avenues for interesting innovation beyond age. It is also like picking an apple from the tree, you use that whisky when it is at its peak, and playing the right role in a vatting rather than just meeting an age statement criteria."

Across the world, we are seeing a greater number of whisky releases from all quarters, be it Scotch, the Grand Old Man of Whisky, through to innovative Irish products, interesting additions from the focused and diligent Japanese, or crazy new world products from Australia, Taiwan and even England.

Yet as whisky drinkers we all know that, even with the most creative of innovations, there is a quality level attached to any bottle labelled 'whisky', any bottled labelled Irish whiskey, and any bottle that reads that famous word, 'Scotch'.

Whisky is not a watch. It is not a suitcase, a car, a pair of headphones. It is a product with real connection, real emotion and real passionate. And we all know that despite, or most likely because of innovation, the future of whisky being the same today, tomorrow and forevermore, is safe.

Joel Harrison is an award winning author, communicator and industry consultant, whose work has been published in over 20 countries, across 16 different languages. His writing work can be seen in publications such as The Wall Street Journal India and The Daily Telegraph in the UK. Harrison also appears regularly on British television across a number of shows as a whisky specialist. He sits as a judge for the International Wine and Spirits Competition (IWSC) where today he holds the role of a Trophy Judge and Chairman across Scotch whisky and other spirits. In 2013 Harrison was made a Keeper of the Quaich.

Talisker Distillery

Malt distilleries

Including the subsections:
Scottish distilleries | New distilleries | Closed distilleries
Japanese distilleries | Distilleries around the globe

Explanations

Owner: Name of the owning company, sometimes with the parent company within brackets.

Region/district: There are five protected whisky regions or localities in Scotland today; Highlands, Lowlands, Speyside, Islay and Campbeltown. Where useful we mention a location within a region e.g. Orkney, Northern Highlands etc.

Founded: The year in which the distillery was founded is usually considered as when construction began. The year is rarely the same year in which the distillery was licensed.

Status: The status of the distillery's production. Active, mothballed (temporarily closed), closed (but most of the equipment still present), dismantled (the equipment is gone but part of or all of the buildings remain even if they are used for other purposes) and demolished.

Visitor centre: The letters (vc) after status indicate that the distillery has a visitor centre. Many distilleries accept visitors despite not having a visitor centre. It can be worthwhile making an enquiry.

Address: The distillery´s address.

Tel: This is generally to the visitor centre, but can also be to the main office.

Website: The distillery's (or in some cases the owner's) website.

Capacity: The current production capacity expressed in litres of pure alcohol (LPA).

History: The chronology focuses on the official history of the distillery and independent bottlings are only listed in exceptional cases.

Tasting notes: For all the Scottish distilleries that are not permanently closed we present tasting notes of what, in most cases, can be called the core expression (mainly their best selling 10 or 12 year old).

We have tried to provide notes for official bottlings but in those cases where we have not been able to obtain them, we have turned to independent bottlers.

The whiskies have been tasted either by Gavin D Smith (GS), a well-known whisky authority and author of 20 books on the subject or by Ingvar Ronde (IR). All notes have been prepared especially for Malt Whisky Yearbook 2020.

Aberfeldy

[ah•bur•**fell**•dee]

Owner:
John Dewar & Sons
(Bacardi)

Region/district:
Southern Highlands

Founded: 1896
Status: Active (vc)
Capacity: 3 400 000 litres

Address: Aberfeldy, Perthshire PH15 2EB

Website: aberfeldy.com
Tel: 01887 822010 (vc)

Aberfeldy has increasingly carved out a place in the single malt world but it still remains an important part of one of the top blends of the world – Dewar´s which is the best selling Scotch whisky in the American market.

The core expression from Dewar´s, White Label, is complemented by 12, 15 and 18 year olds. In spring 2019 the owners launched Dewar's Double Double blend. A four step, marrying process built on the routine introduced in 1901 by Dewar´s first master blender, AJ Cameron, has been used. The limited range features three expressions aged 21 to 32 years and finished in three types of sherry casks; oloroso, palo cortado and Pedro Ximenez.

The equipment consists of a 7.5 ton stainless steel, full lauter mash tun, eight washbacks made of larch (two of them replaced in February 2019) and three made of stainless steel with an average fermentation time of 70 hours, and four stills. With an additional washback, installed in 2014, production has now escalated to 23 mashes per week (working a seven-day week) and 3.4 million litres of alcohol. The owners have also invested £1.2m in a biomass boiler and are currently working on an upgrade of the effluent treatments plant.

Aberfeldy single malt has always been a significant part of the Dewars blended Scotch but since the relaunch in 2014, it has also been the fastest growing single malt in the duty free segment. The distillery also boasts an excellent visitor experience which attracted more than 40,000 visitors last year.

The core range consists of **12, 16 and 21 years old**. An 18 year old, destined for duty free, was recently replaced by a **16 year old** and a **21 year old** both of them finished for up to 12 months in madeira casks (ex-Bual and ex-Malvasia Malmsey respectively). Three recent travel retail bottlings, part of the new Exceptional Cask Series exclusive to the Asian market, were an **18 year old port finish** (two years finish), a **33 year old single cask** and a **Vintage 1999**.

History:

1896 John and Tommy Dewar embark on the construction of the distillery, a stone's throw from the old Pitilie distillery which was active from 1825 to 1867. Their objective is to produce a single malt for their blended whisky - White Label.

1898 Production starts in November.

1917 The distillery closes.

1919 The distillery re-opens.

1925 Distillers Company Limited (DCL) takes over.

1972 Reconstruction takes place, the floor maltings is closed and the two stills are increased to four.

1991 The first official bottling is a 15 year old in the Flora & Fauna series.

1998 Bacardi buys John Dewar & Sons from Diageo at a price of £1,150 million.

2000 A visitor centre opens and a 25 year old is released.

2005 A 21 year old is launched in October, replacing the 25 year old.

2009 Two 18 year old single casks are released.

2010 A 19 year old single cask, exclusive to France, is released.

2011 A 14 year old single cask is released.

2014 The whole range is revamped and an 18 year old for duty free is released.

2015 A 16 year old is released.

2018 A 16 year old and a 21 year old madeira finish are released for duty free.

Tasting notes Aberfeldy 12 years old:

GS – Sweet, with honeycombs, breakfast cereal and stewed fruits on the nose. Inviting and warming. Mouth-coating and full-bodied on the palate. Sweet, malty, balanced and elegant. The finish is long and complex, becoming progressively more spicy and drying.

12 years old

Aberlour

[ah•bur•<u>lower</u>]

Owner:
Chivas Brothers Ltd
(Pernod Ricard)

Region/district:
Speyside

Founded: 1879

Status: Active (vc)

Capacity: 3 800 000 litres

Address: Aberlour, Banffshire AB38 9PJ

Website: aberlour.com

Tel: 01340 881249

The first distillery named Aberlour was built in 1826 but was destroyed by a fire in 1879. Enter local banker James Fleming who decided to build a new Aberlour a couple of kilometres upstream the Spey River.

Fleming was a well-respected person in the Aberlour community and was also known as a philanthropist. He financed the construction of a town hall and made sure Aberlour's streets were lighted up. When he passed away, in 1895, he had established a trust that financed the construction of a school and a hospital. Three years later Aberlour distillery went up in flames again and the famous distillery architect Charles Doig was called in to construct a new distillery.

The distillery is equipped with a 12 ton semi-lauter mash tun, six stainless steel washbacks and two pairs of large and wide stills in a spacious still room. To achieve the desired character of the newmake, which is fruity, the operators run a very slow distillation. With a 7.5 hour spirit cycle, the middle cut (73-63%) takes two hours to complete.

The core range of Aberlour includes **12, 16** and **18 year olds** – all matured in a combination of ex-bourbon and ex-sherry casks. Another core expression is **Aberlour a'bunadh**, matured in ex-Oloroso casks. It is always bottled at cask strength and up to 63 different batches have been released by spring 2019. A new expression was added to the core range in May 2018 - **Aberlour Casg Annamh**. Matured in ex-oloroso casks (both European and American oak) as well as ex-bourbon, this is the first in a new series. For select markets (mainly France) another four expression are available; **10 year old, 12 year old un chill-filtered, 15 year old Select Cask Reserve** and **White Oak Millennium 2004**. Two exclusives are available for duty free – a **12 year old Sherry Cask** and a **15 year old Double Cask**. There are also five cask strength bottlings in the Distillery Reserve Collection, available at all Chivas' visitor centres – from **13 to 20 years old**.

History:

1879 The local banker James Fleming founds the distillery.

1892 The distillery is sold to Robert Thorne & Sons Ltd who expands it.

1898 Another fire rages and almost totally destroys the distillery. The architect Charles Doig is called in to design the new facilities.

1921 Robert Thorne & Sons Ltd sells Aberlour to a brewery, W. H. Holt & Sons.

1945 S. Campbell & Sons Ltd buys the distillery.

1962 Aberlour terminates floor malting.

1973 Number of stills are increased from two to four.

1974 Pernod Ricard buys Campbell Distilleries.

2000 Aberlour a'bunadh is launched.

2001 Pernod Ricard buys Chivas Brothers and merges Chivas Brothers and Campbell Distilleries under the brand Chivas Brothers.

2002 A new, modernized visitor centre is inaugurated in August.

2008 The 18 year old is also introduced outside France.

2013 Aberlour 2001 White Oak is released.

2014 White Oak Millenium 2004 is released.

2018 Casg Annamh is released.

Tasting notes Aberlour 12 year old:

GS – The nose offers brown sugar, honey and sherry, with a hint of grapefruit citrus. The palate is sweet, with buttery caramel, maple syrup and eating apples. Liquorice, peppery oak and mild smoke in the finish.

12 years old

Allt-a-Bhainne

[alt a•<u>vain</u>]

Owner:
Chivas Brothers Ltd
(Pernod Ricard)

Region/district:
Speyside

Founded: **Status:** **Capacity:**
1975 Active 4 200 000 litres

Address: Glenrinnes, Dufftown, Banffshire AB55 4DB

Website: **Tel:**
- 01542 783200

Arguably one of the least known distilleries in Scotland, Allt-a-Bhainne was built to provide malt whisky for Seagram´s range of blends. It soon became an integral part of 100 Pipers, the name taken from the 100 pipers that led Bonnie Prince Charlie into battle during the Jacobite uprising in 1745.

The 100 Pipers brand was launched in 1965. It was the brainchild of Seagram´s owner at the time, Sam Bronfman, and the men making it come true were Jimmy Lang and Alan Bailie. In terms of global sales, the brand has peaked. Sales have halved in the last decade and it is virtually unknown in Europe. But, in the USA it is quite popular being sold since 30 years back by major distilled spirits supplier Heaven Hill Brands. The biggest market for the brand is in Asia and, especially, Thailand and India. In fact, it is so popular in India that in 2018 it became the first ever Scotch whisky brand to break the 1 million case (12 million bottles) in a year mark – an astonishing achievement! The 100 Pipers (which also comes with a 12 year old statement) has a soft, yet distinctive smoky, note which comes from the fact that Allt-a-Bhainne for a number of years have been distilling peated whisky for part of the year. Usually it constitutes 30-50% of the total production and with a phenol specification in the barley between 10 and 20ppm.

In 2015, a new, modern, lauter mash gear was fitted into the existing traditional 9 ton mash tun which had previously been equipped with rakes and ploughs. The rest of the equipment consists of eight stainless steel washbacks with a fermentation time of 48-50 hours and two pairs of stills. The distillery is currently working 7 days a week with 25 mashes resulting in 4 million litres of alcohol.

Historically, the owners have shown no interest in releasing official bottlings. Instead we have had to rely on independent bottlers. In autumn 2018, however, a lightly peated **Allt-a-Bhainne NAS** was released and there is also a **21 year old cask strength** bottling in the Distillery Reserve Collection, available at all Chivas´ visitor centres.

History:

1975 The distillery is founded by Chivas Brothers, a subsidiary of Seagrams, in order to secure malt whisky for its blended whiskies. The total cost amounts to £2.7 million.

1989 Production has doubled.

2001 Pernod Ricard takes over Chivas Brothers from Seagrams.

2002 Mothballed in October.

2005 Production restarts in May.

2018 An official, lightly peated bottling is released.

Allt-a-Bhainne NAS

Tasting notes Allt-a-Bhainne NAS:

IR – Subtle smokiness is mixed with butterscotch, honey, apples and a touch of pepper. Sweet peat on the palate, oranges, ginger, melon, more pepper and vanilla.

Ardbeg

[ard•beg]

Owner:
The Glenmorangie Co
(Moët Hennessy)

Region/district:
Islay

Founded: 1815
Status: Active (vc)
Capacity: 2 400 000 litres

Address: Port Ellen, Islay, Argyll PA42 7EA

Website: ardbeg.com
Tel: 01496 302244 (vc)

History:

1794 First record of a distillery at Ardbeg. It was founded by Alexander Stewart.

1798 The MacDougalls, later to become licensees of Ardbeg, are active on the site through Duncan MacDougall.

1815 The current distillery is founded by John MacDougall, son of Duncan MacDougall.

1853 Alexander MacDougall, John's son, dies and sisters Margaret and Flora MacDougall, assisted by Colin Hay, continue the running of the distillery. Colin Hay takes over the licence when the sisters die.

1888 Colin Elliot Hay and Alexander Wilson Gray Buchanan renew their license.

1900 Colin Hay's son takes over the license.

1959 Ardbeg Distillery Ltd is founded.

1973 Hiram Walker and Distillers Company Ltd jointly purchase the distillery for £300,000 through Ardbeg Distillery Trust.

1977 Hiram Walker assumes single control of the distillery. Ardbeg closes its maltings.

1979 Kildalton, a less peated malt, is produced over a number of years.

1981 The distillery closes in March.

1987 Allied Lyons takes over Hiram Walker and thereby Ardbeg.

Producing an iconic and increasingly popular brand, Ardbeg has struggled in recent years to keep up with demand. From early 2018 until late 2019, the distillery has been an incredibly busy site with lots of new equipment being added.

To a whisky enthusiast, expansion may be about adding a few stills and washbacks but usually there´s much more to it than that. To cope with a larger production at Ardbeg, the malt storage had to be increased from 60 to 120 tonnes, another boiler (there are two now) was added in August 2018 and a new drainage under the mash tun was installed. Add to that a completely new still house with an incredible view towards the sea. With all this happening at the distillery it is important to be able to make a smooth transition from old to new without losing valuable production time.

With everything in place in late 2019 the distillery will be equipped with the following; the current 5 ton stainless steel semi lauter mash tun, eleven washbacks (an additional five) made of Oregon pine with five in the old tun room, four in the old still house and two in the old fuel store and four stills. The two old ones were replaced in summer 2019 and an additional two were installed at the same time. The fermentation time will remain 60 hours and both spirits stills will have purifiers as before to help create the special fruity character of the spirit. The old pair of stills will be used until the expansion has been completed which means 16-17 mashes per week and a production of around 1.4 million litres in 2019.

The core range, all of them non-chill filtered, consists of the **10 year old**, a mix of first and re-fill bourbon casks, **Uigeadail**, a marriage of bourbon and sherry casks and bottled at cask strength, **Corryvreckan**, also a cask strength and a combination of bourbon casks and new French oak and **An Oa**, a vatting of whiskies matured in several types of casks that have been married together for a minimum of three months in three huge vats (14,000 and 30,000 litres respectively). In September 2019, a new expression was added to the range. The 19 year old **Traigh Bhan** has been matured in a combination of American oak and ex-oloroso sherry casks and is bottled at 46,2%. This was the first time in 19 years that Ardbeg released an aged whisky as part of the permanent expressions.

The Ardbeg Day expression for 2019 was **Drum**, matured in ex-bourbon barrels and then finished in ex-rum casks. As usual, a Committee version was launched in March, bottled at 52% while the general release followed in early June, bottled at 46%.

Recent limited expressions include a **21 year old** and **Ardbeg Twenty Something** – a 23 year old made from spirit distilled in the mid-nineties when Allied Distillers used to own the distillery. It was matured in a combination of ex-bourbon and ex-oloroso sherry. It was then followed up in autumn 2018 by a second edition, **22 years old** and matured in ex-bourbon casks.

History continued:

1989 Production is restored. All malt is taken from Port Ellen.

1996 The distillery closes in July and Allied Distillers decides to put it up for sale.

1997 Glenmorangie plc buys the distillery for £7 million. Ardbeg 17 years old and Provenance are launched.

1998 A new visitor centre opens.

2000 Ardbeg 10 years is introduced and the Ardbeg Committee is launched.

2001 Lord of the Isles 25 years and Ardbeg 1977 are launched.

2002 Ardbeg Committee Reserve and Ardbeg 1974 are launched.

2003 Uigeadail is launched.

2004 Very Young Ardbeg (6 years) and a limited edition of Ardbeg Kildalton are launched.

2005 Serendipity is launched.

2006 Ardbeg 1965 and Still Young are launched. Almost There (9 years old) and Airigh Nam Beist are released.

2007 Ardbeg Mor, a 10 year old in 4.5 litre bottles is released.

2008 The new 10 year old, Corryvreckan, Rennaissance, Blasda and Mor II are released.

2009 Supernova is released, the peatiest expression from Ardbeg ever.

2010 Rollercoaster and Supernova 2010 are released.

2011 Ardbeg Alligator is released.

2012 Ardbeg Day and Galileo are released.

2013 Ardbog is released.

2014 Auriverdes and Kildalton are released.

2015 Perpetuum and Supernova 2015 are released.

2016 Dark Cove and a Twenty Something 21 year old are released.

2017 An Oa, Kelpie and Twenty Something 23 year old are released.

2018 Grooves and Twenty Something 22 year old are released.

2019 Drum and Traigh Bhan are released.

Tasting notes Ardbeg 10 year old:

GS – Quite sweet on the nose, with soft peat, carbolic soap and Arbroath smokies. Burning peats and dried fruit, followed by sweeter notes of malt and a touch of liquorice in the mouth. Extremely long and smoky in the finish, with a fine balance of cereal sweetness and dry peat notes.

Traigh Bhan

Uigeadail

An Oa

10 years old

Drum

Corryvreckan

Ardmore

[ard•moor]

Owner:
Beam Suntory

Region/district:
Highland

Founded: 1898 **Status:** Active **Capacity:** 5 550 000 litres

Address: Kennethmont, Aberdeenshire AB54 4NH

Website: ardmorewhisky.com **Tel:** 01464 831213

An excellent example of how intricate the chain of whisky-making is could recently be seen at Ardmore distillery. The introduction of two steps, with the aim to reduce the environmental impact of residues from the distillation, changed the production altogether.

In 2015/2016, the distillery installed a pot ale evaporator and an effluent treatment plant. For them to work properly, it was vital that no anti-foaming agent was used during fermentation. With the distillery actively moving to a clearer wort and thereby standing the risk of foaming in the washbacks, something had to be done. The solution was to reduce the mash charge by 0.5 tonnes which subsequently led to a reduction in washback fill level. Whatever foaming occurred, could then be handled by installing switchers.

The distillery is equipped with a 12 ton, cast iron, semi-lauter mash tun with a copper dome, 14 Douglas fir washbacks (four large and ten smaller ones), as well as four pairs of stills. At the moment, Ardmore is working a seven-day week with 23 mashes per week resulting in 4.3 million litres of alcohol. Traditionally, Ardmore has been the only distillery in the region consistently producing peated whisky with a phenol specification of the barley at 12-14 ppm. The earthy Highland peat is locally sourced from St Fergus. For blending purposes, they also produce the unpeated Ardlair (around half of the yearly output) which is usually produced for the first 6-7 months of the year. The fermentation time for Ardlair is longer than for regular Ardmore – 70 hours compared to 55 hours.

Ardmore serves as the signature malt in Teacher's blended Scotch but is also released as a single malt. The core range is made up of **Legacy**, a mix of 80% peated and 20% unpeated malt, and a **12 year old Port Finish** with four years in port pipes. In 2015, **Tradition** was released as a duty free exclusive together with **Triple Wood** with no age statement and matured in bourbon barrels, quarter casks and sherry puncheons. A new release in 2017 was a **20 year old**, double matured in a mix of first- and second-fill bourbon casks and the second batch was launched in 2018 together with a new **30 year old**.

History:

1898 Adam Teacher, son of William Teacher, starts the construction of Ardmore Distillery which eventually becomes William Teacher & Sons' first distillery. Adam Teacher passes away before it is completed.

1955 Stills are increased from two to four.

1974 Another four stills are added, increasing the total to eight.

1976 Allied Breweries takes over William Teacher & Sons and thereby also Ardmore. The own maltings (Saladin box) is terminated.

1999 A 12 year old is released to commemorate the distillery's 100th anniversary. A 21 year old is launched in a limited edition.

2002 Ardmore is one of the last distilleries to abandon direct heating (by coal) of the stills in favour of indirect heating through steam.

2005 Jim Beam Brands becomes new owner when it takes over some 20 spirits and wine brands from Allied Domecq for five billion dollars.

2007 Ardmore Traditional Cask is launched.

2008 A 25 and a 30 year old are launched.

2014 Beam and Suntory merge. Legacy is released.

2015 Traditional is re-launched as Tradition and a Triple Wood and a 12 year old port finish are released.

2017 A 20 year old, double matured is released.

2018 A 30 year old is released.

Tasting notes Ardmore Legacy:

GS – Vanilla, caramel and sweet peat smoke on the nose, while on the palate vanilla and honey contrast with quite dry peat notes, plus ginger and dark berries. The finish is medium to long, spicy, with persistently drying smoke.

Legacy

Arran

[ar•ran]

Owner:
Isle of Arran Distillers

Region/district:
Highlands (Arran)

Founded: **Status:**   **Capacity:**
1993   Active (vc)   1 200 000 litres

Address: Lochranza, Isle of Arran KA27 8HJ

Website: **Tel:**
arranwhisky.com   01770 830264

Since March 2019, the Island of Arran Distillers operate two distilleries on the island. One in Lochranza, which was opened already in 1995, and a new one situated at Lagg on the south part of Arran.

The distillery in Lochranza is by far the most visited distillery in Scotland. Well over 100,000 people come here every year. With Lagg opened, the owners expect the combined number to rise to more than 200,000 in 2020. Arran distillery is equipped with a 2.5 ton semi-lauter mash tun, six Oregon pine washbacks with an average fermentation time of 60 hours and four new stills. The owners aim to produce 450,000 litres of pure alcohol during 2019. This is 150,000 litres less than the year before and is due to the fact that 250,000 litres of peated newmake will instead be made at the Lagg distillery instead of at Arran.

The Arran range was revamped in autumn 2019 with new bottlings, some name changes of existing expressions and, not least, a completely new and exciting design of both bottles and labels. The new core range consists of **10, 18** and **21 year old** and **Barrel Reserve** (formerly known as Lochranza Reserve). Also included in the core range are **Quarter Cask The Bothy** bottled at 56,2% and **Sherry Cask The Bodega** bottled at 55,8%. Finally, the peated side of Arran is represented by **Machrie Moor** and **Machrie Moor Cask Strength**. A 14 year old which has been a part of the range for many years was temporarily removed in autumn 2019 but is due for a come-back in the future. A range of limited wood finishes include **Amarone, Port, Sauternes** and **Marsala**, and every year a number of **single casks**, matured either in ex-bourbon or ex-sherry, are released. The third and final release in the limited Smuggler's Edition range, **The Exciseman**, was launched in July 2017 and in October the same year, a special bottling was released to celebrate the distillery manager's James MacTaggart 10th anniversary with the distillery. In June 2018, the 20 year old **Brodick Bay**, the first in the new Explorer's Series, was launched. This was followed in spring 2019 by the 21 year old **Lochranza Castle**, matured in sherry hogsheads and finished in amontillado casks.

History:

1993 Harold Currie founds the distillery.

1995 Production starts in full on 17th August.

1998 The first release is a 3 year old.

1999 The Arran 4 years old is released.

2002 Single Cask 1995 is launched.

2003 Single Cask 1997, non-chill filtered and Calvados finish is launched.

2004 Cognac finish, Marsala finish, Port finish and Arran First Distillation 1995 are launched.

2005 Arran 1996 and two finishes, Ch. Margaux and Grand Cru Champagne, are launched.

2006 After an unofficial launch in 2005, Arran 10 years old is released as well as a couple of new wood finishes.

2007 Four new wood finishes and Gordon´s Dram are released.

2008 The first 12 year old is released as well as four new wood finishes.

2009 Peated single casks, two wood finishes and 1996 Vintage are released.

2010 A 14 year old, Rowan Tree, three cask finishes and Machrie Moor (peated) are released.

2011 The Westie, Sleeping Warrior and a 12 year old cask strength are released.

2012 The Eagle and The Devil´s Punch Bowl are released.

2013 A 16 year old and a new edition of Machrie Moor and released.

2014 A 17 year old and Machrie Moor cask strength are released.

2015 A 18 year old and The Illicit Stills are released.

2017 The Exciseman is released.

2018 A 21 year old and Brodick Bay are released.

2019 The core range is revamped and the limited Lochranza Castle is released.

10 years old

Tasting notes Arran 14 year old:

GS – Very fragrant and perfumed on the nose, with peaches, brandy and ginger snaps. Smooth and creamy on the palate, with spicy summer fruits, apricots and nuts. The lingering finish is nutty and slowly drying.

Auchentoshan

[ock•en•tosh•an]

Owner:
Beam Suntory

Region/district:
Lowlands

Founded: **Status:**
1823 Active (vc)

Capacity:
2 000 000 litres

Address: Dalmuir, Clydebank, Glasgow G81 4SJ

Website:
auchentoshan.com

Tel:
01389 878561

Springbank has its triple-distilled Hazelburn and Benriach is making small, yearly runs as well but Auchentoshan is unique in that it is the only distillery in Scotland that triple distill the entire production.

It is a myth that all Lowland whiskies used to be triple-distilled. In fact when Alfred Barnard visited Auchentoshan in 1885, there were only two stills and nobody really knew when they switched to triple distillation. Whenever it was, the inspiration may have come from Ireland where triple distillation was much more common in the 19th century.

Using three stills produces a lighter, purer and more consistent spirit which was advantageous for the Irish compared with their cousins in Scotland who were known for producing a heavier spirit. In fact, Irish whiskey was far more popular than Scotch in Great Britain in the late 1800s and early 1900s. When Auchentoshan practise triple distillation, they use a very narrow middle cut (82%-80%), ending up with a delicate newmake with sweetness and notes of citrus and malt. The implication of this is that wood management is essential. You can easily ruin the future whisky by putting it into very active and flavourful wood. Ex-bourbon casks are perfect but it does not exclude a finish in other types of casks.

The equipment consists of a semi-lauter mash tun with a 6.8 ton mash charge, four Oregon Pine washbacks and three made of stainless steel, all 38,000 litres and with a fermentation time of 50 to 120 hours. There are three stills; wash still (17,500 litres), intermediate still (8,200 litres) and spirit still (11,500 litres). The plan for 2019 is to make 10 to 15 mashes per week and 1.5 million litres of alcohol.

The core range consists of **American Oak**, a first fill bourbon maturation without age statement, **12 years, Three Woods, 18 years** and **21 years**. The former duty free range was replaced in 2015 by **Blood Oak**, matured in a combination of bourbon and red wine casks and the 24 year old **Noble Oak**, a vatting of bourbon and oloroso casks. Two of the previous expressions, **Heartwood** and **Springwood**, have recently been relaunched as well. A limited version named **Bartender's Malt** was launched in summer 2017 with a second version released in 2018.

History:

- **1817** First mention of the distillery Duntocher, which may be identical to Auchentoshan.
- **1823** The distillery is founded by John Bulloch.
- **1823** The distillery is sold to Alexander Filshie.
- **1878** C.H. Curtis & Co. takes over.
- **1903** The distillery is purchased by John Maclachlan.
- **1941** The distillery is severely damaged by a German bomb raid.
- **1960** Maclachlans Ltd is purchased by the brewery J. & R. Tennent Brewers.
- **1969** Auchentoshan is bought by Eadie Cairns Ltd who starts major modernizations.
- **1984** Stanley P. Morrison, eventually becoming Morrison Bowmore, becomes new owner.
- **1994** Suntory buys Morrison Bowmore.
- **2002** Auchentoshan Three Wood is launched.
- **2004** More than a £1 million is spent on a new, refurbished visitor centre. The oldest Auchentoshan ever, 42 years, is released.
- **2006** Auchentoshan 18 year old is released.
- **2007** A 50 year old, the oldest ever Auchentoshan to be bottled, was released.
- **2008** New packaging as well as new expressions - Classic, 18 year old and 1988.
- **2010** Two vintages, 1977 and 1998, are released.
- **2011** Two vintages, 1975 and 1999, and Valinch are released.
- **2012** Six new expressions are launched for the Duty Free market.
- **2013** Virgin Oak is released.
- **2014** American Oak replaces Classic.
- **2015** Blood Oak and Noble Oak are released for duty free.
- **2017** Bartender's Malt is launched.
- **2018** Bartender's Malt 2 and 1988 PX Cask are released.

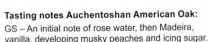

Tasting notes Auchentoshan American Oak:
GS – An initial note of rose water, then Madeira, vanilla, developing musky peaches and icing sugar. Spicy fresh fruit on the palate, chilli notes and more Madeira and vanilla. The finish is medium in length, and spicy to the end.

American Oak

Auchroisk

[ar•thrusk]

Owner: Diageo
Region/district: Speyside

Founded: 1974
Status: Active
Capacity: 5 900 000 litres

Address: Mulben, Banffshire AB55 6XS

Website: malts.com
Tel: 01542 885000

Auchroisk was the fourth distillery to be built by the well-known firm, Justerini & Brooks. The others were Knockando, Strathmill and Glen Spey.

A water source, Dories Well, had been discovered at the distillery location and since the idea was to imitate the style of Glen Spey single malt in the new distillery, water from the well was tankered to Glen Spey in Rothes to see if the expectations were met. The outcome was favourable and Auchroisk came on stream in 1974.

Travelling in Speyside with its more than 50 distilleries, Auchroisk is one that you tend to stumble on more or less by coincidence. If you make a detour off the A95 halfway between Keith and Craigellachie, you will see it, situated right at road B9103. The modern look, however, without any pagoda roofs doesn't really give it away as a distillery.

The equipment consists of a 12 ton stainless steel semi-lauter mash tun, eight stainless steel washbacks with a fermentation time of 53 hours and four pairs of stills. The spacious still house served as a model for the still house of Diageo's latest distillery, Roseisle. Auchroisk is working 24/7 with 24 mashes per week, producing 5.8 million litres of alcohol per year, currently with a nutty/malty character. This has changed over the years though and not so long ago, the style was green/grassy which is not unusual for distilleries that produce malt mainly for blends. It all depends on what the owner predicts they will need for the coming 5 years.

The first, widely available release of Auchroisk single malt was in 1986 under the name Singleton. That particular name is now reserved for another three distilleries in the Diageo range - Dufftown, Glendullan and Glen Ord. In 2001, it was replaced by a **10 year old** in the Flora & Fauna range. Limited bottlings over the years have included a **20 year old** from 1990 and a **30 year old**, both launched as part of the Special Releases. In October 2016, it was time for yet another limited Auchroisk in the Special Releases series; a **25 year old**, distilled in 1990 and bottled at 51.2%.

History:

1972 Building of the distillery commences by Justerini & Brooks (which, together with W. A. Gilbey, make up the group IDV) in order to produce blending whisky. In February the same year IDV is purchased by the brewery Watney Mann which, in July, merges into Grand Metropolitan.

1974 The distillery is completed and, despite the intention of producing malt for blending, the first year's production is sold 12 years later as single malt thanks to the high quality.

1986 The first whisky is marketed under the name Singleton.

1997 Grand Metropolitan and Guinness merge into the conglomerate Diageo. Simultaneously, the subsidiaries United Distillers (to Guinness) and International Distillers & Vintners (to Grand Metropolitan) form the new company United Distillers & Vintners (UDV).

2001 The name Singleton is abandoned and the whisky is now marketed under the name of Auchroisk in the Flora & Fauna series.

2003 Apart from the 10 year old in the Flora & Fauna series, a 28 year old from 1974, the distillery's first year, is launched in the Rare Malt series.

2010 A Manager's Choice single cask and a limited 20 year old are released.

2012 A 30 year old from 1982 is released.

2016 A 25 year old from 1990 is released.

10 years old

Tasting notes Auchroisk 10 year old:

GS – Malt and spice on the light nose, with developing nuts and floral notes. Quite voluptuous on the palate, with fresh fruit and milk chocolate. Raisins in the finish.

Aultmore

[ault•moor]

Owner:
John Dewar & Sons
(Bacardi)

Region/district:
Speyside

Founded: | **Status:** | **Capacity:**
1896 | Active | 3 200 000 litres

Address: Keith, Banffshire AB55 6QY

Website:
aultmore.com

Tel:
01542 881800

Although easily spotted as a "white distillery" on the right hand side of the road when travelling the A96 from Keith towards Fochabers, Aultmore is not a distillery that caters to visitors.

For those of you still keen to have a walk around the premises, there is however a solution. For the last couple of years, the distillery has been part of the "open distillery" theme during the Speyside Festival in May. Even though the distillery was completely rebuilt at the beginning of the 1970s and nothing is left of the old buildings from 1896, it is still a rare opportunity to visit what by many is considered as a hidden gem.

The distillery is equipped with a 10 ton Steinecker full lauter mash tun, six washbacks made of larch with a minimum fermentation time of 56 hours and two pairs of stills. Since 2008 production has been running seven-days a week, which for 2019 means 16 mashes per week and just over 3 million litres of alcohol.

During the UDV/Diageo ownership, Aultmore was part of the Flora & Fauna range as well as the Rare Malt series. When Bacardi/Dewars took over in 1998, they released a 12 year old official bottling but stopped at that. It was still very much a single malt under the radar (except for occasional independent bottlings) until 2014/15 when Bacardi relaunched all their single malts under the name the Last Great Malts.

The core range consists of a **12 year old** and an **18 year old**. The 25 year old, released a few years back, has now been discontinued. For the duty free market there is a **21 year old** and this was also recently released for the domestic US market. In summer 2018, Dewars launched a collection of rare single malts, The Exceptional Cask Series, destined to be sold in the duty free market. Part of that range were three bottlings released at Heathrow in spring 2019; all three were **22 years old** with the final eleven years having been extra matured in casks that had held **Super Tuscan** wine, **Châteauneuf-du-Pape** and **Moscatel** wine respectively. Later in the year a **17 year old** with a finish in several types of casks, i.a. palo cortado, was released.

History:

1896 Alexander Edward, owner of Benrinnes and co-founder of Craigellachie Distillery, builds Aultmore.

1897 Production starts.

1898 Production is doubled; the company Oban & Aultmore Glenlivet Distilleries Ltd manages Aultmore.

1923 Alexander Edward sells Aultmore for £20,000 to John Dewar & Sons.

1925 Dewar's becomes part of Distillers Company Limited (DCL).

1930 The administration is transferred to Scottish Malt Distillers (SMD).

1971 The stills are increased from two to four.

1991 United Distillers launches a 12-year old Aultmore in the Flora & Fauna series.

1996 A 21 year old cask strength is marketed as a Rare Malt.

1998 Diageo sells Dewar's and Bombay Gin to Bacardi for £1,150 million.

2004 A new official bottling is launched
(12 years old).

2014 Three new expressions are released – 12, 25 and 21 year old for duty free.

2015 An 18 year old is released.

2019 Three 22 year old single casks with different second maturations are released for duty free.

Aultmore 12 years old:

GS – A nose of peaches and lemonade, freshly-mown grass, linseed and milky coffee. Very fruity on the palate, mildly herbal, with toffee and light spices. The finish is medium in length, with lingering spices, fudge, and finally more milky coffee.

12 years old

Balblair

[bal•blair]

Owner: | **Region/district:**
Inver House Distillers | Northern Highlands
(Thai Beverages plc) |

Founded: | **Status:** | **Capacity:**
1790 | Active (vc) | 1 800 000 litres

Address: Edderton, Tain, Ross-shire IV19 1LB

Website: | **Tel:**
balblair.com | 01862 821273

Two of Inver House's single malt brands were revamped with new expressions, bottles and labels a few years apart – anCnoc in 2003 and Balblair in 2007. For anCnoc this meant a boost in sales and the introduction of a peated sub-range.

Balblair, on the other hand, has only seen a modest increase in sales in the last decade. When the range was relaunched, the owners decided to sell it by vintage and without age statement. It is possible that this decision made it difficult to raise the volumes. Consumers are not used to vintages in the whisky world and the only other distillery with a similar strategy, Glenrothes, gave up last year and went back to age statements. It was therefore no surprise when, in April 2019, an entire new range with ages on the labels was launched by Balblair. The style of the whisky remains the same but the owners now hope that their range will be easier to navigate for the consumers. In 2012, the distillery opened a small but excellent visitor centre which now attracts around 4,000 visitors yearly.

The distillery is equipped with a stainless steel, 4.4 ton semi lauter mash tun, six Oregon pine washbacks and one pair of stills. The distillery has increased the production substantially since last year moving from a five-day week to a seven-day week. With 21 mashes per week, this means a target of 1.53 million litres for the full year. It also means that fermentation during 2019 will be 60 hours instead of having short (60 hours) and long (90 hours) fermentations. On site, there are also eight dunnage warehouses with a total capacity of 22,500 casks. In 2015 the distillery converted from using heavy fuel oil to gas, thereby reducing the emission of greenhouse gases significantly.

The new core range from Balblair, carrying age statements, consists of **12 year old** matured in ex-bourbon and double-fired American oak, the **15 year old** is matured in ex-bourbon casks followed by time in Spanish oak sherry butts, the **18 year old** has the same maturation as the previous and the **25 year old** starts in ex-bourbon casks and is finished in ex-oloroso sherry casks.

History:
1790 The distillery is founded by James McKeddy.

1790 John Ross takes over

1836 John Ross dies and his son Andrew Ross takes over with the help of his sons.

1872 The distillery is moved to the present location.

1873 Andrew Ross dies and his son James takes over.

1894 Alexander Cowan takes over and rebuilds the distillery

1911 Cowan is forced to cease payments and the distillery closes.

1941 The distillery is put up for sale.

1948 Robert Cumming buys Balblair for £48,000.

1949 Production restarts.

1970 Cumming sells Balblair to Hiram Walker.

1988 Allied Distillers becomes the new owner through the merger between Hiram Walker and Allied Vintners.

1996 The distillery is sold to Inver House Distillers.

2000 Balblair Elements and the first version of Balblair 33 years are launched.

2001 Thai company Pacific Spirits (part of the Great Oriole Group) takes over Inver House.

2004 Balblair 38 years is launched.

2005 12 year old Peaty Cask, 1979 (26 years) and 1970 (35 years) are launched.

2006 International Beverage Holdings acquires Pacific Spirits UK.

2007 Three new vintages replace the former range.

2008 Vintage 1975 and 1965 are released.

2009 Vintage 1991 and 1990 are released.

2011 Vintage 1995 and 1993 are released.

2012 Vintage 1975, 2001 and 2002 are released. A visitor centre is opened.

2013 Vintage 1983, 1990 and 2003 are released.

2014 Vintage 1999 and 2004 are released for duty free.

2016 Vintage 2005 is released.

2019 A new range with age statements is launched.

Tasting notes Balblair 12 year old:
IR – Sugary and malty on the nose with herbal and earthy notes coming through. Rich, creamy and sweet on the palate, grilled corn cobs, caramel, honey and some bitter, oaky notes.

12 years old

Balmenach

[bal•<u>may</u>•nack]

Owner:
Inver House Distillers
(Thai Beverages plc)

Region/district:
Speyside

Founded: 1824
Status: Active
Capacity: 2 800 000 litres

Address: Cromdale, Moray PH26 3PF

Website: inverhouse.com
Tel: 01479 872569

Inver House owns five distilleries in Scotland but only bottles a single malt from four of them. The odd one out is Balmenach but for how long will the owners refrain themselves from offering also this to the consumers?

In recent years, most Scottish malt whisky producers have started bottling the single malts that previously were reserved only as part of a blend. Dewar´s is one example when they releaunched their entire range of malts as The Last of the Great Malts in 2014/15 and even a giant such as Pernod Ricard have official bottlings from all of their distilleries (except the recently opened Dalmunach). There used to be a 12 year old in the Flora & Fauna range from the previous owners, Diageo but the last time an official bottling of Balmenach turned up was in 2002 when a 25 year old was launched to celebrate the Queen´s Golden Jubilee. Independent bottlers on the other hand are keen to release whiskies from this underrated distillery.

Balmenach is equipped with an eight ton stainless steel semi-lauter mash tun, with an old copper canopy. There are six washbacks made of Douglas Fir with a 52 hour fermentation period, and three pairs of stills connected to worm tubs. In 2019, the distillery will be making 15 mashes per week which translates to 1.9 million litres of alcohol. A new biogas plant was installed at Balmenach this summer to treat whisky co-products such as pot ale and spent lees. About 130 m³ of these will each day be processed by the plant and through anaerobic digestion it will be turned into 2,000 m³ of biogas.

For the past decade, Caorunn gin has also been part of the production at Balmenach. Purchased neutral spirit is pumped through a vaporiser and then to a copper berry chamber where the vapours travel upwards passing five trays with different kinds of botanicals.

There is no official bottling of Balmenach single malt. Aberko though, has been working with the distillery for a long time and, over the years, has released Balmenach under the name Deerstalker. The current expression is a 12 year old.

History:

1824 The distillery is licensed to James MacGregor who operated a small farm distillery by the name of Balminoch.

1897 Balmenach Glenlivet Distillery Company is founded.

1922 The MacGregor family sells to a consortium consisting of MacDonald Green, Peter Dawson and James Watson.

1925 The consortium becomes part of Distillers Company Limited (DCL).

1930 Production is transferred to Scottish Malt Distillers (SMD).

1962 The number of stills is increased to six.

1964 Floor maltings replaced with Saladin box.

1992 The first official bottling is a 12 year old.

1993 The distillery is mothballed in May.

1997 Inver House Distillers buys Balmenach from United Distillers.

1998 Production recommences.

2001 Thai company Pacific Spirits takes over Inver House at the price of £56 million. The new owner launches a 27 and a 28 year old.

2002 To commemorate the Queen's Golden Jubilee a 25-year old Balmenach is launched.

2006 International Beverage Holdings acquires Pacific Spirits UK.

2009 Gin production commences.

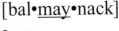

Tasting notes Deerstalker 12 years old:

IR – The nose is sweet and fruity, with green garden notes and sweet liquorice coming through. Sweet, fruity barley on the palate with notes of honey, custard, apricots, peaches and slightly bitter notes from the oak.

Deerstalker 12 years old

Balvenie

[bal•ven•ee]

Owner: **Region/district:**
William Grant & Sons   Speyside

Founded: **Status:**   **Capacity:**
1892   Active (vc)   7 000 000 litres

Address: Dufftown, Keith, Banffshire AB55 4DH

Website: **Tel:**
thebalvenie.com   01340 820373

The appointment of a new master blender is never taken lightly in the Scotch whisky industry. It doesn´t happen overnight but is rather a five to ten year transition where the blender-to-be works as an apprentice alongside the current malt master.

David Stewart is the longest-serving malt master in Scotch whisky and has worked in the industry for 56 years. The potential succesor to this legend is Kelsey McKechnie, a 26 year old Glaswegian with a degree in Biology and Biological Sciences. Alongside her studies, she has been working for William Grants for the last four years.

The distillery is equipped with an 11.8 ton full lauter mash tun, nine wooden and five stainless steel washbacks with a fermentation time of 68 hours, five wash stills and six spirit stills. For 2019, the production plan is 29 mashes per week and 7 million litres of alcohol. The main part is unpeated but each year one week of production comes from peated barley (20-40 ppm). The distillery also has their own floor maltings producing 15% of their needs and there is also a coppersmith and a cooperage on site.

The core range consists of **Doublewood 12 years, Doublewood 17 years, Caribbean Cask 14 years, Single Barrel 12 years First Fill, Single Barrel 15 years Sherry Cask, Single Barrel 25 years Traditional Oak, Portwood 21 years, 30 years, 40 years** and **50 years old**. Recent limited releases include batch 6 of **Tun 1509**, batch 7 of the **Tun 1858** and chapter five of **The Balvenie DCS Compendium**. A new, limited range named **The Balvenie Stories** was launched in May 2019. It compriss of three expressions; the **12 year old The Sweet Toast of American Oak**, matured in first fill bourbon barrels and with a 12 week finish in double toasted American oak, the **14 year old**, heavily peated **A Week of Peat** and a **26 year old** called **A Day of Dark Barley**. For Duty Free there is the **Triple Cask series (12, 16** and **25 years old)** as well as the **21 year old Madeira Cask** and the **14 year old Peated Triple Cask**.

History:

1892 William Grant rebuilds Balvenie New House to Balvenie Distillery (Glen Gordon was the name originally intended). Part of the equipment is brought in from Lagavulin and Glen Albyn.

1893 The first distillation takes place in May.

1957 The two stills are increased by another two.

1965 Two new stills are installed.

1971 Another two stills are installed and eight stills are now running.

1973 The first official bottling appears.

1982 Founder's Reserve is launched.

1996 Two vintage bottlings and a Port wood finish are launched.

2001 The Balvenie Islay Cask is released.

2002 A 50 year old is released.

2004 The Balvenie Thirty is released.

2005 The Balvenie Rum Wood Finish 14 years old is released.

2006 The Balvenie New Wood 17 years old, Roasted Malt 14 years old and Portwood 1993 are released.

2007 Vintage Cask 1974 and Sherry Oak 17 years old are released.

2008 Signature, Vintage 1976, Balvenie Rose and Rum Cask 17 year old are released.

2009 Vintage 1978, 17 year old Madeira finish, 14 year old rum finish and Golden Cask 14 years old are released.

2010 A 40 year old, Peated Cask and Carribean Cask are released.

2011 Second batch of Tun 1401 is released.

2012 A 50 year old and Doublewood 17 years old are released.

2013 Triple Cask 12, 16 and 25 years are launched for duty free.

2014 Single Barrel 15 and 25 years, Tun 1509 and two new 50 year olds are launched.

2015 The Balvenie DCS Compendium is launched.

2016 A 21 year old madeira finish is released.

2017 The Balvenie Peat Week 2002 and Peated Triple Cask are released.

2018 A limited 25 year old is relased.

2019 The Balvenie Stories is released.

Doublewood 12 years old

Tasting notes Balvenie Doublewood 12 years:

GS – Nuts and spicy malt on the nose, full-bodied, with soft fruit, vanilla, sherry and a hint of peat. Dry and spicy in a luxurious, lengthy finish.

Ben Nevis

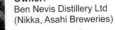
[ben nev•iss]

Owner:
Ben Nevis Distillery Ltd
(Nikka, Asahi Breweries)

Region/district:
Western Highlands

Founded: | **Status:** | **Capacity:**
1825 | Active (vc) | 2 000 000 litres

Address: Lochy Bridge, Fort William PH33 6TJ

Website:
bennevisdistillery.com

Tel:
01397 702476

In December 2018, a rumour that Ben Nevis was about to close within the next couple of years due to a land lease dispute started to spread. It turned out to be just a rumour, but perhaps there was an ounce of truth in it.

The distillery lies on land owned by an aluminium smelter established in 1929 and a couple of years ago sold by Rio Tinto Alcan Inc to GFG Alliance. A 99 year lease, dating back to 1950, exists which means that Ben Nevis could continue producing for at least another 30 years. What happens then remains to be seen. Apart from releasing small amounts of single malt the owners since 1989, Nikka, use the distillery as a supplier of newmake for their hugely popular blend Black Nikka. Up to 75% of the production is shipped to Japan every year. The current whisky legislation in Japan allows foreign whisky in a product that still can be labelled Japanese whisky.

Ben Nevis is equipped with a ten ton full lauter mash tun made of stainless steel, six stainless steel washbacks and two made of Oregon pine with a 48 hour fermentation as well as two pairs of stills. The plan for 2019 is to do 13 mashes in a five-day week and 2 million litres of alcohol. Around 50,000 litres of this will be heavily peated with a phenol specification of 40ppm in the barley. Early 2018, the distillery installed an LPG boiler (gas) instead of the old one which was fired by heavy fuel oil.

The core range consists of **MacDonald's Ben Nevis 10 year old** and the peated **MacDonald's Traditional Ben Nevis**. The latter is an attempt to replicate the style of Ben Nevis single malt from the 1880s. In 2018, a limited version of the 10 year old was released under the name **Ben Nevis 10 years old Batch No. 1**. Bottled at 62,4% the whisky had been matured in a combination of first fill bourbon and sherry and wine casks. Very old versions of Ben Nevis have occurred in recent years with three special bottlings distilled in **1966**, **1967** and **1968** respectively going to Taiwan during 2019.

History:

1825 The distillery is founded by 'Long' John McDonald.

1856 Long John dies and his son Donald P. McDonald takes over.

1878 Demand is so great that another distillery, Nevis Distillery, is built nearby.

1908 Both distilleries merge into one.

1941 D. P. McDonald & Sons sells the distillery to Ben Nevis Distillery Ltd headed by the Canadian millionaire Joseph W. Hobbs.

1955 Hobbs installs a Coffey still which makes it possible to produce both grain and malt whisky.

1964 Joseph Hobbs dies.

1978 Production is stopped.

1981 Joseph Hobbs Jr sells the distillery back to Long John Distillers and Whitbread.

1984 After restoration and reconstruction totalling £2 million, Ben Nevis opens up again.

1986 The distillery closes again.

1989 Whitbread sells the distillery to Nikka Whisky Distilling Company Ltd.

1990 The distillery opens up again.

1991 A visitor centre is inaugurated.

1996 Ben Nevis 10 years old is launched.

2006 A 13 year old port finish is released.

2010 A 25 year old is released.

2011 McDonald's Traditional Ben Nevis is released.

2014 Forgotten Bottlings are introduced.

2015 A 40 year old "Blended at Birth" single blend is released.

2018 Ben Nevis 10 years old Batch No. 1 is released.

10 years old

Tasting notes Ben Nevis 10 years old:

GS – The nose is initially quite green, with developing nutty, orange notes. Coffee, brittle toffee and peat are present on the slightly oily palate, along with chewy oak, which persists to the finish, together with more coffee and a hint of dark chocolate.

Benriach

[ben•ree•ack]

Owner:
BenRiach Distillery Company
(Brown Forman)

Region/district:
Speyside

Founded: 1897
Status: Active (vc)
Capacity: 2 800 000 litres

Address: Longmorn, Elgin, Morayshire IV30 8SJ

Website: benriachdistillery.com
Tel: 01343 862888

With 27 years in the whisky industry, Rachel Barrie is an experienced master blender. Over the years, she has been working with great brands such as Glenmorangie, Ardbeg, Bowmore and Laphroaig.

In early 2017 she took over from Billy Walker as blender for BenRiach, following the company take-over by Brown Forman. The following year was exceptionally successful for Rachel Barrie when she became the first female master blender to receive an honorary doctorate from the University of Edinburgh. This was followed later in the same year with her induction into Whisky Magazine's Hall of Fame.

The distillery is located just by the A941, a few miles south of Elgin. It is impossible to miss when driving by but despite the excellent location, BenRiach doesn't have a proper visitor centre. However, with many people stopping by and asking to see the distillery, the owners decided to offer tours on Tuesdays and Thursdays with the possibility to buy bottles in a small shop.

BenRiach distillery is equipped with a 5.8 ton traditional cast iron mash tun with a stainless steel shell, eight washbacks made of stainless steel with short (55 hours) and long fermentations (+100 hours) and two pairs of stills. The production for 2019 will be 1.8 million litres which includes peated spirit at 35ppm as well as 15,000 litres of triple-distilled spirit. In 2013, the owners revamped the malting floor but it has only been used sporadically, most recently in April 2019..

The core range consists of **Heart of Speyside**, **Cask Strength**, **10, 12, 21, 25** and **30** year old. Peated varieties include **Birnie Moss, Curiositas 10 year old, Temporis 21 year old** (aged in virgin oak, ex-bourbon, ex-oloroso and ex-PX casks) and **Authenticus 30 year old**. Currently, there are three different wood finishes, all of them 22 years old - **Dark Rum, Dunder** and the peated **Albariza**. Exclusive to the travel retail segment are **10 year old Triple-distilled** as well as duty-free versions of **Classic Quarter Cask** and **Peated Quarter Cask**. Finally, batch 16 of the **Cask Bottlings** (single casks) was launched in July 2019 with no less than 24 different single casks!

History:

1897 John Duff & Co founds the distillery.

1900 The distillery is closed.

1965 The distillery is reopened by the new owner, The Glenlivet Distillers Ltd.

1972 Production of peated Benriach starts.

1978 Seagram Distillers takes over.

1985 The number of stills is increased to four.

1998 The maltings is decommissioned.

2002 The distillery is mothballed in October.

2004 Intra Trading, buys Benriach together with the former Director at Burn Stewart, Billy Walker.

2004 Standard, Curiositas and 12, 16 and 20 year olds are released.

2005 Four different vintages are released.

2006 Sixteen new releases, i.a. a 25 year old, a 30 year old and 8 different vintages.

2007 A 40 year old and three new heavily peated expressions are released.

2008 New expressions include a peated Madeira finish, a 15 year old Sauternes finish and nine single casks.

2009 Two wood finishes (Moscatel and Gaja Barolo) and nine single casks are released.

2010 Triple distilled Horizons and heavily peated Solstice are released.

2011 A 45 year old and 12 vintages are released.

2012 Septendecim 17 years is released.

2013 Vestige 46 years is released. The maltings are working again.

2015 Dunder, Albariza, Latada and a 10 year old are released.

2016 Brown Forman buys the company for £285m. BenRiach cask strength and Peated Quarter Cask are launched.

2017 10 year old Triple Distilled and Peated Cask Strength are released.

2018 A 12 and a 21 year old as well as Temporis 21 year old and Authenticus 30 year old are released.

Tasting notes BenRiach 10 year old:

GS – Earthy and nutty on the early nose, with apples, ginger and vanilla. Smooth and rounded on the palate, with oranges, apricots, mild spice and hazelnuts. The finish is medium in length, nutty and spicy.

10 years old

2

Benrinnes

[ben rin•ess]

Owner:		Region/district:
Diageo		Speyside

Founded:	Status:	Capacity:
1826	Active	3 500 000 litres

Address: Aberlour, Banffshire AB38 9NN

Website:	Tel:
malts.com	01340 872600

For those travelling the southern parts of Speyside, the mountain of Benrinnes is a well-known feature in the landscape. From its 840 metres above sea level peak, one can see eight different Scottish counties.

With its location, right at the foot of the mountain, it is also appropriate that Benrinnes distillery has taken the same name. It draws its process water from wells on the mountain in the same way as four other distilleries located a little bit further away do – namely Aberlour, Allt-a-Bhainne, Dailuaine and Glenfarclas.

There are no visible remains of the first Benrinnes distillery dating back to 1826. In fact it was situated a few kilometres from where the current one is. Nor can the second distillery be found, instead you will see a creation from the 1950s.

A major upgrade took place in autumn 2012 which included a full automation of the process, as well as a new control room where one operator can handle all the machinery. The equipment consists of an 8.5 ton semi-lauter mash tun, eight washbacks made of Oregon pine with a fermentation time ranging from 65 to 100 hours. There are also two wash stills and four spirit stills. From 1966 until a few years ago, these were run three and three with a partial triple distillation. This system has now been abandoned and one wash still now serves two spirit stills. The spirit vapours are cooled using cast iron worm tubs which contribute to the character of Benrinnes' newmake, which is light sulphury. The wide spirit cut (73%-58%) also plays its part in creating a robust and meaty spirit. In the last couple of years, Benrinnes has been alternating between a seven-day production week and a five-day with either 21 or 15 mashes per week.

Most of the production goes into blended whiskies – J&B, Johnnie Walker and Crawford's 3 Star – and there is currently only one official single malt, the **Flora & Fauna 15 year old**. In 2010 a **Manager's Choice** from 1996 was released and in autumn 2014 it was time for a **21 year old Special Release** bottled at 57%.

History:

1826 Lyne of Ruthrie distillery is built at Whitehouse Farm by Peter McKenzie.

1829 A flood destroys the distillery and a new distillery is constructed by John Innes a few kilometres from the first one.

1834 John Innes files for bankruptcy and William Smith & Co takes over.

1864 William Smith & Co goes bankrupt and David Edward becomes the new owner.

1896 Benrinnes is ravaged by fire which prompts major refurbishment. Alexander Edward takes over.

1922 John Dewar & Sons takes over ownership.

1925 John Dewar & Sons becomes part of Distillers Company Limited (DCL).

1956 The distillery is completely rebuilt.

1964 Floor maltings is replaced by a Saladin box.

1966 The number of stills doubles to six.

1984 The Saladin box is taken out of service and the malt is purchased centrally.

1991 The first official bottling from Benrinnes is a 15 year old in the Flora & Fauna series.

1996 United Distillers releases a 21 year old cask strength in their Rare Malts series.

2009 A 23 year old is launched as a part of this year´s Special Releases.

2010 A Manager´s Choice 1996 is released.

2014 A limited 21 year old is released.

15 years old

Tasting notes Benrinnes 15 years old:

GS – A brief flash of caramel shortcake on the initial nose, soon becoming more peppery and leathery, with some sherry. Ultimately savoury and burnt rubber notes. Big-bodied, viscous, with gravy, dark chocolate and more pepper. A medium-length finish features mild smoke and lively spices.

Benromach

[ben•ro•mack]

Owner:
Gordon & MacPhail

Region/district:
Speyside

Founded: 1898
Status: Active (vc)
Capacity: 700 000 litres

Address: Invererne Road, Forres,
Morayshire IV36 3EB

Website:
benromach.com

Tel:
01309 675968

To get to Benromach, you only need to look for the signs along the A96 between Elgin and Inverness. It is set in the small town of Forres, known for its connection with Macbeth.

Macbeth was a Scottish king from 1040 to 1057 and he was made immortal in William Shakespeares famous play with the same name. Tourists may come to Forres for that reason but in recent years also in increasing numbers to visit Benromach. Last year 13,000 people toured the distillery and in 2018 it achieved a Five Star Visitor Attraction status from Visit Scotland. There are four different tours to choose from and you can also take part in a gin tour as Red Door gin is produced at the distillery as well.

The goal at Benromach is to produce a Speyside whisky, just like it used to taste back in the 1950s. This is achieved by predominantly using medium peated barley (12ppm). The distillery is equipped with a 1.5 ton semi-lauter mash tun with a copper dome and 13 washbacks made of larch (nine of which were commissioned in early 2017) with a fermentation time of 67-115 hours. There is also one pair of stills with the condensers outside. In 2019 the production will be 14 mashes per week and 400,000 litres of pure alcohol including two weeks of peated production. On site there are five dunnage warehouses, with another two coming in by the end of 2019.

The core range consists of **10** and **15 year old** and the new **Cask Strength Vintage 2008** (which replaces the 100° Proof). Recent limited releases include a **1978 single cask**, a **2009 Triple Distilled** and a **20th Anniversary bottling**. In spring 2019, two vintage expressions (**1972** and **1977**) were released as part of the Heritage Collection. There are also special editions; **Organic 2011** and **Peat Smoke 2008**. A special version of the latter is **Peat Smoke Sherry Cask Matured** which was released in March 2019. Recent wood finishes include **Chateau Cissac 2010** and **Sassicaia 2011**. In October 2018, the very first cask that was filled after the re-opening in 1998, was launched. All profits from **Benromach Cask No. 1** were donated to charitable causes

History:

1898 Benromach Distillery Company starts the distillery.

1911 Harvey McNair & Co buys the distillery.

1919 John Joseph Calder buys Benromach and sells it to recently founded Benromach Distillery Ltd owned by several breweries.

1931 Benromach is mothballed.

1937 The distillery reopens.

1938 Joseph Hobbs buys Benromach and sells it on to National Distillers of America (NDA).

1953 NDA sells Benromach to Distillers Company Limited (DCL).

1966 The distillery is refurbished.

1968 Floor maltings is abolished.

1983 Benromach is mothballed.

1993 Gordon & McPhail buys Benromach.

1998 The distillery is once again in operation.

1999 A visitor centre is opened.

2004 The first bottle distilled by the new owner is 'Benromach Traditional'.

2005 A Port Wood finish (22 years old) and a Vintage 1968 are released together with the Benromach Classic 55 years.

2006 Benromach Organic is released.

2007 Peat Smoke, the first heavily peated whisky from the distillery, is released.

2008 Benromach Origins Golden Promise is released.

2009 Benromach 10 years old is released.

2011 New edition of Peatsmoke, a 2001 Hermitage finish and a 30 year old are released.

2014 Three new bottlings are launched; a 5 year old, 100 Proof and Traveller's Edition.

2015 A 15 year old and two wood finishes (Hermitage and Sassicaia) are released.

2016 A 35 year old and 1974 single cask are released.

2017 A 1976 single cask and a 2009 Triple Distilled are released.

2018 A 20th Anniversary bottling and a Sassicaia 2010 are released.

2019 Peat Smoke Sherry Cask Matured is released.

Tasting notes Benromach 10 year old:

GS – A nose that is initially quite smoky, with wet grass, butter, ginger and brittle toffee. Mouth-coating, spicy, malty and nutty on the palate, with developing citrus fruits, raisins and soft wood smoke. The finish is warming, with lingering barbecue notes.

10 years old

Bladnoch

[blad•nock]

Owner: David Prior   **Region/district:** Lowlands

Founded: 1817   **Status:** Active (vc)   **Capacity:** 1 500 000 litres

Address: Bladnoch, Wigtown, Wigtonshire DG8 9AB

Website: bladnoch.com   **Tel:** 01988 402605

The "new" Bladnoch has only been running for close to three years but it's already time for a change of the guards. When Australian entrepreneur David Prior bought the distillery he managed to engage one of the stalwarts of the industry – Ian MacMillan.

Ian has been working with whisky for more than 45 years and before he joined the Bladnoch team, he was responsible for Burn Stewart's three distilleries as well as their range of whiskies. In January 2019 he started up his own whisky consultancy firm and his successor as master distiller is Nick Savage. Prior to joining the Bladnoch team, Nick was the Macallan master distiller for three years. The owners have also been working hard on the construction of a visitor centre and in July 2019 it was finally opened to the public.

The distillery is equipped with a five ton stainless steel semi-lauter mash tun, six Douglas fir washbacks with a combination of short (76 hours) and long (100 hours) fermentations and two pairs of stills. The plan for 2019 is to do 6-7 mashes per week and 600,000 litres of pure alcohol. A small part of heavily peated production started already in 2017 and will continue in 2019 with 45,000 litres.

When the new owners took over, the deal included several thousand casks of whisky dating back to the 1980s. Where needed, MacMillan re-filled whiskies from inferior casks into quality wood and in November 2016, the first new whiskies were released. The range today includes **Samsara** with no age statement and matured in ex-bourbon and casks that had contained Californian red wine, the 15 year old **Adela** matured in ex-oloroso casks and the 25 year old **Talia**. Two more Talia expressions have since been released – a **25 year old port finish** and a **27 year old bourbon cask finish**. For the 200th anniversary, a **Vintage 1988** finished in moscatel casks was released as well as **10 year old** matured in ex-bourbon. The latest limited release is a **17 year old California red wine finish**.

History:

1817 Founded by Thomas and John McClelland.

1878 John McClelland's son Charlie reconstructs and refurbishes the distillery.

1905 Production stops.

1911 Dunville & Co. buys T. & A. McClelland Ltd. Production is intermittent until 1936.

1937 Dunville & Co. is liquidated and Bladnoch is wound up. Ross & Coulter from Glasgow buys the distillery after the war. The equipment is dismantled and shipped to Sweden.

1956 A. B. Grant (Bladnoch Distillery Ltd.) takes over and restarts production with four new stills.

1964 McGown and Cameron becomes new owners.

1973 Inver House Distillers buys Bladnoch.

1983 Arthur Bell and Sons take over.

1985 Guiness Group buys Arthur Bell & Sons which, from 1989, are included in United Distillers.

1988 A visitor centre is built.

1993 United Distillers mothballs Bladnoch in June.

1994 Raymond Armstrong buys Bladnoch in October.

2000 Production commences in December.

2003 The first bottles from Armstrong are launched, a 15 year old cask strength from UD casks.

2008 First release of whisky produced after the take-over in 2000 - three 6 year olds.

2009 An 8 year old of own production and a 19 year old are released.

2014 The distillery is liquidated.

2015 The distillery is bought by David Prior.

2016 Samsara, Adela and Talia are released.

2017 Production starts again and a Vintage 1988 is released.

2018 A 10 year old is released.

2019 A visitor centre is opened.

Samsara

Tasting notes Bladnoch Samsara:

GS – The nose is slightly savoury, with soft spices and tinned peaches with cream. Sweet on the palate, with passion fruit and vanilla. Long in the finish with ripe pears, and increasingly dry spices, plus a suggestion of tannins.

Blair Athol

[blair ath•ull]

Owner:	**Region/district:**
Diageo	Southern Highlands
Founded: **Status:**	**Capacity:**
1798 Active (vc)	2 800 000 litres

Address: Perth Road, Pitlochry, Perthshire PH16 5LY

Website:	**Tel:**
malts.com	01796 482003

In the last five years alone, the number of visitors to Scottish distilleries has increased by almost 100% and in 2018, around two million people travelled to a distillery to find out more about Scotch.

One would have thought that being well-established and situated in close proximity to a major road would be a prerequisite for receiving high visitor numbers, but that's not the fact. Three of the five most visited distilleries are on islands – Arran, Skye and Harris. Perhaps the biggest surprise is Harris distillery in the distant Outer Hebrides which attracted more than 90,000 visitors in 2018 – their third year of opening!

Blair Athol on the other hand has always been able to rely on the busy traffic of the A9 with people going from Edinburgh north to the Highlands. It is the most visited of all Diageo distilleries, attracting 89,000 visitors in 2018. A new feature since last year is a used mash tun from Clynelish which has been turned into a whisky tasting bar! Blair Athol is the spiritual home of Bell's blended whisky. Connected from the very start in 1896 when Bell's was introduced, the relation became even stronger when Bell's acquired the distillery in 1933. The independence of Arthur Bell & Sons lasted much longer than most of the old family companies and it wasn't until 1985 that they were sold to Guinness and later became a part of Diageo.

The equipment of Blair Athol distillery consists of an 8.2 ton semi-lauter mash tun, six washbacks made of stainless steel and two pairs of stills. The part of the spirit which goes into Bell's is matured mainly in bourbon casks, while the rest is matured in sherry casks. The last couple of years, the distillery has been working a five-day week with 12 mashes per week and around two million litres of alcohol. This also means a scheme of short (46 hours) and long (104 hours) fermentations. A very cloudy wort gives Blair Athol newmake a nutty and malty character.

The only official bottling is the **12 year old Flora & Fauna**. In autumn 2017, however, a **23 year old**, matured in ex-bodega European oak butts was released as part of the Special Releases.

History:

1798 John Stewart and Robert Robertson found Aldour Distillery, the predecessor to Blair Athol. The name is taken from the adjacent river Allt Dour.

1825 The distillery is expanded by John Robertson and takes the name Blair Athol Distillery.

1826 The Duke of Atholl leases the distillery to Alexander Connacher & Co.

1860 Elizabeth Connacher runs the distillery.

1882 Peter Mackenzie & Company Distillers Ltd of Edinburgh (future founder of Dufftown Distillery) buys Blair Athol and expands it.

1932 The distillery is mothballed.

1933 Arthur Bell & Sons takes over by acquiring Peter Mackenzie & Company.

1949 Production restarts.

1973 Stills are expanded from two to four.

1985 Guinness Group buys Arthur Bell & Sons.

1987 A visitor centre is built.

2003 A 27 year old cask strength from 1975 is launched in Diageo's Rare Malts series.

2010 A distillery exclusive with no age statement and a single cask from 1995 are released.

2016 A distillery exclusive without age statement is released.

2017 A 23 year old is released as part of the Special Releases.

12 years old

Tasting notes Blair Athol 12 years old:

GS – The nose is mellow and sherried, with brittle toffee. Sweet and fragrant. Relatively rich on the palate, with malt, raisins, sultanas and sherry. The finish is lengthy, elegant and slowly drying.

Bowmore

[bow•moor]

Owner: | **Region/district:**
Beam Suntory | Islay

Founded: | **Status:** | **Capacity:**
1779 | Active (vc) | 2 000 000 litres

Address: School Street, Bowmore, Islay, Argyll PA43 7GS

Website: | **Tel:**
bowmore.com | 01496 810441

One of Bowmore distillery's impressive features is the oldest existing warehouse in Scotland – Vault No. 1 – in which part of the warehouse walls are actually situated below sea level.

The Vaults have been highlighted recently in the way of a special series of whiskies and, in 2019, a special tasting room will also be opened in the Vaults. The distillery is one of only a few Scottish distilleries with its own floor maltings (three floors), with 30% of the malt requirement being produced in-house. The remaining part is bought from Simpson's. Both parts have a phenol specification of 25-30 ppm and are always mixed on a ratio of 2.5 tons in house malt and 5.5 tons of malt from Simpsons before mashing. At least until 2016. That year (and the following two) 7 mashes were made using only malt from their own floors. In 2019, this increased to 28 mashes (two weeks of production).

Bowmore is equipped with an eight ton stainless steel semi-lauter mash tun with a copper lid. It was previously in use at Jura distillery. There are also two magnificent and unusual hot water tanks made of copper to feed the mash tun. The six washbacks are made of Oregon pine, with both short (60 hours) and long (90 hours) fermentations and have all been named after previous owners of the distillery. One washback will feed one of the two wash stills and is then in turn split between the two spirit stills. The 27,000 casks are stored in two dunnage and one racked warehouse. In 2019, they will be doing 14 mashes per week, seven short and seven long fermentations, which amounts to 1.8 million litres of alcohol.

The core range for domestic markets includes **No. 1, 12 years, 15 years** (with the name Darkest having been dropped), **18 years** and **25 years**. A limited release was made in autumn 2016, highlighting the influence from Vault No. 1. The first expression, bottled at 51.5% was called Bowmore Vault Edition with the addition "Atlantic Sea Salt". A second instalment named **Peat Smoke** was released in 2018. In autumn 2017 a **50 year old**, distilled in 1966, was released which was followed by a **Vintage 1965** in 2018. A new range, highlighting how wine casks interact with Bowmore single malt, was introduced in autumn 2017. The first two bottlings in the Vintner's Trilogy were the **18 year old Double Matured Manzanilla** and a **26 year old** which had received a second maturation (for 13 years) in ex-wine barriques. The third expression, a **27 year old port finish**, was released in summer 2018. The **36 year old Dragon Edition**, an exclusive to China and the first in a range of four, was launched in August 2019. The duty free line-up was completely revamped in spring 2017. The new range consists of **10 year old** (Dark and Intense), **15 year old** (Golden and Elegant) and **18 year old** (Deep and Complex). Limited releases for duty free also occur, the latest (in 2019) being a **21 year old**. Finally, there were two bottlings for Feis Ile 2019; a **23 year old Vintage 1995** (55.2%) matured in sherry casks and a **15 year old** (51.7%) matured in ex-bourbon casks.

History:

1779 Bowmore Distillery is founded by David Simpson and becomes the oldest Islay distillery.

1837 The distillery is sold to James and William Mutter of Glasgow.

1892 After additional construction, the distillery is sold to Bowmore Distillery Company Ltd, a consortium of English businessmen.

1925 J. B. Sheriff and Company takes over.

1929 Distillers Company Limited (DCL) takes over.

1950 William Grigor & Son takes over.

1963 Stanley P. Morrison buys the distillery and forms Morrison Bowmore Distillers Ltd.

1989 Japanese Suntory buys a 35% stake in Morrison Bowmore.

1993 The legendary Black Bowmore is launched.

1994 Suntory now controls all of Morrison Bowmore.

1996 A Bowmore 1957 (38 years) is bottled at 40.1% but is not released until 2000.

1999 Bowmore Darkest with three years finish on Oloroso barrels is launched.

2000 Bowmore Dusk with two years finish in Bordeaux barrels is launched.

2001 Bowmore Dawn with two years finish on Port pipes is launched.

2002 A 37 year old Bowmore from 1964 and matured in fino casks is launched in a limited edition of 300 bottles (recommended price £1,500).

2003 Another two expressions complete the wood trilogy which started with 1964 Fino - 1964 Bourbon and 1964 Oloroso.

2005 Bowmore 1989 Bourbon (16 years) and 1971 (34 years) are launched.

History continued:

2006 Bowmore 1990 Oloroso (16 years) and 1968 (37 years) are launched. A new and upgraded visitor centre is opened.

2007 An 18 year old is introduced. 1991 (16yo) Port and Black Bowmore are released.

2008 White Bowmore and a 1992 Vintage with Bordeaux finish are launched.

2009 Gold Bowmore, Maltmen's Selection, Laimrig and Bowmore Tempest are released.

2010 A 40 year old and Vintage 1981 are released.

2011 Vintage 1982 and new batches of Tempest and Laimrig are released.

2012 100 Degrees Proof, Springtide and Vintage 1983 are released for duty free.

2013 The Devil´s Casks, a 23 year old Port Cask Matured and Vintage 1984 are released.

2014 Black Rock, Gold Reef and White Sands are released for duty free.

2015 New editions of Devil´s Cask, Tempest and the 50 year old are released as well as Mizunara Cask Finish.

2016 A 9 year old, a 10 year old travel retail exclusive and Bowmore Vault Edit1on are released as well as the final batch of Black Bowmore.

2017 No. 1 is released together with three new expressions for travel retail.

2018 Vintner´s Trilogy is launched.

2019 Vault Edit1on Peat Smoke, a 21 year old for duty free and the 36 year old Dragon Edition are released.

Tasting notes Bowmore 12 year old:

GS – An enticing nose of lemon and gentle brine leads into a smoky, citric palate, with notes of cocoa and boiled sweets appearing in the lengthy, complex finish.

Vintner´s Trilogy
27 year old Port Cask

18 years old
Travel Retail

Vault Edition
Peat Smoke

No. 1

12 years old

25 years old

3

Braeval

[bre•vaal]

Owner: | **Region/district:**
Chivas Brothers (Pernod Ricard) | Speyside

Founded: | **Status:** | **Capacity:**
1973 | Active | 4 200 000 litres

Address: Chapeltown of Glenlivet, Ballindalloch, Banffshire AB37 9JS

Website: | **Tel:**
- | 01542 783042

The largest surge in demand from Scotch whiskies are currently in single malts and companies are keen to release bottlings from distilleries that have until now been reserved for producing malt for blends.

Chivas Brothers started doing this a little later than others but have been very active in the last couple of years. The final distillery in their range (except for Dalmunach which started just four years ago) to receive such attention was Braeval. An official cask strength available only at the Chivas distilleries had been around for a while but since summer 2019 there are no less than three, widely available bottlings from the owners.

They are all part of a new series called The Secret Speyside Collection with Caperdonich, Glen Keith and Longmorn being the other three included. In this range, Braeval is called "the remote distillery" which is not an understatement. Situated in the isolated Braes south of Glenlivet, it is not a distillery you stumble across by accident. When you drive on the B9008, you reach the small hamlet, Auchnarrow, which is just a mile south of Tamnavulin. From there you follow the narrow road east towards Chapeltown and drive until the road ends. Set in an idyllic and pastoral environment, lies the impressive Braeval. This area was regarded as a haven for illicit distillers from the 1780s to the early 1800s and the whisky was smuggled out of the valley along narrow paths.

The equipment at Braeval consists of a 9 ton stainless steel, full lauter mash tun, 13 stainless steel washbacks with a fermentation time of 70 hours and six stills. There are two wash stills with aftercoolers and four spirit stills, and with the possibility of doing 26 mashes per week, the distillery can now produce 4.2 million litres per year.

In 2017, a **16 year old single cask**, available at Chivas' visitor centres was released. This was followed in July 2019 with three bottlings, initially reserved for duty-free but will be avaliable in domestic retail in 2020; **25, 27** and **30 year old**, all matured in ex-bourbon barrels and hogsheads. The first two are bottled at 48% and the 30 year old is a cask strength (50.3%).

History:

1973 The distillery is founded by Chivas Brothers (Seagram's) and production starts in October.

1975 Three stills are increased to five.

1978 Five stills are further expanded to six.

1994 The distillery changes name to Braeval.

2001 Pernod Ricard takes over Chivas Brothers.

2002 Braeval is mothballed in October.

2008 The distillery starts producing again in July.

2017 The first official bottling, a 16 year old single cask, is released.

2019 Three new official bottlings in a new range, The Secret Speyside Collection, are released.

25 years old

Tasting notes Braeval 16 year old:

GS – Marzipan, milk chocolate-coated Turkish Delight and orange peel on the nose. The palate is sweet and fruity, with stewed apples, sugared almonds, nutmeg and ginger. Medium to long in the finish, consistently sugary and spicy.

Pioneers of Whisky

Annabel Thomas
Founder
Ncnéan Distillery, Scotland

It is a long and winding road before you reach Ncnéan Distillery on Morvern peninsula in western Scotland but it is well worth the effort once you arrive. Not only is the old farmstead, now turned into a distillery, beautiful; the view towards Mull and Tobermory is also truly stunning.

One may wonder why someone would build a distillery in such a remote place but the answer is quite simple: Derek and Louise Lewis bought the 7,000 acre Drimnin Estate in 2001 and their daughter Annabel is the founder of the distillery, so the land for the distillery site was already in the family. But Annabel admits there are other, more emotional reasons.

"I've always had links with Scotland – and I got married in the tiny chapel that sits on the shore just below the distillery."

She spent some years in London working as a management consultant for a large company but also had a two year spell with Innocent drinks where she obtained an insight into how to run a responsible but commercial enterprise. Her relationship with whisky is quite recent.

"Until about 2012 I didn't like whisky. I'd only ever tried it neat when friends or family had been drinking it, and probably also frequently tried peated whisky and I honestly thought it was disgusting. I finally had a 'light bulb' moment just before my first trip to Islay when i fell in love with it – and sadly I can't actually remember which whisky it was that converted me."

Her reasons for building Ncnéan (which by the way is pronounced nook-knee-anne) was twofold. Together with her parents there had always been the vision of creating sustainable employment in the area based on the farm and the land. But even more fundamental was Annabel's belief that there was a space within Scotch to look at whisky with a new pair of eyes. Asked to elaborate on that she says "For me, this means creating as sustainable a whisky as possible and challenging the 'norms' of how whisky is made and drunk."

"My observation was that there was a lot more going on to interest the next generation of consumers in whisky outside Scotland, and in other categories like gin, than from within Scotland. And with such great heritage and quality, that seemed a shame. So I took inspiration more from outside the Scotch industry – Sipsmith, Kavalan, Starward – than within."

The foundation of the distillery was funded by a number of individuals with one core investor and there are now 44 shareholders with the Lewis family still owning a significant minority. The actual construction of the distillery was challenging, not least due to the location. Some of the issues were lack of accommodation for the builders, having a three-hour round trip to the nearest building suppliers and power supply when you are at the end of a single phase line. While these issues were more or less anticipated, there were other things that turned out to be more or less difficult as the work proceeded, such as how to operate the biomass boiler so that it would work with the very up and down demand for steam or how to operate and maintain the chipper so it had a reliable source of wood.

Annabel explains; "Perhaps it would be fair to generalise that it's our efforts to operate the distillery as sustainably as possible that seem to have caused most of the headaches!"

This brings us to one of two key things when it comes to the direction to take for the distillery. Sustainability on all levels is fundamental and this spans sourcing organic raw materials, equipment and production techniques for example using on-site timber for the biomass boiler, recycling the cooling water or having a closed loop on site process for waste disposal. But it doesn't stop at that. Ncnéan Distillery is constantly working on how to reduce their chemical and plastic usage and make their packaging as sustainable as possible.

The second thing has more to do with the actual product. "… an attitude or approach that gives us leeway to experiment, to think about flavour before we think about rules. That is completely fundamental to us, but not something we particularly designed the physical distillery around – just something that is central to who we are."

All the whisky currently produced at Ncnéan is un-peated and while a couple of core recipes are followed there is still plenty of space for experiments, for example with different kinds of yeast. One spirit that wasn't even in the business plan from the beginning is their Botanical Spirit, which basically is their newmake flavoured with various botanicals, some of them sourced locally. So far that is the only product released from the distillery.

While Annabel is fine with the basic, wider regulations that define a whisky, she is more hesitant when it comes to the detailed rules around Scotch.

"I think they can stifle innovation, cause consumer confusion (by defining these rather complex categories – single malts, vatted malts, grain etc) and also appear to go against modern consumer demands (e.g. the regulations surrounding what you can and cannot disclose about the components that make up each bottle of Scotch fly in the face of a move towards more and more transparency in business)."

Obviously the next step for the distillery, which has been producing now for a little more than two years, is to start thinking about sales and distribution.

Without getting into any specifics, Annabel says "What we do know is that we are not planning to take a traditional route – for example in how we recommend the whisky is drunk, the distributors we partner with and the bars, restaurants and shops we target to sell."

Annabel's final advice
to someone wanting to start a distillery of their own;

" Really challenge yourself on why you want to do it – you must get clear on that – both from a business and personal point of view – and then test it on other people – do they think the world needs it? The reason this is so important is that a) you will probably have to raise a lot of money and if other people aren't excited by your 'why' then it will be difficult to achieve and b) it is going to take a long time and a lot of effort, and unless you are 100% committed to your idea, it's going to be hard to stick with it."

Bruichladdich

[brook•lad•dee]

Owner:
Rémy Cointreau

Region/district:
Islay

Founded: 1881
Status: Active (vc)
Capacity: 1 500 000 litres

Address: Bruichladdich, Islay, Argyll PA49 7UN

Website:
bruichladdich.com

Tel:
01496 850221

Bruichladdich has always been concerned with how different barley varieties affect the taste of the whisky and, not least, if the soil has an impact as well.

At the same, they want to make a 100% Islay whisky. The only problem is that few farmers on the island used to grow barley. One of the issues, and an unusual one, is the tens of thousands of protected Barnacle Geese wintering on the island and feasting on crops. Together with farmers, the distillery has been working on developing barley varieties that can be sown later and harvested earlier when the geese are not around. Bruichladdich was one of the first distilleries in Scotland in modern times to produce rye whisky. In 2017, 30 casks made of three types of American oak were filled with spirit with a mashbill of 55% unmalted rye and 45% malted barley. Following a pause in 2018, trials of rye whisky production was recommenced in 2019.

The distillery is equipped with a 7 ton cast iron, open mash tun with rakes, six washbacks of Oregon pine with a fermentation time between 60 and 100 hours and two pairs of stills. All whisky produced is based on Scottish barley, 50% of which comes from Islay and with 5% being organically grown. During 2019, they will be doing 9-10 mashes per week and one million litres of alcohol. The breakdown of the three whisky varieties during 2019 are 50% Bruichladdich, 40% Port Charlotte and 10% Octomore.

The malting floors at Bruichladdich were closed in 1961 but the decision has now been taken to start up own malting again within the next three to five years. Instead of traditional floor malting, Saladin boxes will be installed. The main reason for this is to keep up the consistency of the spirit since Baird's, their current malt supplier, are using that same technique. The malting will be located in one of the current warehouses.

Bruichladdich has three product lines; unpeated Bruichladdich, heavily peated Port Charlotte and ultra-heavily peated Octomore. The only core expressions, in the sense that they are widely available and not released by vintage, are **The Classic Laddie** and **Port Charlotte 10 year old**. The following appear every year with new batches/vintages; for Bruichladdich there are **Islay Barley 2011, Bere Barley 2010** and **The Organic 2010** and for **Port Charlotte, Islay Barley 2011** and **MRC:01 2010**. The duty free range is made up of **The Laddie Eight, Bruichladdich 1990, Port Charlotte MC:01** and **Octomore 09.2**. Recent limited expressions include the **Rare Cask** series which is made up of three Bruichladdich bottlings distilled in **1984-1986**. **Black Art 7** was released in September 2019 and was followed in October by three **Octomores**; the 5 year old, bourbonmatured **10.1**, the 8 year old **10.2** with a double maturation (bourbon/sauternes) and the 5 year old **10.3** where the barley comes from Octomore farm on Islay. January 2020 will see the release of the 3 year old Octomore **10.4** matured in virgin Limousin oak. The special bottlings for Feis Ile 2019 were the 12 year old **Octomore Event Horizon** and the 13 year old, sherry matured **Bruichladdich Valinch**.

History:

1881 Barnett Harvey builds the distillery with money left by his brother William III to his three sons William IV, Robert and John Gourlay.

1886 Bruichladdich Distillery Company Ltd is founded and reconstruction commences.

1929 Temporary closure.

1936 The distillery reopens.

1938 Joseph Hobbs, Hatim Attari and Alexander Tolmie purchase the distillery through the company Train & McIntyre.

1952 The distillery is sold to Ross & Coulter.

1960 A. B. Grant buys Ross & Coulter.

1961 Own maltings ceases.

1968 Invergordon Distillers take over.

1975 The number of stills increases to four.

1983 Temporary closure.

1993 Whyte & Mackay buys Invergordon Distillers.

1995 The distillery is mothballed in January.

1998 In production again for a few months.

2000 Murray McDavid buys the distillery from JBB Greater Europe for £6.5 million.

2001 The first distillations of Port Charlotte and Bruichladdich starts in July.

2002 Octomore, the world's most heavily peated whisky (80ppm) is distilled.

2004 Second edition of the 20 year old (nick-named Flirtation) and 3D, also called The Peat Proposal, are launched.

2005 Several new expressions are launched - the second edition of 3D, Infinity, Rocks, Legacy Series IV, The Yellow Submarine and The Twenty 'Islands'.

2006 The first official bottling of Port Charlotte; PC5.

2007 New releases include Redder Still, Legacy 6, PC6 and an 18 year old.

History continued:

2008 New expressions include the first Octomore, Bruichladdich 2001, PC7 and Golder Still.

2009 New releases include Classic, Organic, Black Art, Infinity 3, PC8, Octomore 2 and X4+3 - the first quadruple distilled single malt.

2010 PC Multi Vintage, Organic MV, Octomore/3_152, Bruichladdich 40 year old are released.

2011 The first 10 year old from own production is released as well as PC9 and Octomore 4_167.

2012 Ten year old versions of Port Charlotte and Octomore are released as well as Laddie 16 and 22, Bere Barley 2006, Black Art 3 and DNA4. Rémy Cointreau buys the distillery.

2013 Scottish Barley, Islay Barley Rockside Farm, Bere Barley 2nd edition, Black Art 4, Port Charlotte Scottish Barley, Octomore 06.1 and 06.2 are released.

2014 PC11 and Octomore Scottish Barley are released.

2015 PC12, Octomore 7.1 and High Noon 134 are released.

2016 The Laddie Eight, Octomore 7.4 and Port Charlotte 2007 CC:01 are released.

2017 Black Art 5 and 25 year old sherry cask are launched. The limited Rare Cask series is launched.

2018 The Port Charlotte range is revamped and a 10 year old and Islay Barley 2011 are released.

2019 Bere Barley 10, Organic 10, Black Art 7 and Octomore 10.1, 10.2, 10.3 and 10.4 are released.

Bere Barley 2010   Black Art 7   The Organic 2010

Tasting notes Bruichladdich Scottish Barley:

GS – Mildly metallic on the early nose, then cooked apple aromas develop, with a touch of linseed. Initially very fruity on the gently oily palate. Ripe peaches and apricots, with vanilla, brittle toffee, lots of spice and sea salt. The finish is drying, with breakfast tea.

Tasting notes Port Charlotte Scottish Barley:

GS – Wood smoke and contrasting bonbons on the nose. Warm Tarmac develops, with white pepper. Finally, fragrant pipe tobacco. Peppery peat and treacle toffee on the palate, with a maritime note. Long in the finish, with black pepper and oak.

Tasting notes Octomore Scottish Barley:

GS – A big hit of sweet peat on the nose; ozone and rock pools, supple leather, damp tweed. Peat on the palate is balanced by allspice, vanilla and fruitiness. Very long in the finish, with chilli, dry roasted nuts and bonfire smoke.

Port Charlotte
Islay Barley 2011

The Classic Laddie Scottish Barley

Port Charlotte
10 year old

Bunnahabhain

[buh•nah•<u>hav</u>•enn]

Owner:
Distell International Ltd.

Region/district:
Islay

Founded: **Status:**
1881   Active (vc)

Capacity:
3 200 000 litres

Address: Port Askaig, Islay, Argyll PA46 7RP

Website:
bunnahabhain.com

Tel:
01496 840646

The distillery is currently going through an intensive time of much needed upgrading. Or to quote a company official - Bunnahabhain has been "underinvested in". Four warehouses have been demolished and another 15 houses have been knocked down.

Part of the refurbishment will entail a new visitor centre which will be situated on the left side of the road, approaching the distillery. Hopefully, this will be ready in late 2019/early 2020. On the right hand side of the road where the warehouses used to be, a new filling store will be built as well as more space for parking. Most of the produce will from now on be matured on the mainland. The production side of the distillery also gets its share of the investment where the manual mashhouse will become computer controlled. The entire project will cost £10.5m.

The distillery is equipped with a 12.5 ton traditional stainless steel mash tun with a copper lid, six washbacks made of Oregon pine and two pairs of stills. Two of the washbacks were replaced in early 2018 and another two will be replaced in late 2019. The fermentation time varies between 55 and 110 hours. The production for 2019 will be 2.1 million litres, split between 40% peated and 60% unpeated. The peated volume has increased during the last years and the peating level has also increased to 35-45ppm.

The core range consists of **12, 18** and **25 year olds** as well as a **40 year old**. The peated side of Bunnahabhain is represented by **Toiteach a Dha** without age statement and matured in both bourbon and sherry casks and **Stiùireadair**, matured in first and re-fill sherry casks. Recent limited bottlings (released in July 2019) include a **2007 port pipe finish**, a **2007 French brandy finish** and a **1988 Marsala finish**. There are three travel retail exclusives – **Cruach-Mhòna** which comprises of young, heavily peated Bunnahabhain, **Eirigh Na Greine**, a vatting of whisky from bourbon, sherry and red wine casks and the sherry-matured **An Cladach**. Two special bottlings were released for Feis Ile 2019, a peated **2008 Moine French oak finish** and a **2001 sauternes finish**.

History:

1881   William Robertson of Robertson & Baxter, founds the distillery together with the brothers William and James Greenless, owners of Islay Distillers Company Ltd.

1883   Production starts in earnest in January.

1887   Islay Distillers Company Ltd merges with William Grant & Co. in order to form Highland Distilleries Company Limited.

1963   The two stills are augmented by two more.

1982   The distillery closes.

1984   The distillery reopens. A 21 year old is released to commemorate the 100th anniversary.

1999   Edrington takes over Highland Distillers and mothballs Bunnahabhain but allows for a few weeks of production a year.

2001   A 35 year old from 1965 is released during Islay Whisky Festival.

2002   A 35 year old from 1965 is released during Islay Whisky Festival. Auld Acquaintance 1968 is launched at the Islay Jazz Festival.

2003   Edrington sells Bunnahabhain and Black Bottle to Burn Stewart Distilleries for £10 million. A 40 year old from 1963 is launched.

2004   The first limited edition of the peated version is a 6 year old called Moine.

2005   Three limited editions are released - 34 years old,18 years old and 25 years old.

2006   14 year old Pedro Ximenez and 35 years old are launched.

2008   Darach Ur is released for the travel retail market and Toiteach (a peated 10 year old) is launched on a few selected markets.

2009   Moine Cask Strength is released.

2010   The peated Cruach-Mhòna and a limited 30 year old are released.

2013   A 40 year old is released.

2014   Eirigh Na Greine and Ceobanach are released.

2017   Moine Oloroso, Stiùireadair and An Cladach are released.

2018   Toiteach a Dha and a 20 year old Palo Cortado are released.

2019   A 2007 brandy finish and a 1988 marsala finish are released.

Tasting notes Bunnahabhain 12 years old:

GS – The nose is fresh, with light peat and discreet smoke. More overt peat on the nutty and fruity palate, but still restrained for an Islay. The finish is full-bodied and lingering, with a hint of vanilla and some smoke.

12 years old

Caol Ila

[cull eel•a]

Owner:	**Region/district:**
Diageo	Islay

Founded:	**Status:**	**Capacity:**
1846	Active (vc)	6 500 000 litres

Address: Port Askaig, Islay, Argyll PA46 7RL

Website:	**Tel:**
malts.com	01496 302760

Following the first appearances in the Flora & Fauna and Rare Malts series, Caol Ila was properly introduced by the owners in 2002 as a single malt with a dedicated core range.

Today, approximately 15% of the production is reserved for single malts while the rest is an important part of the Johnnie Walker blend, not least Black Label. That is also why the distillery was chosen to be one of four Diageo distilleries that would have their visitor centres completely transformed within the next couple of years. This is in line with the huge investment in a Johnnie Walker experience in Edinburgh where distilleries in the four corners of Scotland representing the four pillars of the Johnnie Walker character will be highlighted. Caol Ila will of course represent the smoky side of the famous blend. The visitor centre is planned to be ready in 2020 and it will include a bar looking out across the Sound of Islay and a footbridge connecting the car park and the roof of one of the warehouses leading directly in to the visitor centre.

Caol Ila is equipped with a 12.5 ton full lauter mash tun, eight wooden washbacks (two of which will be replaced in November 2019) and two made of stainless steel and three pairs of stills. In recent years, the distillery has either been working a seven-day week with 26 mashes or a five-day week with 16 mashes. On a 5-day week production, there is a mix of short (55 hours) and long (120 hours) fermentations. Caol Ila is known for its peated whisky but unpeated new-make is also produced. The cut points for the two versions are the same at 72-63%.

The core range consists of **Moch** without age statement, **12, 18** and **25 year old, Distiller's Edition** with a moscatel finish and **Cask Strength**. The release for Feis Ile 2019 was a **10 year old** that had been matured in refill American oak hogsheads and rejuvenated European oak butts. As usual, Caol Ila was also represented by an unpeated expression in the Special Releases 2018. This time it was a **15 year old**. But that was not all – in the range there could also be found an unusually old Caol Ila, a **35 year old** bottling of the peated version.

History:
1846 Hector Henderson founds Caol Ila.

1852 Henderson, Lamont & Co. is subjected to financial difficulties and Henderson is forced to sell Caol Ila to Norman Buchanan.

1863 Norman Buchanan sells to the blending company Bulloch, Lade & Co. from Glasgow.

1879 The distillery is rebuilt and expanded.

1920 Bulloch, Lade & Co. is liquidated and the distillery is taken over by Caol Ila Distillery.

1927 DCL becomes sole owners.

1972 All the buildings, except for the warehouses, are demolished and rebuilt.

1974 The renovation, which totals £1 million, is complete and six new stills are installed.

1999 Experiments with unpeated malt.

2002 The first official bottlings since Flora & Fauna/Rare Malt appear; 12 years, 18 years and Cask Strength (c. 10 years).

2003 A 25 year old cask strength is released.

2006 Unpeated 8 year old and 1993 Moscatel finish are released.

2007 Second edition of unpeated 8 year old.

2009 The fourth edition of the unpeated version (10 year old) is released.

2010 A 25 year old, a 1999 Feis Isle bottling and a 1997 Manager´s Choice are released.

2011 An unpeated 12 year old and the unaged Moch are released.

2012 An unpeated 14 year old is released.

2013 Unpeated Stitchell Reserve is released.

2014 A 15 year old unpeated and a 30 year old are released.

2016 A 15 year old unpeated is released.

2017 An 18 year old unpeated is released.

2018 Two bottlings in the Special Releases - a 15 year old and a 35 year old.

Tasting notes Caol Ila 12 year old:
GS – Iodine, fresh fish and smoked bacon feature on the nose, along with more delicate, floral notes. Smoke, malt, lemon and peat on the slightly oily palate. Peppery peat in the drying finish.

12 years old

Cardhu

[car•<u>doo</u>]

Owner:
Diageo

Region/district:
Speyside

Founded: 1824
Status: Active (vc)
Capacity: 3 400 000 litres

Address: Knockando, Aberlour, Moray AB38 7RY

Website:
malts.com

Tel:
01479 874635

The connection between Cardhu single malt and the Johnnie Walker blended Scotch dates back to 1893 when the distillery was the first to be bought by John Walker & Sons for the sum of £20,500.

The blending company had been a customer of the distillery for many years and when Johnnie Walker finally took over the distillery, they decided to let the previous owners continue operations. It couldn't have been in safer hands: Helen Cumming was the widow of the founder John Cumming and, together with her daughter-in-law, Elizabeth, the two became legends in Scotch whisky history. They skilfully managed and expanded the distillery and Elizabeth was dubbed "The Queen of the Whisky Trade".

Both women will have a central part in the new distillery visitor centre that is about to be built. A planning application was submitted to the Council in April 2019 and when finished it will be a part of the Johnnie Walker Four Corners project, a £150m investment with a major visitor experience in central Edinburgh and four "hubs" at Cardhu, Clynelish, Caol Ila and Glenkinchie. One of the new features at Cardhu will be a warehouse tasting experience.

Cardhu distillery is equipped with an eight ton stainless steel full lauter mash tun with a copper top, ten washbacks (eight made of Douglas fir and two of stainless steel in a separate room), all with a fermentation time of 75 hours, and three pairs of stills. Four of the wooden washbacks are new, having replaced four old ones made of larch. In 2019, Cardhu will be working a seven-day week with 21 mashes per week and a production of 3.4 million litres of alcohol.

Cardhu is Diageo's third best selling single malt after The Singleton and Talisker with close to 3 million bottles sold in 2018. The core range from the distillery is **12, 15** and **18 year old** and two expressions without age statement – **Amber Rock** and **Gold Reserve**. There is also a **Special Cask Reserve** matured in rejuvenated bourbon casks. In spring 2019, **Cardhu Gold Reserve House Targaryen** was released in the Game of Thrones series and this was followed in autumn by a **14 year old** Special Releases.

History:

1824 John Cumming applies for and obtains a licence for Cardhu Distillery.

1846 John Cumming dies and his wife Helen and son Lewis takes over.

1872 Lewis dies and his wife Elizabeth takes over.

1884 A new distillery is built to replace the old.

1893 John Walker & Sons purchases Cardhu for £20,500.

1908 The name reverts to Cardow.

1960 Reconstruction and expansion of stills from four to six.

1981 The name changes to Cardhu.

1998 A visitor centre is constructed.

2002 Diageo changes Cardhu single malt to a vatted malt with contributions from other distilleries in it.

2003 The whisky industry protests sharply against Diageo's plans.

2004 Diageo withdraws Cardhu Pure Malt.

2005 The 12 year old Cardhu Single Malt is relaunched and a 22 year old is released.

2009 Cardhu 1997, a single cask in the new Manager´s Choice range is released.

2011 A 15 year old and an 18 year old are released.

2013 A 21 year old is released.

2014 Amber Rock and Gold Reserve are launched.

2016 A distillery exclusive is released.

2019 A 14 year old appears in the Special Releases.

Tasting notes Cardhu 12 years old:

GS – The nose is relatively light and floral, quite sweet, with pears, nuts and a whiff of distant peat. Medium-bodied, malty and sweet in the mouth. Medium-length in the finish, with sweet smoke, malt and a hint of peat.

Game of Thrones
House Targaryen

Websites to watch

scotchwhisky.com
Without a doubt the best whiskysite there is! It covers every possible angle of the subject in an absolutely brilliant way.

whiskyfun.com
Serge Valentin, one of the Malt Maniacs, is almost always first with well written tasting notes on new releases.

Ralfydotcom on Youtube
Ralfy does this video blog with tastings and field reports in an educational yet easy-going and entertaining way.

malt-whisky-madness.com
A new site from Johannes van den Heuvel but he´s also working on the reconstruction of legendary maltmadness.com.

edinburghwhiskyblog.com
Mainly run by Chris White these days it´s about reviews of new releases and news from the whisky world.

whiskycast.com
The best whisky-related podcast on the internet and one that sets the standard for podcasts in other genres as well.

whiskyintelligence.com
The best site on all kinds of whisky news. The first whisky website you should log into every morning!

whisky-news.com
Apart from daily news, this site contains tasting notes, distillery portraits, lists of retailers, events etc.

thewhiskylady.net
Anne-Sophie Bigot´s mission is "to remove the dusty cliché that whisky is only an old man´s drink" and she does it so well!

malt-review.com
Mark, Jason and Abby provide very honest and comprehensive reviews on whiskies as well as well-written features.

meleklerinpayi.com
One of the most visited whisky blogs in the world and soon Burkay Adalig will publish the first whisky book in Turkish.

whiskynotes.be
This blog is almost entirely about tasting notes (and lots of them, not least independent bottlings) plus some news.

whiskyforeveryone.com
Educational site, perfect for beginners, with a blog where both new releases and affordable standards are reviewed.

blog.thewhiskyexchange.com
A knowledgeable team from The Whisky Exchange write about new bottlings and the whisky industry in general.

whisky-distilleries.net
Ernie Scheiner describes more than 130 distilleries in both text and photos and we are talking lots of great images!

whiskyandwisdom.com
Excellent blog by Andrew Derbidge. Reviews but also distillery features, news, opinions and insightful articles.

canadianwhisky.org
Davin de Kergommeaux presents reviews, news and views on all things Canadian whisky. High quality content.

whiskyisrael.co.il
Gal Granov is definitely one of the most active of all bloggers. Well worth checking out daily!

spiritsjournal.klwines.com
Reviews about whiskies and the whisky industry in general from the US retailer K&L Wines.

thewhiskywire.com
Steve Rush mixes reviews of the latest bottlings with presentations of classics plus news, interviews etc.

bestshotwhiskyreviews.com
Jan van den Ende presents his honest opinions on everything from cheap blends to rare single cask bottlings.

scotch-whisky.org.uk
The official site of SWA (Scotch Whisky Association) with i.a. press releases and publications about the industry.

whiskysaga.com
Brilliant blog by Norwegian whisky enthusiast Thomas Öhrbom - not least on every detail relating to Nordic whiskies.

whiskysponge.com
The brilliantly sarcastic blog by Angus Macraild is where everyone in the business secretely wants to be mentioned.

speller.nl
Thomas and Ansgar Speller explore the world of whisky and they´ve been to more distilleries than most people.

thewhiskeywash.com
Since 2015, this great team of whisky writers has brought us initiated reviews and recently they started a podcast as well.

whiskeyreviewer.com
This is a web magazine covering the world of whisky through news, reviews, features and interviews.

thewhiskyphiles.com
An incredibly comprehensive blog filled no just with well written reviews but also news, comments, distillery profiles...

spiritedmatters.com
When Billy Abbott, known from The Whisky Exchange website, writes something it´s always thoughtful and interesting.

whiskywaffle.com
Enjoy the no-nonsense reviews and comments about affordable whiskies and not just the latest unobtainable single cask.

allthingswhisky.com
Even though postings sometimes occur intermittently, these reviews and musings are always an interesting read.

thedramble.com
Loads of (currently around 900) extensive and well-written reviews paired with well-researched articles

Clynelish

[cline•leash]

Owner: Diageo

Region/district: Northern Highlands

Founded: 1967

Status: Active (vc)

Capacity: 4 800 000 litres

Address: Brora, Sutherland KW9 6LR

Website: malts.com

Tel: 01408 623003 (vc)

Clynelish malt is a vital part of several Johnnie Walker expressions, not least the Gold Label version. Responsible for the quality and consistency of the world's most sold Scotch is Dr. Jim Beveridge who has been with Diageo for almost forty years.

Even though his team consists of twelve, Dr. Beveridge is ultimately responsible as the Master Blender and he is only the sixth person to hold that title since the brand was established in the 1800s. Most blenders have favourite malts that they find more useful in blending than others. If pressed for an answer, Jim's favourites are Talisker and Clynelish. For his long time of service to the Scotch whisky industry, he was awarded an OBE (Officer of the Most Excellent Order of the British Empire) in June 2019. The accolade couldn't have come at a better time with Johnnie Walker celebrating its 200th anniversary in 2020.

Following a year-long upgrade which was completed in June 2017, the distillery is now equipped with a 12.5 ton full lauter mash tun, eight wooden washbacks and two made of stainless steel. The still room, with its three pairs of stills, has stunning views towards the village of Brora and the North Sea. Clynelish is usually operational seven-days a week, producing around 4.8 million litres of alcohol. Approximately 6,000 casks of Clynelish are stored in the two old Brora warehouses next door, but most of the production is matured elsewhere.

A completely new visitor centre is currently being built at Clynelish and at the same time, the second distillery on site, Brora, will open up again in 2020 after a substantial upgrade and renovation.

Official bottlings include a **14 year old** and a **Distiller's Edition**, with an Oloroso Seco finish. Recent limited bottlings include **Clynelish Select Reserve** which was launched two years in a row as part of the annual Special Releases. In spring 2019, Clynelish was also part of the series of eight single malts named after houses in the popular Tv series Game of Thrones. **Clynelish House of Tyrell** was launched without age statement and bottled at 51.6%.

History:

1819 The 1st Duke of Sutherland founds a distillery called Clynelish Distillery.

1827 The first licensed distiller, James Harper, files for bankruptcy and John Matheson takes over.

1846 George Lawson & Sons become new licensees.

1896 James Ainslie & Heilbron takes over.

1912 James Ainslie & Co. narrowly escapes bankruptcy and Distillers Company Limited (DCL) takes over together with James Risk.

1916 John Walker & Sons buys a stake of James Risk's stocks.

1931 The distillery is mothballed.

1939 Production restarts.

1960 The distillery becomes electrified.

1967 A new distillery, also named Clynelish, is built adjacent to the first one.

1968 'Old' Clynelish is mothballed in August.

1969 'Old' Clynelish is reopened as Brora and starts using a very peaty malt.

1983 Brora is closed in March.

2002 A 14 year old is released.

2006 A Distiller's Edition 1991 finished in Oloroso casks is released.

2009 A 12 year old is released for Friends of the Classic Malts.

2010 A 1997 Manager's Choice single cask is released.

2014 Clynelish Select Reserve is released.

2015 Second version of Clynelish Select Reserve is released.

2017 The distillery produces again after a year long closure for refurbishing.

2019 Clynelish House Tyrell is released as part of the Game of Thrones series.

Game of Thrones
House Tyrell

Tasting notes Clynelish 14 year old:

GS – A nose that is fragrant, spicy and complex, with candle wax, malt and a whiff of smoke. Notably smooth in the mouth, with honey and contrasting citric notes, plus spicy peat, before a brine and tropical fruit finish.

Cragganmore

[crag•an•moor]

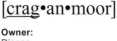

Owner: **Region/district:**
Diageo   Speyside

Founded: **Status:**   **Capacity:**
1869   Active (vc)   2 200 000 litres

Address: Ballindalloch, Moray AB37 9AB

Website: **Tel:**
malts.com   01479 874700

Barely 200 metres from the river Spey, at the end of a narrow road lies Cragganmore distillery. In spite of being one of the six original Classic Malts and situated in the busy and well-visited Speyside area, only 8,000 people find their way here each year.

One can't help thinking that the location has something to do with it. From the busy A95, between Aberlour and Boat of Garten, you have to take a small road to reach the distillery and the sign is easily missed. For those who finally arrive, there are plenty of tours to choose from. Those who select a more expensive one end up with a tasting in the extraordinary Clubroom. With a library, an open fire and Chesterfield furniture it resembles a beautiful drawing room from Victorian days. Once the distillery office, it was from here the founder John Smith ran the distillery in the 1870s.

Cragganmore is equipped with a 6.8 ton stainless steel full lauter mash tun with a copper canopy and six washbacks made of Oregon pine. Since they are working a five-day week in 2019 there will be six short (60 hours) and six long (90 hours) fermentations. There are two large wash stills with sharply descending lyne arms and two considerably smaller spirit stills with boil balls and long, slightly descending lyne arms. The two spirit stills are most peculiar with flat tops, which had already been introduced during the times of the founder, John Smith. The stills are attached to worm tubs on the outside for cooling the spirit vapours. In 2019, the production will amount to 1.65 million litres of alcohol.

Cragganmore single malt plays an important part in Old Parr blended whisky which was first introduced in 1909 and is very popular in Japan and Latin America. Except for a brief surge in the first decade of the new millennium, sales have decreased since its heydays and it sold around 12 million bottles in 2018. The official core range of Cragganmore is made up of a **12 year old** and a **Distiller's Edition** with a finish in port pipes. In 2019, a **12 year old** bottled at cask strength appeared as part of the Special Releases.

History:

1869 John Smith, who already runs Glenfarclas distillery, founds Cragganmore.

1886 John Smith dies and his brother George takes over operations.

1893 John's son Gordon, at 21, is old enough to assume responsibility for operations.

1901 The distillery is refurbished and modernized with help of the famous architect Charles Doig.

1912 Gordon Smith dies and his widow Mary Jane supervises operations.

1917 The distillery closes.

1918 The distillery reopens and Mary Jane installs electric lighting.

1923 The distillery is sold to the newly formed Cragganmore-Glenlivet Distillery Co. where Mackie & Co. and Sir George Macpherson-Grant of Ballindalloch Estate share ownership.

1927 White Horse Distillers is bought by DCL which thus obtains 50% of Cragganmore.

1964 The number of stills is increased from two to four.

1965 DCL buys the remainder of Cragganmore.

1988 Cragganmore 12 years becomes one of six selected for United Distillers' Classic Malts.

1998 Cragganmore Distillers Edition Double Matured (port) is launched for the first time.

2002 A visitor centre opens in May.

2006 A 17 year old from 1988 is released.

2010 Manager's Choice single cask 1997 and a limited 21 year old are released.

2014 A 25 year old is released.

2016 A Special Releases vatting without age statement and a distillery exclusive are released.

2019 A 12 year bottled at cask strength appears in the Special Releases series.

12 years old

Tasting notes Cragganmore 12 years old:

GS – A nose of sherry, brittle toffee, nuts, mild wood smoke, angelica and mixed peel. Elegant on the malty palate, with herbal and fruit notes, notably orange. Medium in length, with a drying, slightly smoky finish.

Craigellachie

Owner:	**Region/district:**
John Dewar & Sons (Bacardi)	Speyside
Founded: **Status:**	**Capacity:**
1891 Active	4 100 000 litres

Address: Aberlour, Banffshire AB38 9ST

Website:	**Tel:**
craigellachie.com	01340 872971

The whisky boom in the late 19th century put pressure on the big, blending companies. Stocks were running low due to the rapidly increased demand and building or buying a distillery, became necessary.

Peter Mackie, admittedly, already owned Lagavulin, but the sales of his White Horse blend increased so rapidly that he needed another distillery and, together with Alexander Edward, he built Craigellachie. He established a sales office in London but became annoyed with the competition from the large producers of grain whisky which, according to him, sold "young, cheap, fiery whisky". During his entire life he was an advocate for quality products and his opinion was crystal clear: "What the public wants is age and plenty of it".

Craigellachie distillery is equipped with a Steinecker full lauter mash tun, installed in 2001, which replaced the old, open cast iron mash tun. There are also eight 47,000 litre washbacks made of larch with a fermentation time of 56-60 hours and two pairs of stills. Both stills are attached to worm tubs. The old cast iron tubs were exchanged for stainless steel in 2014 and the existing copper worms were moved to the new tubs. Production in 2019 will be 21 mashes per week and 3.9 million litres of alcohol.

Apart from a 14 year old, which at times could be hard to get hold of, there was no official bottling of Craigellachie until 2014, when the brand was re-launched. Three new expressions (**13, 17** and **23 year old**) were released for selected domestic markets. This was followed by a range for duty free, **19, 31** and **33 year old**, where the first two now have been discontinued. The 33 year old was later rolled out onto the US domestic market. Two small batch releases were added to the duty free range in early 2018 – a **24 year old** and a **17 year old** with a palo cortado finish. The oldest official bottling of Craigellachie is a **51 year old** which was launched in autumn 2018 in a rather unusual way. Instead of releasing at a premium price, the whisky was offered to whisky enthusiasts for free! Pop up bars in the UK, Australia, South Africa and the US displayed the whisky and to be able to enjoy the rare drops you had to register for a ticket.

History:

1890 The distillery is built by Craigellachie–Glenlivet Distillery Company which has Alexander Edward and Peter Mackie as part-owners.

1891 Production starts.

1916 Mackie & Company Distillers Ltd takes over.

1924 Peter Mackie dies and Mackie & Company changes name to White Horse Distillers.

1927 White Horse Distillers are bought by Distillers Company Limited (DCL).

1930 Administration is transferred to Scottish Malt Distillers (SMD), a subsidiary of DCL.

1964 Refurbishing takes place and two new stills are bought, increasing the number to four.

1998 United Distillers & Vintners (UDV) sells Craigellachie together with Aberfeldy, Brackla and Aultmore and the blending company John Dewar & Sons to Bacardi Martini.

2004 The first bottlings from the new owners are a new 14 year old which replaces UDV's Flora & Fauna and a 21 year old cask strength from 1982 produced for Craigellachie Hotel.

2014 Three new bottlings for domestic markets (13, 17 and 23 years) and one for duty free (19 years) are released.

2015 A 31 year old is released.

2016 A 33 year old and a 1994 Madeira single cask are released.

2018 A 24 year old and and a 17 year old palo cortado -finish are released for duty free and the oldest official Craigellachie so far, 51 years old, is launched.

13 years old

Tasting notes Craigellachie 13 years old:

GS – Savoury on the early nose, with spent matches, green apples and mixed nuts. Malt join the nuts and apples on the palate, with sawdust and very faint smoke. Drying, with cranberries, spice and more subtle smoke.

Dailuaine

[dall•yoo•an]

Owner: **Region/district:**
Diageo   Speyside

Founded: **Status:**   **Capacity:**
1852   Active   5 200 000 litres

Address: Carron, Banffshire AB38 7RE

Website: **Tel:**
malts.com   01340 872500

Reading through Alfred Barnard's "The Whisky Distilleries of the United Kingdom" from the late 1880s makes you realise how much has changed since then and, yet at the same time, how much remains the same.

As for Dailuaine, a distillery that he designates a full seven pages, it is clear that the setting and the surroundings caught him off guard. Barnard, in awe, writes: "No wonder with these surroundings that the pure spirit emerging from such an Eden should be appreciated by mortals all the world over" and adds "The whole scene is dainty enough for a fairy's palace."

Even today, Dailuaine epitomises what a Speyside distillery is all about. You drive along the winding roads, crossing the river Spey on numerous occasions and then, all of a sudden, you come across yet another distillery hidden in the glen. As romantic as it seems, Dailuaine was in those days one of the biggest distilleries in the area, producing close to one million litres per year.

Dailuaine distillery is equipped with a stainless steel, 11.25 ton full lauter mash tun, eight washbacks made of Douglas fir, plus two stainless steel ones placed outside and three pairs of stills. All the condensers are made of copper as usual but, until a few years ago, some were made of stainless steel to help achieve a sulphury style of new make. In 2015, the fermentation time was changed to help achieve a more waxy character to the spirit. The reason for the change was that Clynelish distillery had been closed for refurbishing. That is the only Diageo distillery so far that has accounted for this style which is so important for some blends. During 2019 they will be doing four short (80 hours) and eight long fermentations (107 hours) per week amounting to 2.6 million litres of pure alcohol.

Dailuaine is one of many distilleries whose main task is to produce malt whisky which is to become part of blended Scotch. The only core bottling is the **16 year old Flora & Fauna**. In 2015, a **34 year old** from 1980 was launched as part of the Special Releases.

History:

1852 The distillery is founded by William Mackenzie.

1865 William Mackenzie dies and his widow leases the distillery to James Fleming, a banker from Aberlour.

1879 William Mackenzie's son forms Mackenzie and Company with Fleming.

1891 Dailuaine-Glenlivet Distillery Ltd is founded.

1898 Dailuaine-Glenlivet Distillery Ltd merges with Talisker Distillery Ltd and forms Dailuaine-Talisker Distilleries Ltd.

1915 Thomas Mackenzie dies without heirs.

1916 Dailuaine-Talisker Company Ltd is bought by the previous customers John Dewar & Sons, John Walker & Sons and James Buchanan & Co.

1917 A fire rages and the pagoda roof collapses. The distillery is forced to close.

1920 The distillery reopens.

1925 Distillers Company Limited (DCL) takes over.

1960 Refurbishing. The stills increase from four to six and a Saladin box replaces the floor maltings.

1965 Indirect still heating through steam is installed.

1983 On site maltings is closed down and malt is purchased centrally.

1991 The first official bottling, a 16 year old, is launched in the Flora & Fauna series.

1996 A 22 year old cask strength from 1973 is released as a Rare Malt.

1997 A cask strength version of the 16 year old is launched.

2000 A 17 year old Manager's Dram matured in sherry casks is launched.

2010 A single cask from 1997 is released.

2012 The production capacity is increased by 25%.

2015 A 34 year old is launched as part of the Special Releases.

16 years old

Tasting notes Dailuaine 16 years old:

GS – Barley, sherry and nuts on the substantial nose, developing into maple syrup. Medium-bodied, rich and malty in the mouth, with more sherry and nuts, plus ripe oranges, fruitcake, spice and a little smoke. The finish is lengthy and slightly oily, with almonds, cedar and slightly smoky oak.

Dalmore

[dal•moor]

Owner:
Whyte & Mackay Ltd
(Emperador Inc)

Region/district:
Northern Highlands

Founded: 1839
Status: Active (vc)
Capacity: 4 300 000 litres

Address: Alness, Ross-shire IV17 0UT

Website:
thedalmore.com

Tel:
01349 882362

Dalmore opened up their first visitor centre in 2004 and in 2011 it was refurbished. But with an excellent location next to the busy A9 going from Inverness to Thurso, the distillery should be able to attract more visitors than the 12,000 now coming here annually.

Apparently the owners were of the same opinion and planning permission for a new visitor centre was granted in October 2018. An office building and a filling store were demolished to make way for a completely renovated and expanded brand home which will more than double in size. The plan was to have it ready by August 2019. Added to the distillery's location is the splendid view across the Cromarty Firth, a view which is sadly spoiled by the many oil rigs now resting in the water waiting for scrapping due to the North Sea oil downturn.

The distillery is equipped with a 10.4 ton stainless steel, semi-lauter mash tun, eight washbacks made of Oregon pine with a fermentation time of 50 hours and four pairs of stills. All the wash stills have peculiar flat tops while the spirit stills are equipped with water jackets, which allow cold water to circulate between the reflux bowl and the neck of the stills, thus increasing the reflux. The owners expect to do 22 mashes per week during 2019, producing close to 4 million litres.

The core range consists of **12, 15, 18** and **25 year old, 1263 King Alexander III** and **Cigar Malt Reserve**. A new addition to the range in 2018 was the **Port Wood Reserve**. Initially matured in ex-bourbon casks, half of the stock was transferred into tawny port pipes for additional maturation and then married together with the ex-bourbon casks. **Valour**, exclusive to travel retail, was joined in 2016 by **Regalis** with a finish in Amoroso casks, **Luceo** with a final maturation in Apostoles casks and **Dominium** finished in Matusalem casks. Recent limited bottlings include new versions of the **35** and **40 year old**, a range called **Vintage Port Collection** with three different expressions and, in spring 2018, a **45 year old**. The highlight of limited releases from the distillery appeared in autumn 2019 when a **60 year old** was launched.

History:

1839 Alexander Matheson founds the distillery.
1867 Three Mackenzie brothers run the distillery.
1891 Sir Kenneth Matheson sells the distillery for £14,500 to the Mackenzie brothers.
1917 The Royal Navy moves in to start manufacturing American mines.
1920 The Royal Navy moves out and leaves behind a distillery damaged by an explosion.
1922 The distillery is in production again.
1956 Floor malting replaced by Saladin box.
1960 Mackenzie Brothers (Dalmore) Ltd merges with Whyte & Mackay.
1966 Number of stills is increased to eight.
1982 The Saladin box is abandoned.
1990 American Brands buys Whyte & Mackay.
1996 Whyte & Mackay changes name to JBB (Greater Europe).
2001 Through management buy-out, JBB (Greater Europe) is bought from Fortune Brands and changes name to Kyndal Spirits.
2002 Kyndal Spirits changes name to Whyte & Mackay.
2007 United Spirits buys Whyte & Mackay. A 15 year old, and a 40 year old are released.
2008 1263 King Alexander III is released.
2009 New releases include an 18 year old, a 58 year old and a Vintage 1951.
2010 The Dalmore Mackenzie 1992 is released.
2011 More expressions in the River Collection and 1995 Castle Leod are released.
2012 The visitor centre is upgraded and Constellaton Collection is launched.
2013 Valour is released for duty free.
2014 Emperador Inc buys Whyte & Mackay.
2016 Three new travel retail bottlings are released as well as a 35 year old and Quintessence.
2017 Vintage Port Collection is launched.
2018 The Port Wood Reserve is released.
2019 A 60 year old is released.

Tasting notes Dalmore 12 years old:

GS – The nose offers sweet malt, orange marmalade, sherry and a hint of leather. Full-bodied, with a dry sherry taste though sweeter sherry develops in the mouth along with spice and citrus notes. Lengthy finish with more spices, ginger, Seville oranges and vanilla.

12 years old

Dalwhinnie

[dal•whin•nay]

Owner:
Diageo

Region/district:
Speyside

Founded: **Status:**
1897   Active (vc)

Capacity:
2 200 000 litres

Address: Dalwhinnie, Inverness-shire PH19 1AB

Website:
malts.com

Tel:
01540 672219 (vc)

USA has since many years back been one of the most important markets for Scotch whisky. Over the years, American companies have also approached Scottish distilleries with acquisition offers.

There are several US-owned distilleries today but already as long ago as in 1905 Cook & Bernheimer took over Dalwhinnie. Cook & Bernheimer, established already in 1863, was a spirits wholesaler with outlets all over USA. Their impressive Head Office was situated on Franklin Street, south Manhattan in the area nowadays known as Tribeca. But the company is also testament to how quickly the spirit business changed. At the start of prohibition in 1920 the company was dissolved and Dalwhinne came under Scottish ownership again.

The distillery is equipped with a 7.3 ton full lauter mash tun and six wooden washbacks with the fermentation split into four short sessions of 60 hours and six long, fermenting over the weekend, of 110 hours. There is one pair of stills, replaced in 2018, attached to worm tubs which were replaced with new ones in 2015. In February and March 2018 the distillery was closed when the two stills were exchanged for new ones. A five-day production week for 2019 means 10 mashes per week which gives 1.4 million litres of alcohol in the year. Dalwhinnie is one of Diageo's best selling single malts and comes in at sixth place with 1.2 million bottles. It is also a key malt in two major Scotch blends – Buchanans and Black & White. The latter has increased sales by no less than 93% since 2014 and now sells 32 million bottles, mainly in Brazil, Mexico and Colombia.

The core range is made up of a **15 year old, Distiller's Edition** with a finish in oloroso casks and **Dalwhinnie Winter's Gold**. In 2018, **Lizzie's Dram**, a distillery exclusive was laun- ched. In spring 2019, Dalwhinnie was also part of the series named after the popular TV series Game of Thrones. **Dalwhinnie Winter's Frost**, representing House Stark was matured in ex- bourbon barrels. In autumn 2019, a **30 year old** was launched in the Special Releases range.

History:
1897 John Grant, George Sellar and Alexander Mackenzie commence building the facilities. The first name is Strathspey.
1898 The owner encounters financial troubles and John Somerville & Co and A P Blyth & Sons take over and change the name to Dalwhinnie.
1905 Cook & Bernheimer in New York, buys Dalwhinnie for £1,250 at an auction.
1919 Macdonald Greenlees & Willliams Ltd headed by Sir James Calder buys Dalwhinnie.
1926 Macdonald Greenlees & Williams Ltd is bought by Distillers Company Ltd (DCL) which licences Dalwhinnie to James Buchanan & Co.
1930 Operations are transferred to Scottish Malt Distilleries (SMD).
1934 The distillery is closed after a fire in February.
1938 The distillery opens again.
1968 The maltings is decommissioned.
1987 Dalwhinnie 15 years becomes one of the selected six in United Distillers´ Classic Malts.
1991 A visitor centre is constructed.
1992 The distillery closes and goes through a major refurbishment costing £3.2 million.
1995 The distillery opens in March.
2002 A 36 year old is released.
2006 A 20 year old is released.
2012 A 25 year old is released.
2014 A triple matured bottling without age statement is released for The Friends of the Classic Malts.
2015 Dalwhinnie Winter´s Gold and a 25 year old are released.
2016 A distillery exclusive without age statement is released.
2018 Lizzie´s Dram, a distillery exclusive bottling, is released.
2019 Dalwhinnie Winter´s Frost, part of the Game of Thrones series, is released as well as a 30 year old in the annual Special Releases..

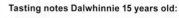

15 years old

Tasting notes Dalwhinnie 15 years old:

GS – The nose is fresh, with pine needles, heather and vanilla. Sweet and balanced on the fruity palate, with honey, malt and a very subtle note of peat. The medium length finish dries elegantly.

Deanston

[deen•stun]

Owner:
Distell International Ltd.

Region/district:
Southern Highlands

Founded: 1965
Status: Active (vc)
Capacity: 3 000 000 litres

Address: Deanston, Perthshire FK16 6AG

Website:
deanstonmalt.com

Tel:
01786 843010

To extract and convert the starch in the barley into sugar, every distillery needs to have a mash tun. Hot water is added to the grist and the wort, the sugary liquid, is drained off and pumped to the washbacks for fermentation.

These days, the vast majority of distilleries use one of two types of mash tuns – semi-lauter or full lauter. In the first type, knives on rotating arms cut through the soaked grist to facilitate the extraction and the drainage. Full lauter is based on the same principle but here the knives can be raised and lowered as well.

The first lauter tun in a Scottish whisky distillery was installed in 1974 at Tomatin. Before that a traditional mash tun with rakes and ploughs was used. They are less efficient than lauter tuns and produce a cloudier wort. Only fourteen distilleries still use traditional mash tuns and most of them are reluctant to exchange them for modern ones for fear of changing spirit character. So when Deanston installed a new mash tun in May 2019, they exchanged it like for like but changed from cast iron to stainless steel.

Deanston is equipped with a new 10.5 ton traditional open top, stainless steel mash tun and eight stainless steel washbacks with a fermentation time of 85 hours. The distillery is working on a seven-day production which means 13 mashes per week and 2.4 million litres in the year. There are also two pairs of stills with ascending lyne arms. Having started in 2000, organic spirit is produced every year. Due to the demand for "traditional" Deanston single malt, the volume of organic spirit has been reduced to 10,000 litres. In 2012 a visitor centre was opened which today is one of the best in the industry.

The core range is **12, 18 year old, Virgin oak** matured in first fill bourbon and with a one to three month finish in virgin oak casks and the 15 year old **Organic**. Recent limited bottlings, released in July 2019, include a **1997 Palo Cortado finish**, a **2002 Organic Oloroso finish**, a **2006 Cream Sherry finish**, a **2006 Fino finish** and, for the US market only, a **2012 Beer finish**. The only available duty-free exclusive is a **10 year old Bordeaux red wine cask finish**.

History:

1965 A weavery from 1785 is transformed into Deanston Distillery by James Finlay & Co. and Brodie Hepburn Ltd Brodie Hepburn also runs Tullibardine Distillery.

1966 Production commences in October.

1971 The first single malt is named Old Bannockburn.

1972 Invergordon Distillers takes over.

1974 The first single malt bearing the name Deanston is produced.

1982 The distillery closes.

1990 Burn Stewart Distillers from Glasgow buys the distillery for £2.1 million.

1991 The distillery resumes production.

1999 C L Financial buys an 18% stake of Burn Stewart.

2002 C L Financial acquires the remaining stake.

2006 Deanston 30 years old is released.

2009 A new version of the 12 year old is released.

2010 Virgin Oak is released.

2012 A visitor centre is opened.

2013 Burn Stewart Distillers is bought by South African Distell Group for £160m

2014 An 18 year old cognac finish is released in the USA.

2015 An 18 year old is released.

2016 Organic Deanston is released.

2017 A 40 year old and Vintage 2008 are released.

2018 A 10 year old Bordeaux finish is released for duty free.

2019 A number of limited releases including 1997 Palo Cortado finish, 2006 Cream Sherry finish and a 2012 Beer finish.

12 years old

Tasting notes Deanston 12 years old:

GS – A fresh, fruity nose with malt and honey. The palate displays cloves, ginger, honey and malt, while the finish is long, quite dry and pleasantly herbal.

Dufftown

[duff•town]

Owner: Diageo
Region/district: Speyside

Founded: 1896
Status: Active
Capacity: 6 000 000 litres

Address: Dufftown, Keith, Banffshire AB55 4BR

Website: malts.com thesingleton.com
Tel: 01340 822100

Modern distilleries, and those that have been revamped in the last couple of decades, usually have spacious still rooms with plenty of space to move around in. Dufftown is not one of them.

They probably have the most cramped still room in the industry and it is almost impossible to pass between the hot copper stills. Until recently, the operators had a screen placed right between the stills (as can be seen from the picture below) where they could monitor the distillation. In 2019, however, they were finally blessed with a separate still house control room.

With water being one of the key prerequisites for making whisky, many distilleries are built next to a river. To be able to control the water intake, dams are often constructed but these can sometimes become a barrier for wild fish. A completely new fish ladder has now been built by the distillery to ensure that the brown trout population of the Dullan river are able to make it up the river beyond the distillery's dam wall.

Dufftown distillery is equipped with a 13 ton full lauter mash tun, 12 stainless steel washbacks and three pairs of stills. Furthermore, all stills have sub coolers. The style of Dufftown single malt is green and grassy which is achieved by a clear wort and long fermentation (75 hours minimum). In a seven day week, no less than 165 still runs are completed (110 in the wash stills and 55 in the spirits stills) which clearly shows what a busy distillery Dufftown is. Dufftown has been working 24/7 since 2007 and during 2019 they will be producing 6 million litres of alcohol.

The core range, which received a substantial upgrade in terms of packaging in 2018, consists of **The Singleton of Dufftown 12, 15** and **18 year old**. At the same time, a new expression was added to the line-up – **Malt Master's Selection**. Without age statement, the whisky has been matured in a combination of bourbon and sherry casks with a high proportion of refill casks. This means that Tailfire, Sunray and Spey Cascade have been discontinued. A range for duty-free was introduced in 2013 and now consists of **Trinité, Liberté** and **Artisan**.

History:

1895 Peter Mackenzie, Richard Stackpole, John Symon and Charles MacPherson build the distillery Dufftown-Glenlivet in an old mill.

1896 Production starts in November.

1897 The distillery is owned by P. Mackenzie & Co., who also owns Blair Athol in Pitlochry.

1933 P. Mackenzie & Co. is bought by Arthur Bell & Sons for £56,000.

1968 The floor maltings is discontinued and malt is bought from outside suppliers. The number of stills is increased from two to four.

1974 The number of stills is increased from four to six.

1979 The stills are increased by a further two to eight.

1985 Guinness buys Arthur Bell & Sons.

1997 Guinness and Grand Metropolitan merge to form Diageo.

2006 The Singleton of Dufftown 12 year old is launched as a special duty free bottling.

2008 The Singleton of Dufftown is made available also in the UK.

2010 A Manager´s Choice 1997 is released.

2013 A 28 year old cask strength and two expressions for duty free - Unité and Trinité - are released.

2014 Tailfire, Sunray and Spey Cascade are released.

2016 Two limited releases are made - a 21 year old and a 25 year old.

2018 Malt Master´s Selection is released.

Singleton of Dufftown 12 year

Tasting notes Dufftown 12 years old:

GS – The nose is sweet, almost violet-like, with underlying malt. Big and bold on the palate, this is an upfront yet very drinkable whisky. The finish is medium to long, warming, spicy, with slowly fading notes of sherry and fudge.

Edradour

[ed•ra•<u>dow</u>•er]

Owner:
Signatory Vintage
Scotch Whisky Co. Ltd

Region/district:
Southern Highland

Founded: **Status:** **Capacity:**
1825 Active (vc) 260 000 litres

Address: Pitlochry, Perthshire PH16 5JP

Website: **Tel:**
edradour.com 01796 472095

The last couple of years have been busy for the good people at Edradour. What was once known as the smallest distillery in Scotland has been expanded, doubling the capacity.

On the other side of the burn, a replica of the original distillery has been built. This includes an open, traditional cast iron mash tun with a mash size of 1.1 tons and also the distillery "trademark" – an unusual Morton refrigerator to cool the wort. There are four washbacks made of Oregon pine with space left to install another two in the future. The dumpy wash still and the spirit still are attached to a worm tub on the outside and new warehouses were erected next to the new still house But if 2018 was busy, the owners didn´t rest on their laurels during 2019. A complete refurbishment of Edradour #1 (the old distillery) was carried through and, with the unusually hot and dry summer of 2018 in mind, they are also contemplating installing cooling towers to mitigate the effect of lack of water in the future.

With 6 mashes per week at each distillery they produced 260,000 litres of alchol during 2018 but with two more washbacks and running double shifts, the theoretical capacity could be 500,000 litres. Due to rapidly increasing demand for the unpeated Edradour, there was no peated production in the last couple of years but recently they have started doing around 25,000 litres per year. Whenever peated production occurs, it will be in the old distillery.

The core range consists of **10 year old, 12 year old Caledonia Selection** (oloroso finish), **Cask Strength Sherry 10 year old**, and **Cask Strength Bourbon 10 year old**. There is also the peated **Ballechin 10 year old**. The Straight From The Cask range is made up of a number of expressions, all fully matured in different casks. Some of the latest include **Edradour 2006 Madeira, 2008 Sherry, 2007 Rum, 2007 Sauternes, 2007 Burgundy** and **2003 Chardonnay** and **Ballechin 2007 Bordeaux, 2007 Oloroso** and **2008 Bourbon**. A limited release is the interesting **8 year old vatting** of sherry matured Edradour and bourbon matured Ballechin.

History:

1825 Probably the year when a distillery called Glenforres is founded by farmers in Perthshire.

1837 The first year Edradour is mentioned.

1841 The farmers form a proprietary company, John MacGlashan & Co.

1886 John McIntosh & Co. acquires Edradour.

1933 William Whiteley & Co. buys the distillery.

1982 Campbell Distilleries (Pernod Ricard) buys Edradour and builds a visitor centre.

1986 The first single malt is released.

2002 Edradour is bought by Andrew Symington from Signatory for £5.4 million. The product range is expanded with a 10 year old and a 13 year old cask strength.

2003 A 30 year old and a 10 year old are released.

2004 A number of wood finishes are launched as cask strength.

2006 The first bottling of peated Ballechin is released.

2007 A Madeira matured Ballechin is released.

2008 A Ballechin matured in Port pipes and a 10 year old Edradour with a Sauternes finish are released.

2009 Fourth edition of Ballechin (Oloroso) is released.

2010 Ballechin #5 Marsala is released.

2011 Ballechin #6 Bourbon and a 26 year old PX sherry finish are relased.

2012 A 1993 Oloroso and a 1993 Sauternes finish as well as the 7th edition of Ballechin (Bordeaux) are released.

2013 Ballechin Sauternes is released.

2014 The first release of a 10 year old Ballechin.

2015 Fairy Flag is released.

2017 New releases include an 8 year old vatting of Edradour and Ballechin.

2018 The new distillery is commissioned.

Tasting notes Edradour 10 years old:

GS – Cider apples, malt, almonds, vanilla and honey ar present on the nose, along with a hint of smoke and sherry. The palate is rich, creamy and malty, with a persistent nuttiness and quite a pronounced kick of slightly leathery sherry. Spices and sherry dominate the medium to long finish.

10 years old

Fettercairn

[fett•er•cairn]

Owner:
Whyte & Mackay (Emperador)

Region/district:
Eastern Highlands

Founded: 1824

Status: Active (vc)

Capacity: 3 200 000 litres

Address: Fettercairn, Laurencekirk, Kincardineshire AB30 1YB

Website:
fettercairnwhisky.com

Tel:
01561 340205

Whyte & Mackay has their four malt distilleries nicely spread out, not only geographically but also in terms of character.

From the rich and complex Dalmore in the North, through the dry and malty Jura in the West via the light, perfect blending malt Tamnavulin in the middle of Speyside to the fruity Fettercairn in the Eastern Highlands. With such big diversity it was no wonder that there were bidders willing to take over the company in 2015 when Diageo put it on the market. The winning bid came from the Philippine brandy conglomerate Emperador.

The Gladstone family were the owners of Fettercairn distillery for more than a century. It was bought by John Gladstone, father of the four times Prime Minister William Gladstone, in 1830. Although never directly involved in the running of the distillery, the Gladstone family remained owners until 1923. The whisky industry of today owes a great deal to William Gladstone. As Chancellor of the Exchequer, he passed laws in 1853 and 1860 that allowed for whisky to be blended and, not least important, to be matured under bonds. The latter meant that taxes could be paid on the amount of whisky left in the cask after maturation and not when they were filled. In other words, no tax on the angel's share and an incentive to make older and better whiskies.

Fettercairn distillery is equipped with a traditional, five ton cast iron mash tun and eleven washbacks with a fermentation time of 60 hours. There are two pairs of stills with a feature making Fettercairn unique among Scottish distilleries (although a similar technique is used at Dalmore): When collecting the middle cut, cooling water is allowed to trickle along the outside of the spirit still necks in order to increase reflux and thereby produce a lighter spirit. The production goal for 2019 is 18 mashes per week and 1.5 million litres of alcohol.

The core range consists of a **12 year old** and a **28 year old**, both matured in ex-bourbon casks. There is also a **40 year old** with a finish in Palo Cortado sherry casks and a **50 year old**, finished in tawny port pipes for five years. In 2019 there was also the release of a **12 year old PX sherry finish** exclusive to duty-free.

History:

1824 Sir Alexander Ramsay founds the distillery.

1830 Sir John Gladstone buys the distillery.

1887 A fire erupts and the distillery is forced to close for repairs.

1890 Thomas Gladstone dies and his son John Robert takes over. The distillery reopens.

1912 The company is close to liquidation and John Gladstone buys out the other investors.

1926 The distillery is mothballed.

1939 The distillery is bought by Associated Scottish Distillers Ltd. Production restarts.

1960 The maltings discontinues.

1966 The stills are increased from two to four.

1971 The distillery is bought by Tomintoul-Glenlivet Distillery Co. Ltd.

1973 Tomintoul-Glenlivet Distillery Co. Ltd is bought by Whyte & Mackay Distillers Ltd.

1974 The mega group of companies Lonrho buys Whyte & Mackay.

1988 Lonrho sells to Brent Walker Group plc.

1989 A visitor centre opens.

1990 American Brands Inc. buys Whyte & Mackay for £160 million.

1996 Whyte & Mackay and Jim Beam Brands merge to become JBB Worldwide.

2001 Kyndal Spirits buys Whyte & Mackay from JBB Worldwide.

2002 The whisky changes name to Fettercairn 1824.

2003 Kyndal Spirits changes name to Whyte & Mackay.

2007 United Spirits buys Whyte & Mackay. A 23 year old single cask is released.

2009 24, 30 and 40 year olds are released.

2010 Fettercairn Fior is launched.

2012 Fettercairn Fasque is released.

2015 Emperador Inc buys Whyte & Mackay.

2018 A new range is launched; 12, 28, 40 and 50 year old.

2019 A 12 year old PX finish is released for duty free.

12 years old

Tasting notes Fettercairn 12 years old:

IR – A delicious combination of pineapple, banana and mango together with coffee beans, cured ham and dried flowers. Still fruity on the palate but also becomes more spicy and malty and with a bit of mint at the end.

Glenallachie

[glen•alla•key]

Owner:
The Glenallachie Distillers Co.

Region/district:
Speyside

Founded: 1967
Status: Active
Capacity: 4 000 000 litres

Address: Aberlour, Banffshire AB38 9LR

Website:
www.theglenallachie.com

Tel:
01236 422120

When whisky veteran Billy Walker is involved, things tend to happen quickly. Glenallachie was taken over in July 2017, early in 2018 a number of single casks were released and a couple of months later an entire new core range was launched.

In May 2019, less than two years after the take-over, a visitor centre with a shop was opened up in connection with the Speyside Festival. The distillery offers two types of tours – "The Wee" with a tour and sampling of three drams and "The Connoisseur"with access to all areas ending with five drams.

The distillery is equipped with a 9.4 ton semi-lauter mash tun, six washbacks made of mild steel, but lined with stainless steel, plus another two washbacks made from stainless steel which were brought in from Caperdonich when that was demolished. There are also two pairs of unusually wide stills with horizontal condensers. Distilling 700,000 litres in a distillery built to make 4 million litres allows for very long fermentations – in GlenAllachie's case up to 160 hours. 100,000 litres of heavily peated spirit with an 80ppm phenol specification is also made. On site, there are 16 warehouses (two dunnage, two palletised and 12 racked). The owners fill the newmake into casks at three different strengths to achieve different flavour profiles – 63.5%, 68% and 72%.

Under Chivas' regime, Glenallachie single malt was a key ingredient in one of the top selling blends – Clan Campbell. With the new owners, a range of single casks was released in March 2018 – distilled in 1978 until 1991. A few months later a core range was unveiled; a **10 year old cask strength, 12, 18** and **25 year old**. All of them are non-chill filtered, without colouring and bottled either at cask strength or at 46-48%. In September 2019, a **15 year old** was added to the range. Recent limited releases appeared in July 2019; **8 year old Koval finish, 10 year old port finish** and a **12 year old PX sherry finish**. A **13 year old single cask** was released for the Spirit of Speyside festival and the owners also have a range of peated, blended malts called MacNair's Lum Reek.

History:

1967 The distillery is founded by Mackinlay, McPherson & Co., a subsidiary of Scottish & Newcastle Breweries Ltd. William Delmé Evans is architect.

1985 Scottish & Newcastle Breweries Ltd sells Charles Mackinlay Ltd to Invergordon Distillers which acquires both Glenallachie and Isle of Jura.

1987 The distillery is decommissioned.

1989 Campbell Distillers (Pernod Ricard) buys the distillery, increases the number of stills from two to four and takes up production again.

2005 The first official bottling for many years is a Cask Strength Edition from 1989.

2017 Glenallachie Distillery Edition is released and the distillery is sold to The Glenallachie Consortium.

2018 A series of single casks is released followed by a core range consisting of 12, 18 and 25 year old.

2019 A range of wood finishes is launched as well as a 15 year old core bottling. A visitor centre is opened.

Tasting notes Glenallachie 12 years old:

IR – Baked apples with almonds and custard, lemon zest and pine needles on the nose. Rich and lively on the palate, sweet spices, ginger, bananas, liquorice, raisins and hints of pepper.

12 years old

Glenburgie

[glen•bur•gee]

Owner:	**Region/district:**
Chivas Brothers	Speyside
(Pernod Ricard)	

Founded:	**Status:**	**Capacity:**
1810	Active	4 250 000 litres

Address: Glenburgie, Forres, Morayshire IV36 2QY

Website:	**Tel:**
-	01343 850258

It doesn't matter if you're a single malt fan or a lover of blended Scotch - it is clear that none of these two would survive without the other. With the blends being the driving force behind Scotch whisky's dominant place in the spirits world, the malts have always been there rendering character to the blends.

So it seems nothing more than fair that the world's second largest producer of spirits, Pernod Ricard, allows attention for some of their more anonymous malts yet at the same time put them in a context that involves a blend and not just any blend. Ballantines is the second most sold Scotch in the world with more than 80 million bottles sold in 2018!

Glenburgie is one of three key malts in Ballantines (Miltonduff and Glentauchers being the other two) and since 2017 all three distilleries are present on the whisky scene with a 15 year old official bottling. In Glenburgie's case there's even an 18 year old which was first released in Sweden in spring 2019 and later rolled out onto other markets.

For slightly more than two decades, 1958-1981, two Lomond stills were operative at Glenburgie. Instead of the traditional swan neck, they had columns with a number of plates inside. The plates were adjustable and the idea with this was to be able to produce different kinds of newmake from the same still.

Glenburgie is equipped with a 7.5 ton full lauter mash tun, 12 stainless steel washbacks and three pairs of stills. In older days, the fermentation time used to be around 70 hours, but has now been reduced to 52. The majority of the production is filled into bourbon casks and a part thereof is matured in four dunnage, two racked and two palletised warehouses.

The new official bottlings are **15 and 18 year old** both aged in ex-bourbon casks and bottled at 40%. A **17 year old cask strength** in the range The Distillery Reserve Collection is also available at Chivas' visitor centres.

History:

1810 William Paul founds Kilnflat Distillery. Official production starts in 1829.

1870 Kilnflat distillery closes.

1878 The distillery reopens under the name Glenburgie-Glenlivet, Charles Hay is licensee.

1884 Alexander Fraser & Co. takes over.

1925 Alexander Fraser & Co. files for bankruptcy and the receiver Donald Mustad assumes control of operations.

1927 James & George Stodart Ltd (owned by James Barclay and R A McKinlay since 1922) buys the distillery which by this time is inactive.

1930 Hiram Walker buys 60% of James & George Stodart Ltd.

1936 Hiram Walker buys Glenburgie Distillery in October. Production restarts.

1958 Lomond stills are installed producing a single malt, Glencraig. Floor malting ceases.

1981 The Lomond stills are replaced by conventional stills.

1987 Allied Lyons buys Hiram Walker.

2002 A 15 year old is released.

2004 A £4.3 million refurbishment and reconstruction takes place.

2005 Chivas Brothers (Pernod Ricard) becomes the new owner through the acquisition of Allied Domecq.

2006 The number of stills are increased from four to six in May.

2017 A 15 year old is released.

2019 An 18 year old is released.

Tasting notes Glenburgie 15 years old:

IR – Very fruity on the nose with notes of pears, apple pie, honey, marzipan and roasted nuts. The palate reveals tropical fruits, white chocolate, marmalade, vanilla and caramel.

15 years old

Glencadam

[glen•ka•dam]

Owner: Angus Dundee Distillers

Region/district: Eastern Highlands

Founded: 1825

Status: Active

Capacity: 1 300 000 litres

Address: Brechin, Angus DD9 7PA

Website: glencadamwhisky.com

Tel: 01356 622217

A few years ago I referred to Glencadam as an "underrated, overlooked and a hidden gem". Since then, the range of whiskies has been hugely expanded and the owners have intensified the promotion of their products.

It comes natural to facilitate wider awareness of the brand so in 2018 a planning application for a visitor centre was submitted to Angus Council and it was finally approved in May 2019. This will incorporate an extension to the current office building where the shop and visitor centre will be on the ground floor and a café and outdoor terrace on the first floor. If everything goes according to plan, the centre could open mid to late 2020.

There was a time when eight distilleries were in operation on the East coast between Aberdeen and Dundee. Following the big downturn for Scotch whisky in the mid 1980s, only two survived, Glencadam and Fettercairn but the number has now increased to three since Arbikie opened in 2014.

Glencadam distillery is equipped with a traditional, 4.9 ton cast iron mash tun, six stainless steel washbacks with a fermentation time of 52 hours and one pair of stills. The external heat exchanger on the wash still is from the fifties and perhaps the first in the business. On site, two dunnage warehouses from 1825, three from the 1950s and one modern racked can be found. The distillery is currently working a 7-day week, which enables 16 mashes per week and 1.3 million litres of alcohol. The owners also produce a large number of blends. These are blended in 16 enourmous steel tanks next to the distillery. From here the spirit is sent to the bottling plant in Coatbridge east of Glasgow.

The core range consists of **Origin 1825, 10, 13, 18, 21** and **25 year old**. Two wood finishes are also in the core range; a **17 year old port finish** and a **19 year old oloroso finish**. Due to the fact that the distillery was silent between 2000 and 2003, the **15 year old** was taken off the range for a couple of years but from 2019 it is now back. Recent limited editions include a **1978 30 year old single sherry cask**.

History:

1825 George Cooper founds the distillery.

1827 David Scott takes over.

1837 The distillery is sold by David Scott.

1852 Alexander Miln Thompson becomes the owner.

1857 Glencadam Distillery Company is formed.

1891 Gilmour, Thompson & Co Ltd takes over.

1954 Hiram Walker takes over.

1959 Refurbishing of the distillery.

1987 Allied Lyons buys Hiram Walker Gooderham & Worts.

1994 Allied Lyons changes name to Allied Domecq.

2000 The distillery is mothballed.

2003 Allied Domecq sells the distillery to Angus Dundee Distillers.

2005 The new owner releases a 15 year old.

2008 A re-designed 15 year old and a new 10 year old are introduced.

2009 A 25 and a 30 year old are released in limited numbers.

2010 A 12 year old port finish, a 14 year old sherry finish, a 21 year old and a 32 year old are released.

2012 A 30 year old is released.

2015 A 25 year old is launched.

2016 Origin 1825, 17 year old port finish, 19 year old oloroso finish, an 18 year old and a 25 year old are released.

2017 A 13 year old is released.

2019 The 15 year old is back in the range and batch two of the 25 year old is released.

10 years old

Tasting notes Glencadam 10 years old:

GS – A light and delicate, floral nose, with tinned pears and fondant cream. Medium-bodied, smooth, with citrus fruits and gently-spiced oak on the palate. The finish is quite long and fruity, with a hint of barley.

GlenDronach

[glen•dro•nack]

Owner:
Benriach Distillery Co
(Brown Forman)

Region/district:
Highlands

Founded: 1826
Status: Active (vc)
Capacity: 1 400 000 litres

Address: Forgue, Aberdeenshire AB54 6DB

Website: glendronachdistillery.co.uk
Tel: 01466 730202

Glendronach's sherried whiskies have had a loyal following of fans for many years. But what happens when one of the top sellers in the range all of a sudden disappears? In autumn 2015, the owners announced that the 15 year old Revival would be discontinued.

Consumers couldn't understand why but there was a perfectly good reason to be found (or actually not found) in the warehouses. There was simply no stock left of 15 year old GlenDronach. The owners at the time, Allied Domecq, closed the distillery in 1996 and did not resume production until 2002. This means that Revival, when introduced in 2009, was indeed 15 years old but in 2015, when it was finally taken off the shelves it had to be at least 20-21 years old but priced as a 15 year old. Since October 2018, Revival is back but the recipe has changed. Both oloroso casks and PX casks are now involved in the maturation process.

The distillery equipment consists of a 3.7 ton cast iron mash tun with rakes, nine washbacks made of larch with a fermentation time of 60 to 90 hours and two pairs of stills. The plan is to produce 1.2 million litres of alcohol in 2019. The distillery has made increasingly larger volumes of peated spirit (38ppm in the barley), but in 2018 and 2019 there was no peated production.

The core range is **The Hielan 8 years, Original 12 years, Revival 15 years, Allardice 18 years, Parliament 21 years** and **Peated** (first introduced in 2015). The latter has been matured in bourbon casks and then finished in a combination of oloroso and PX sherry casks. Recent limited releases include **Cask Strength Batch 7**, the **27 year old Grandeur, Peated Port Wood, Master Vintage 1993** (a combination of oloroso and PX casks) and **GlenDronach Port Wood**. The latter, released in summer 2019, was matured in ex-sherry wood followed by a second maturation in ex-port pipes. In July 2019, batch 17 of the **Cask Bottlings** was launched. The first GlenDronach for duty free appeared in autumn 2018 when **10 year old Forgue** was launched followed in May 2019 by the **16 year old Boynsmill** where port pipes had been used to complement the sherried profile.

History:

1826 The distillery is founded by a consortium with James Allardes as one of the owners.

1837 Parts of the distillery is destroyed in a fire.

1852 Walter Scott (from Teaninich) takes over.

1887 Walter Scott dies and Glendronach is taken over by a consortium from Leith.

1920 Charles Grant buys Glendronach for £9,000.

1960 William Teacher & Sons buys the distillery.

1966 The number of stills is increased to four.

1976 Allied Breweries takes over William Teacher & Sons.

1996 The distillery is mothballed.

2002 Production is resumed on 14th May.

2005 The distillery closes to rebuild from coal to in-direct firing by steam. Reopens in September. Chivas Brothers (Pernod Ricard) becomes new owner through the acquisition of Allied Domecq.

2008 Pernod Ricard sells the distillery to the owners of BenRiach distillery.

2009 Relaunch of the whole range including 12, 15 and 18 year old.

2010 A 31 year old, a 1996 single cask and a total of 11 vintages and four wood finishes are released. A visitor centre is opened.

2011 The 21 year old Parliament and 11 vintages are released.

2012 A number of vintages are released.

2013 Recherché 44 years and a number of new vintages are released.

2014 Nine different single casks are released.

2015 The Hielan, 8 years old, is released.

2016 Brown Forman buys the distillery. Peated GlenDronach and Octaves Classic are released.

2017 A range of new single casks is released.

2018 Two bottlings for duty free are released - 10 year old Forgue and 16 year old Boynsmill.

2019 Port Wood is released.

Tasting notes GlenDronach 12 years old:

GS – A sweet nose of Christmas cake fresh from the oven. Smooth on the palate, with sherry, soft oak, fruit, almonds and spices. The finish is comparatively dry and nutty, ending with bitter chocolate.

12 years old Original

Glendullan

[glen•dull•an]

Owner:
Diageo

Region/district:
Speyside

Founded: **Status:**   **Capacity:**
1897   Active   5 000 000 litres

Address: Dufftown, Keith, Banffshire AB55 4DJ

Website:
www.thesingleton.com

Tel:
01340 822100

Glendullan is one of three Diageo distilleries working under the brand name The Singleton. Glen Ord and Dufftown are the other two. It was built in 1897 by William Williams – a blender from Aberdeen.

The working distillery we see today is of a much later date, built in 1972 and the two plants were operated simultaneously until 1985 when the old distillery was closed. These buildings are now used by Diageo's engineering team which delivers maintenance to 23 Diageo distilleries, bio plants and dark grain facilities across Scotland. The old distillery, with one pair of stills, had a capacity of one million litres a year.

Glendullan distillery is situated just one minute drive east of Glenfiddich near a river which, in spite of the distillery's name isn't Dullan, but Fiddich. The confluence of the two rivers lies a mile to the south of Glendullan. The distillery is equipped with a 12 ton full lauter stainless steel mash tun, 8 washbacks made of larch and two made of stainless steel with a fermentation time of 75 hours to promote a green/grassy character of the whisky, as well as three pairs of stills. In 2019 the distillery will be doing 21 mashes per week, producing 5 million litres of alcohol.

The recently re-packaged core range of Singleton of Glendullan consists of **12, 15** and **18 year old**. There is also a range exclusive to duty free, The Singleton Reserve Collection with **Classic** (matured in American oak), **Double Matured** (matured separately in American and European oak and then married together) and **Master's Art** (with a finish in Muscat casks). The Forgotten Drops Series was created in 2017 for Glen Ord, Dufftown and Glendullan with the aim to present old and limited expressions. The first for Glendullan, in spring 2018, was a **40 year old,** the oldest whisky ever released from the distillery. This was followed up in spring 2019 with a second edition, **41 years old**. Glendullan was also a part of the Game of Thrones series launched in early 2019. Representing **House Tully**, the whisky was matured in ex-bourbon casks and bottled at 40%.

History:

1896   William Williams & Sons, a blending company with Three Stars and Strahdon among its brands, founds the distillery.

1902   Glendullan is delivered to the Royal Court and becomes the favourite whisky of Edward VII.

1919   Macdonald Greenlees buys a share of the company and Macdonald Greenlees & Williams Distillers is formed.

1926   Distillers Company Limited (DCL) buys Glendullan.

1930   Glendullan is transferred to Scottish Malt Distillers (SMD).

1962   Major refurbishing and reconstruction.

1972   A brand new distillery is constructed next to the old one and both operate simultaneously during a few years.

1985   The oldest of the two distilleries is mothballed.

1995   The first launch of Glendullan in the Rare Malts series is a 22 year old from 1972.

2005   A 26 year old from 1978 is launched in the Rare Malts series.

2007   Singleton of Glendullan is launched in the USA.

2013   Singleton of Glendullan Liberty and Trinity are released for duty free.

2014   A 38 year old is released.

2015   Classic, Double Matured and Master's Art are released.

2018   The Forgotten Drops 40 years old is released.

2019   House of Tully, part of the Game of Thrones series, as well as a 41 year old are released

Tasting notes Singleton of Glendullan 12 years:

GS – The nose is spicy, with brittle toffee, vanilla, new leather and hazelnuts. Spicy and sweet on the smooth palate, with citrus fruits, more vanilla and fresh oak. Drying and pleasingly peppery in the finish.

Game of Thrones House Tully

Glen Elgin

[glen el•gin]

Owner: Diageo

Region/district: Speyside

Founded: 1898

Status: Active

Capacity: 2 700 000 litres

Address: Longmorn, Morayshire IV30 8SL

Website: malts.com

Tel: 01343 862100

There are three distilleries that are intricably linked to the blended Scotch White Horse. First and foremost is Lagavulin which Peter Mackie, the creator of the brand, inherited from his father.

The second is Craigellachie which Peter himself built in 1891 in order to secure malts for the blend. And the third is Glen Elgin, a distillery that was never owned by Mackie but a part of the mighty DCL who acquired the White Horse Distillers in 1927. All three single malts still contribute to the blend which was created in 1890, and during the first half of the 20th century, was one of the top selling whiskies of the world. Today, it has dropped to 15th place but, in parts of the world, it is still the most popular standard Scotch and a total of 19 million bottles were sold in 2018.

Even though Glen Elgin was never one of the six original Classic Malts when that concept was launched in 1988, it was actually sold as a single malt already in 1977. It has always had a solid reputation amongst the blenders and apparently the owners saw a potential with the consumers as well. Briefly it appeared as one of many in the Flora & Fauna range but in 2002, Diageo decided to give four of their brands some more attention and the Hidden Malts range was launched comprising Glen Elgin, Caol Ila, Clynelish and Glen Ord. Hidden Malts existed for several years but today three of the whiskies are included in the extended Classic Malts range while Glen Ord has become part of the hugely successful Singleton family.

The distillery is equipped with an 8.4 ton Steinecker full lauter mash tun from 2001, nine washbacks made of larch (two of them replaced in 2018) and six small stills. The distillery alternates between 12 and 16 mashes per week. The stills are connected to six wooden worm tubs where the spirit vapours are condensed. The production plan for 2019 is a five-day week, producing 1.8 million litres of alcohol.

The only official bottling is a **12 year old**, but a limited **18 year old**, matured in ex-bodega European oak butts was one of the Special Releases in 2017.

History:

1898 The former manager of Glenfarclas, William Simpson and banker James Carle found Glen Elgin.

1900 Production starts in May but the distillery closes just five months later.

1901 The distillery is auctioned for £4,000 to the Glen Elgin-Glenlivet Distillery Co. and is mothballed.

1906 The wine producer J. J. Blanche & Co. buys the distillery for £7,000 and production resumes.

1929 J. J. Blanche dies and the distillery is put up for sale again.

1930 Scottish Malt Distillers (SMD) buys it and the license goes to White Horse Distillers.

1964 Expansion from two to six stills plus other refurbishing takes place.

1992 The distillery closes for refurbishing and installation of new stills.

1995 Production resumes in September.

2001 A 12 year old is launched in the Flora & Fauna series.

2002 The Flora & Fauna series malt is replaced by Hidden Malt 12 years.

2003 A 32 year old cask strength from 1971 is released.

2008 A 16 year old is launched as a Special Release.

2009 Glen Elgin 1998, a single cask in the new Manager's Choice range is released.

2017 An 18 year old is launched as part of the Special Releases.

Tasting notes Glen Elgin 12 years old:

GS – A nose of rich, fruity sherry, figs and fragrant spice. Full-bodied, soft, malty and honeyed in the mouth. The finish is lengthy, slightly perfumed, with spicy oak.

12 years old

Glenfarclas

[glen•fark•lass]

Owner: **Region/district:**
J. & G. Grant   Speyside

Founded: **Status:**   **Capacity:**
1836   Active (vc)   3 500 000 litres

Address: Ballindalloch, Banffshire AB37 9BD

Website: **Tel:**
glenfarclas.com   01807 500257

Glenfarclas is known for its devotion to sherry casks, an expensive way to mature whisky nowadays as ex-sherry casks can be hard to come by. Glenfarclas way of solving this has been to build a 25 year long relationship with José y Miguel Martin bodega which supplies all their casks.

However, Glenfarclas has always been more than "sherry bombs" with over-powering taste of dried fruit. It is equally important to the owners that the DNA of Glenfarclas newmake is present in the bottle. To showcase that sherry maturation is not just about very active first fill butts, the family recently released three vintages that had been matured in 2nd, 3rd and 4th filled oloroso sherry casks respectively. Over the years there have been plenty of offers from other companies to buy Glenfarclas distillery, but none that has reached the public ear. Every offer has been politely turned down by the owners since 150 years ago, the six generations of the Grant family.

The distillery is equipped with a 16.5 ton semi-lauter mash tun and twelve stainless steel washbacks with a minimum fermentation time of 60 hours but a current average of 102 hours. There are three pairs of directly fired stills and the wash stills are equipped with rummagers. This is a copper chain rotating at the bottom of the still to prevent solids from sticking to the copper. During 2019, the distillery will do around 9 mashes per week which means 2.3 million litres of pure alcohol. A number of new dunnage warehouses have recently been built and the owners can now store 105,000 casks on site in 38 warehouses.

The Glenfarclas core range consists of **8, 10, 12, 15, 21** and **25 year old**, as well as the lightly sherried **Glenfarclas Heritage** which comes without an age statement and the **105 Cask Strength**. The latter was the first commercially available cask strength single malt in the industry. There is also a **17 year old** destined for the USA, Japan and Sweden. The **30 and 40 year olds** are limited but new editions occur regularly. An **18 year old** exclusive to travel retail was launched in 2014 and in 2019, a **2003 single cask** was released for the Speyside Festival.

The owners quite often make spectacular limited releases and the rarity of the expressions clearly show the impressive selection that they have available in their warehouses. In the past five years they have released a series of bottlings called The Generations Range which included whiskies made in the 1950s! The owners also continue to release bottlings in their **Family Casks** series with vintages ranging from 1954 to 2004. Another spectacular release occurred in late 2018 when **Glenfarclas Family Trunk** appeared. The collection contains 50 200ml bottles covering every single malt in the Family Cask collection going back to 1954. The bottles are packaged in a trunk resembling old shipping luggage. Sixty trunks were made available at a price of £100,000. Finally, in May 2019 the **Glenfarclas Trilogy** was released with three bottlings aged 14, 20 and 27 years, matured in ex-oloroso casks.

History:

1836 Robert Hay founds the distillery on the original site since 1797.

1865 Robert Hay passes away and John Grant and his son George buy the distillery. They lease it to John Smith at The Glenlivet Distillery.

1870 John Smith resigns in order to start Cragganmore and J. & G. Grant Ltd takes over.

1889 John Grant dies and George Grant takes over.

1890 George Grant dies and his widow Elsie takes over the license while sons John and George control operations.

1895 John and George Grant take over and form The Glenfarclas-Glenlivet Distillery Co. Ltd with the infamous Pattison, Elder & Co.

1898 Pattison becomes bankrupt. Glenfarclas encounters financial problems after a major overhaul of the distillery but survives by mortgaging and selling stored whisky to R. I. Cameron, a whisky broker from Elgin.

1914 John Grant leaves due to ill health and George continues alone.

1948 The Grant family celebrates the distillery's 100th anniversary, a century of active licensing. It is 9 years late, as the actual anniversary coincided with WW2.

1949 George Grant senior dies and sons George Scott and John Peter inherit the distillery.

1960 Stills are increased from two to four.

1968 Glenfarclas is first to launch a cask-strength single malt. It is later named Glenfarclas 105.

1972 Floor maltings is abandoned and malt is purchased centrally.

1973 A visitor centre is opened.

1976 Enlargement from four stills to six.

History continued:

2002 George S Grant dies and is succeeded as company chairman by his son John L S Grant

2003 Two new gift tins are released (10 years old and 105 cask strength).

2005 A 50 year old is released to commemorate the bi-centenary of John Grant´s birth.

2006 Ten new vintages are released.

2007 Family Casks, a series of single cask bottlings from 43 consecutive years, is released.

2008 New releases in the Family Cask range. Glenfarclas 105 40 years old is released.

2009 A third release in the Family Casks series.

2010 A 40 year old and new vintages from Family Casks are released.

2011 Chairman´s Reserve and 175th Anniversary are released.

2012 A 58 year old and a 43 year old are released.

2013 An 18 year old for duty free is released as well as a 25 year old quarter cask.

2014 A 60 year old and a 1966 single fino sherry cask are released.

2015 A 1956 Sherry Cask and Family Reserve are released.

2016 40 year old, 50 year old, 1981 Port and 1986 cask strength are released.

2018 A 22 year old version of the 105 Cask Strength is released.

2019 Glenfarclas Trilogy is released.

Tasting notes Glenfarclas 10 year old:

GS – Full and richly sherried on the nose, with nuts, fruit cake and a hint of citrus fruit. The palate is big, with ripe fruit, brittle toffee, some peat and oak. Medium length and gingery in the finish.

105 Cask Strength   50 years old

18 years old   Family Cask 1959

12 years old   21 years old   40 years old

Glenfiddich

[glen•fidd•ick]

Owner:
William Grant & Sons

Region/district:
Speyside

Founded: 1886

Status: Active (vc)

Capacity: 13 700 000 litres

Address: Dufftown, Keith, Banffshire AB55 4DH

Website: glenfiddich.com

Tel: 01340 820373 (vc)

Most of William Grants previous distillery expansions have been performed according to plans and often very much under the radar. One good example is the building of Ailsa Bay distillery and before that (in the 1960s) Girvan grain distillery.

The ongoing expansion of Glenfiddich distillery is a bit different. Planning approval was granted in December 2015. The first indications from the company said the new distillery would be up and running by the end of 2018. This was later changed to mid 2019 and the latest bid is mid 2020. For many whisky companies such a delay would raise questions but not so for William Grants. The family owned company is in a good position with the latest figures showing profits of £304m with Glenfiddich, Balvenie and Hendricks gin being the main contributors. The main reason for the delay is probably the fact that Glenfiddich is the number one single malt in the world (16,8 million bottles sold in 2018) and any change or expansion of production must be carried out so that the character and the quality of the spirit remains the same.

Currently, Glenfiddich is equipped with two, stainless steel full lauter mash tuns – both with a ten ton mash. There are 32 washbacks made of Douglas fir with a minimum fermentation time of 68 hours but typically 72. Two still rooms hold a total of 11 wash stills and 20 spirit stills. All the stills in still house number 2 are directly fired using gas. The production for 2018 will be 68 mashes per week and 13.65 million litres of pure alcohol. The ongoing expansion includes a new tun room and a new still house. In terms of equipment (details not yet confirmed by the owners) it looks as there will be one more mash tun, 16 washbacks, five wash stills and ten spirit stills. The new capacity for the combined distilleries will be in the vicinity of 20 million litres.

The core range consists of **12, 15, 18, 21, Excellence 26, 30** and **40 year old**. The **Rich Oak 14 year old** is an exclusive to Canada and a **14 year old** bourbon matured is available only in the USA and Canada. A new addition to the core range appeared in September 2019 with **Grand Cru 23 year old**, finished in champagne casks. Recent limited releases include **Glenfiddich The Original**, the **38 year old Glenfiddich Ultimate** and a **50 year old**. A special range is called Experimental Series where **IPA Experiment** and **Project XX** are ongoing items. A third bottling, the **21 year old Winter Storm**, was released in spring 2018 followed by **Fire & Cane**. The latter is a smoky whisky finished in rum casks. In 2014, the **Glenfiddich Gallery** was introduced where Malt Master, Brian Kinsman, has selected special casks. Included in the duty free range is the Cask Collection with **Select Cask, Reserve Cask, Vintage Cask** and **Finest Solera**. Another duty free exclusive is **Glenfiddich Rare Oak 25 years**. Finally, a **15 year old Distillery Edition** is available at the distillery as well as **Rare Collection Cask No. 20050**, released in 2019. Two bottlings exclusive to Taiwan were released in August 2019, **Black Queen** and **Ice Breaker**, both finished in Taiwan wine casks.

History:

1886 The distillery is founded by William Grant, 47 years old, who had learned the trade at Mortlach Distillery. The equipment is bought from Mrs. Cummings of Cardow Distillery. The construction totals £800.

1887 The first distilling takes place on Christmas Day.

1892 William Grant builds Balvenie.

1898 The blending company Pattisons, largest customer of Glenfiddich, files for bankruptcy and Grant decides to blend their own whisky. Standfast becomes one of their major brands.

1903 William Grant & Sons is formed.

1957 The famous, three-cornered bottle is introduced.

1958 The floor maltings is closed.

1963 Glennfiddich becomes the first whisky to be marketed as single malt in the UK and the rest of the world.

1964 A version of Standfast's three-cornered bottle is launched for Glenfiddich in green glass.

1969 Glenfiddich becomes the first distillery in Scotland to open a visitor centre.

1974 16 new stills are installed.

2001 1965 Vintage Reserve is launched in a limited edition of 480 bottles. Glenfiddich 1937 is bottled (61 bottles).

2002 Glenfiddich Gran Reserva 21 years old, Caoran Reserve 12 years and Glenfiddich Rare Collection 1937 (61 bottles) are launched.

2003 1973 Vintage Reserve (440 bottles) is launched.

2004 1991 Vintage Reserve (13 years) and 1972 Vintage Reserve (519 bottles) are launched.

2005 Circa £1.7 million is invested in a new visitor centre.

History continued:

2006 1973 Vintage Reserve, 33 years (861 bottles) and 12 year old Toasted Oak are released.

2007 1976 Vintage Reserve, 31 years is released.

2008 1977 Vintage Reserve is released.

2009 A 50 year old and 1975 Vintage Reserve are released.

2010 Rich Oak, 1978 Vintage Reserve, the 6th edition of 40 year old and Snow Phoenix are released.

2011 1974 Vintage Reserve and a 19 year old Madeira finish are released.

2012 Cask of Dreams and Millenium Vintage are released.

2013 A 19 year old red wine finish and 1987 Anniversary Vintage are released. Cask Collection with three different expressions is released for duty free.

2014 The 26 year old Glenfiddich Excellence, Rare Oak 25 years and Glenfiddich The Original are released.

2015 A 14 year old for the US market is released.

2016 Finest Solera is released for travel retail. Two expressions in the Experimental Series are launched; Project XX and IPA Experiment.

2017 Winter Storm is released.

2018 A new expression in the Experimental Series is released - Fire & Cane.

2019 Grand Cru 23 year old and Rare Collection Cask No. 20050 are released.

Tasting notes Glenfiddich 12 year old:

GS – Delicate, floral and slightly fruity on the nose. Well mannered in the mouth, malty, elegant and soft. Rich, fruit flavours dominate the palate, with a developing nuttiness and an elusive whiff of peat smoke in the fragrant finish.

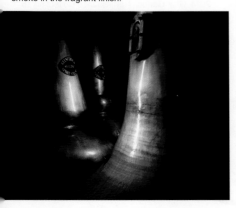

Project XX   Fire & Cane   IPA Experiment

Reserve Cask   Vintage Cask

Our Original Twelve

Our Solera Fifteen

Winter Storm

Glen Garioch

[glen gee•ree]

Owner:
Beam Suntory

Region/district:
Eastern Highlands

Founded: 1797

Status: Active (vc)

Capacity: 1 370 000 litres

Address: Oldmeldrum, Inverurie, Aberdeenshire AB51 0ES

Website: glengarioch.com

Tel: 01651 873450

Every distillery, or region of distilleries, in Scotland these days organise some kind of event on a yearly basis where visitors have a chance to experience the distillery in depth.

Glen Garioch is no exception. For three years now, in December, they celebrate the Rare Fayre for one day which includes tours, masterclasses, new whiskies and food and craft stalls. The archaic spelling of the "fair" is well placed in Glen Garioch's case. It is one of the oldest distilleries in Scotland. Officially it was founded in 1797 but is probably older. An article in the Aberdeen Journal from December 1785 noted that distillation had now started at Glen Garioch in Oldmeldrum.

Glen Garioch single malt is typically unpeated, but smoky notes can easily be detected in expressions distilled before 1994 when their own floor maltings closed. Until then, the malt was peated with a phenol specification of 8-10ppm. The distillery is equipped with a four ton full lauter mash tun, eight stainless steel wash-backs with a fermentation time of 72 hours, one wash still and one spirit still (replaced in 2016). There is also a third still, which has not been used for a long time. The spirit is tankered to Glasgow, filled into casks and returned to the distillery's four warehouses. During 2019 the production will be seven mashes per week and around 450,000 litres in the year.

The core range is the **1797 Founder's Reserve** (without age statement) and a **12 year old**, both of them bottled at 48%. A limited **15 year old** matured in oloroso casks was launched, mainly for the duty free market, in autumn 2018. Since 2013 there is also **Virgin Oak**, fully matured in virgin American white oak. Recent limited releases include a selection of **Vintage** expressions with **1978, 1990, 1994, 1997** and **1998**. For the Rare Fayre there was also a **Vintage 2007**. The first chapter of a new range of cask strength bottlings called Glen Garioch Renaissance Collection was released in 2014 with a 15 year old and the fourth and final instalment came in 2018 by way of an **18 year old**.

History:

1797 John Manson founds the distillery.

1798 Thomas Simpson becomes licensee.

1825 Ingram, Lamb & Co. bcome new owners.

1837 The distillery is bought by John Manson & Co.

1884 The distillery is sold to J. G. Thomson & Co.

1908 William Sanderson buys the distillery.

1933 Sanderson & Son merges with the gin maker Booth's Distilleries Ltd.

1937 Booth´s Distilleries Ltd is acquired by Distillers Company Limited (DCL).

1968 Glen Garioch is decommissioned.

1970 It is sold to Stanley P. Morrison Ltd.

1973 Production starts again.

1978 Stills are increased from two to three.

1994 Suntory controls all of Morrison Bowmore Distillers Ltd.

1995 The distillery is mothballed in October.

1997 The distillery reopens in August and from now on, it is using unpeated malt.

2004 Glen Garioch 46 year old is released.

2005 15 year old Bordeaux Cask Finish is launched. A visitor centre opens in October.

2006 An 8 year old is released.

2009 Complete revamp of the range - 1979 Founders Reserve (unaged), 12 year old, Vintage 1978 and 1990 are released.

2010 1991 vintage is released.

2011 Vintage 1986 and 1994 are released.

2012 Vintage 1995 and 1997 are released.

2013 Virgin Oak, Vintage 1999 and 11 single casks are released.

2014 Glen Garioch Renaissance Collection 15 years is released.

2018 The fourth and final installment of the Rennaisance Collection is released.

Tasting notes Glen Garioch 12 years old:

GS – Luscious and sweet on the nose, peaches and pineapple, vanilla, malt and a hint of sherry. Full-bodied and nicely textured, with more fresh fruit on the palate, along with spice, brittle toffee and finally dry oak notes.

12 years old

Pioneers of Whisky

David Vitale
Founder
Starward Distillery, Australia

In terms of new distilleries being opened, Australia is definitely one of the hottest places right now and has been for more than a decade. From the first whisky distillery founded in modern days in 1992, there are now more than fifty. It takes some hard work and skills to become heard of in such a context. David Vitale and his Starward Distillery managed to do just that and did it loud enough for such a giant player as Diageo to invest in his business.

But let's start from the beginning. Around the turn of the millennium, when David was running a tech business developing an e-learning platform for financial advisers, he also had a secret passion – craft beer! Looking for a career change, he decided to set up a craft brewery. Whisky wasn't really on his radar at that time. As he himself puts it,

"Johnnie Walker Black was what my Greek friend's dads drank when playing cards, and Chivas Regal 12 yo was what we gave our accountant for Christmas every year. My passion was for craft beer and wine."

Before the brewery idea had taken off, David discovered that good beer doesn't travel well and shelved the idea. Through a mutual contact, he was introduced to Bill Lark at Lark Distillery in Hobart, Tasmania and quickly realized that the world of single malt whisky was as diverse and interesting (if not even more interesting) than craft brewing. Bill Lark, who started his distillery in 1992, is often referred to as the godfather of Australian whisky and apart from running his own distillery he has been involved as a consultant in plenty set-ups in Australia. David joined Lark's team for three years to learn about whisky production.

At the same time he met Chris Middleton who had just returned to Australia after having worked for Jack Daniels in USA. Chris convinced David that there was space for a distinctly Australian whisky in the global whisky market. With his encyclopedic knowledge of whisky, Chris was a great sounding board on what was and wasn't innovative and authentic to a new, world whisky.

"Without Bill Lark and Chris Middleton, there would be no Starward." says David.

After having spent three years at Lark distillery, David decided it was time to start his own enterprise and he investigated possible funding. This was in autumn 2008 and the timing couldn't have been worse. A major financial crisis swept round the globe like a wildfire. David remembers;

" The fact we were able to raise capital during that period of time, particularly given the long lead times for cashflow in whisky, was a testament to the vision we have for Starward and the patience and belief of those early round investors."

In order to make a whisky that stood out from the crowd, David used brewing barley with a dark and rich character and also a hybrid yeast strain to provide more of the fruity esters. But his wood policy really differed from everything else:

" Our use of Australian Red Wine Barrels is critical to our point of difference. Not only because they are local, but they deliver such an immense part of the aroma, flavour and mouthfeel in Starward." David continues; " There were plenty of wine cask finished whiskies, but when we started, full maturation in Red Wine barrels was really unique – particularly in the nascent new world whisky category."

But what sets them apart is also the most complicated aspect to manage. Sourcing barrels from over 50 wineries and handling them according to oak source, size, toasting and type of wine and then maturing the whisky in the reactive climate of Melbourne is quite challenging. When asked about the style of whisky, David keeps coming back to "distinctively Australian".

"What that actually meant when we started was a bit of a soul searching exercise. We didn't have a formula as such, it was very much "I'll know it when I see it". Distinctive flavours can be at the extremes of your palate – think peat – and not so approachable. To come up with something that is truly different, delicious and approachable for a new whisky drinker - and at a sharp price point - was the essence of our vision. And we think we've done a pretty good job of that with our whiskies."

The rules and regulations surrounding production and sales of whisky vary a lot around the world. In some markets that is the real challenge for a start-up but in Australia the situation is different.

"The great thing about Australian whisky regulations is that they are so broad you can shoot a canon through them. Even though they were written in 1901, they got the important things right in an elegantly simple fashion. I'm paraphrasing but the definition is "Whisky is a distilled spirit from a fermented grain mash that spends a minimum of two years in wood and has all the attributes of whisky." This definition stimulates innovation while still keeping distillers honest."

Starward moved to a new location one mile from Melbourne in 2016 but now they have outgrown storage capacity at the new distillery. Being located in the Melbourne area with a 20°C temperature range in a single day, gives a profound impact on the whisky. Depth and complexity develops much quicker in that climate.

While focusing on expansion, Starward it at the same time in a phase where distribution and sales are equally important. The first expressions were released in 2013 and today the brand is ranged in (apart from the home market) the UK, France, six states in the US, and selected Duty Free markets around the world. Asked about the route to market strategy, David says,

"I often think of sales and building the brand like a game of Risk. For us, we have a strategy of deep, not wide markets. There is no such thing as a "light-touch" territory. We fully invest with our trade partners to roll out a sales and marketing plan that makes sense for the brand in that particular market."

David's final advice
to someone wanting to start a distillery of their own:

"Be interesting, dream big, and find a team that believes in your vision of what whisky means to you. On finances, take your first budget, add 20%... and then double it!"

Glenglassaugh

[glen•gla•ssa]

Owner:
Glenglassaugh Distillery Co
(BenRiach Distillery Co.)

Region/district:
Highlands

Founded: 1875
Status: Active (vc)
Capacity: 1 100 000 litres

Address: Portsoy, Banffshire AB45 2SQ

Website:
glenglassaugh.com

Tel:
01261 842367

The history of Glenglassaugh distillery clearly follows the ebb and flow of Scotch whisky. It opened during whisky's first golden era, around the end of the 1800s.

Blending of whisky had become a fashion and the malt was used for famous brands such as Teachers. It was then closed for more or less 50 years when two World Wars and the prohibition in the USA slowed down sales of Scotch. When it was re-opened in 1960, the industry was in the midst of its second golden era. Blended Scotch ruled the world and Glenglassaugh could be found in Famous Grouse and Cutty Sark. The market deflated in the mid 1980s and Glenglassaugh, together with around 20 other distilleries, was forced to close again. After 22 years, and well into the third golden era, new owners took over and the emphasis is now on single malt. Since the resurrection in 2008, the distillery has changed hands on two occasions and is now run, together with BenRiach and GlenDronach, by Brown Forman.

The equipment of the distillery consists of a 5.2 ton Porteus cast iron mash tun with rakes, four wooden washbacks and two stainless steel ones with a fermentation time between 54 and 80 hours and one pair of stills. The production is 800,000 litres of pure alcohol, of which 40,000 litres is peated (30ppm). The main part (85%) is filled to be used as single malt while the rest is sold externally.

The core range is **Revival**, finished in oloroso casks, **Evolution**, matured in American oak and the peated **Torfa** without age statement. Limited releases include **30, 40** and **51 year old**, as well as single casks in the **Rare Cask Series** where batch three was released in autumn 2018. The second release of **Octaves Classic** and **Octaves Peated**, matured around seven years in small casks holding around 65 litres, appeared at the same time. Autumn 2017 saw the first release of a wood finish series from the distillery – **Port, PX Sherry, Peated Port** and **Peated Virgin oak**. All expressions started out in ex-bourbon casks and were bottled at 46%.

History:

1873 The distillery is founded by James Moir.

1887 Alexander Morrison embarks on renovation work.

1892 Morrison sells the distillery to Robertson & Baxter. They in turn sell it on to Highland Distilleries Company for £15,000.

1908 The distillery closes.

1931 The distillery reopens.

1936 The distillery closes.

1957 Reconstruction takes place.

1960 The distillery reopens.

1986 Glenglassaugh is mothballed.

2005 A 22 year old is released.

2006 Three limited editions are released - 19 years old, 38 years old and 44 years old.

2008 The distillery is bought by the Scaent Group for £5m. Three bottlings are released - 21, 30 and 40 year old.

2009 New make spirit and 6 months old are released.

2010 A 26 year old replaces the 21 year old.

2011 A 35 year old and the first bottling from the new owners production, a 3 year old, are released.

2012 A visitor centre is inaugurated and Glenglassaugh Revival is released.

2013 BenRiach Distillery Co buys the distillery and Glenglassaugh Evolution and a 30 year old are released.

2014 The peated Torfa is released as well as eight different single casks and Massandra Connection (35 and 41 years old).

2015 The second batch of single casks is released.

2016 Octaves Classic and Octaves Peated are released.

2017 Three wood finishes are released.

2018 Batch three in the Rare Cask series and the second release of Octaves are released.

Tasting notes Glenglassaugh Evolution:

GS – Peaches and gingerbread on the nose, with brittle toffee, icing sugar, and vanilla. Luscious soft fruits dipped in caramel figure on the palate, with coconut and background stem ginger. The finish is medium in length, with spicy toffee.

Evolution

Glengoyne

[glen•goyn]

Owner:	**Region/district:**
Ian Macleod Distillers	Southern Highlands
Founded: **Status:**	**Capacity:**
1833 Active (vc)	1 100 000 litres

Address: Dumgoyne by Killearn, Glasgow G63 9LB

Website:	**Tel:**
glengoyne.com	01360 550254 (vc)

Since 2010, the number of visitors to Glengoyne has gone up 100% and in 2018, more than 90,000 people travelled here to do a tour.

Apart from being a beautiful distillery with a variety of different tours to choose from, the proximity to Glasgow plays a vital part. If you don't have a car, you can go by bus from Buchannan Station in Glasgow. It takes an hour and the bus stops right at the distillery which is situated in the lovely Trossachs national park, immortalised by Walter Scott in his novel Rob Roy.

Even if some of Glengoyne's newmake is filled into ex-bourbon wood, the owners rely heavily on sherry cask maturation. Air-dried European oak from Galicia as well as American oak is used for the butts that are then filled with sherry in Jerez to season the casks. The entire process from cutting down a tree to the stage when the cask is filled at Glengoyne can take up to six years.

The distillery is equipped with a 3.84 ton semi lauter mash tun. There are also six Oregon pine washbacks, as well as the rather unusual combination of one wash still and two spirit stills. Both short (56 hours) and long (110 hours) fermentations are practised. In 2019, the production will be 920,000 litres of alcohol.

The core range consists of **10, 12, 18, 21** and **25 year old**. There is also batch seven of the **Cask Strength** which for the first time has a small proportion of bourbon cask matured whisky in the recipe. Recent limited releases include the first instalment in a new series called **Glengoyne Legacy**. Released in February 2019, it was a no-age-statement whisky matured initially in first fill European oak oloroso sherry casks and then finished in casks that had already held Glengoyne single malt. Other limited releases were batch 2 of the **30 year old** and batch 7 of the popular **The Teapot Dram**, available only at the distillery. The line-up for duty free was completely revamped in April 2018. The 15 year old Distiller's Gold was replaced by a range called Spirit of Oak with no less than four expressions — all of them heavily influenced by sherry casks; **Cuartillo** (American oak oloroso), **Balbaine** (European oak oloroso), **28 year old** (a combination of American and European oak oloroso) and **Glengoyne PX** (American and European oak with a finish in PX casks).

History:

1833 The distillery is licensed under the name Burnfoot Distilleries by the Edmonstone family.

1876 Lang Brothers buys the distillery and changes the name to Glenguin.

1905 The name changes to Glengoyne.

1965 Robertson & Baxter takes over Lang Brothers and the distillery is refurbished. The stills are increased from two to three.

2001 Glengoyne Scottish Oak Finish (16 years old) is launched.

2003 Ian MacLeod Distillers Ltd buys the distillery plus the brand Langs from the Edrington Group for £7.2 million.

2005 A 19 year old, a 32 year old and a 37 year old cask strength are launched.

2006 Nine "choices" from Stillmen, Mashmen and Manager are released.

2007 A new version of the 21 year old, two Warehousemen´s Choice, Vintage 1972 and two single casks are released.

2008 A 16 year old Shiraz cask finish, three single casks and Heritage Gold are released.

2009 A 40 year old, two single casks and a new 12 year old are launched.

2010 Two single casks, 1987 and 1997, released.

2011 A 24 year old single cask is released.

2012 A 15 and an 18 year old are released as well as a Cask Strength with no age statement.

2013 A limited 35 year old is launched.

2014 A 25 year old is released.

2018 A new range for duty free is released – Cuartillo, Balbaine, a 28 year old and Glengoyne PX.

2019 Glengoyne Legacy is launched.

12 years old

Tasting notes Glengoyne 12 years old:

GS – Slightly earthy on the nose, with nutty malt, ripe apples, and a hint of honey. The palate is full and fruity, with milk chocolate, ginger and vanilla. The finish is medium in length, with milky coffee and soft spices.

Glen Grant

[glen grant]

Owner: **Region/district:**
Campari Group   Speyside

Founded: **Status:**   **Capacity:**
1840   Active (vc)   6 200 000 litres

Address: Elgin Road, Rothes, Banffshire AB38 7BS

Website: **Tel:**
glengrant.com   01340 832118

Campari took over Glen Grant from Chivas Brothers in 2006. At that time it had slipped from being the second best selling single malt in the world to place number four. Twelve years later, the famous whisky is on the verge of slipping out of the Top 10 list

The former success of Glen Grant was largely based on selling young whiskies up to five years of age and had Italy as the main market. Accounting for just 1% of Campari's sales and dropping almost 6% in volumes in 2018, the owners have now decided to buck the declining trend by doing just the opposite. According to Campari's CEO, the brand is now in transition. The idea is to move it up the value chain by focusing on aged whisky. Unaged Glen Grant will be put on allocation and by laying down more stocks to mature the company will hopefully have more of the 12 and 18 year old to sell in the future. America will be an important market in this context and distribution in USA, especially in the duty-free segment, is about to be expanded.

The major person behind Glen Grant brand is without doubt Dennis Malcolm who started at the distillery as an apprentice cooper in 1961. Still, 58 years later he is active as Master Distiller which means he is responsible for the development of new expressions.

The distillery is equipped with a 12.3 ton semi-lauter mash tun, ten Oregon pine washbacks with a minimum fermentation time of 48 hours and four pairs of stills. The wash stills are peculiar in that they have vertical sides at the base of the neck and all eight stills are fitted with purifiers. This gives an increased reflux and creates a light and delicate whisky. A new, extremely efficient £5m bottling hall was inaugurated in 2013. It has a capacity of 12,000 bottles an hour and Glen Grant is alone among the larger distillers bottling the entire production on site. In 2015 a second line for the premium range was installed. During 2019, the operators will split their time between the distillery and the bottling plant, making 20 mashes per week for 20 weeks which will amount to 2 million litres of pure alcohol.

The Glen Grant core range consists of **Major's Reserve** with no age statement, a **5 year old** sold in Italy only, a **10 year old**, a **12 year old** matured in both bourbon and sherry casks and an **18 year old** bourbon matured. Summer 2016 also saw the introduction of a **12 year old non chill-filtered** expression for the duty free market. In early 2018, a **15 year old**, matured in first fill bourbon and bottled at 50% was launched for the American market as well as for duty free. The current distillery exclusive bottling, available only at the distillery, is a **10 year port finish**. Older expressions are rarely released by the owners but can from time to time be found in the range from Gordon & MacPhail. Recently a collection of six bottlings distilled from 1950 to 1955 was launched by the independent bottler.

History:

1840   The brothers James and John Grant, managers of Dandelaith Distillery, found the distillery.

1861   The distillery becomes the first to install electric lighting.

1864   John Grant dies.

1872   James Grant passes away and the distillery is inherited by his son, James junior (Major James Grant).

1897   James Grant decides to build another distillery across the road; it is named Glen Grant No. 2.

1902   Glen Grant No. 2 is mothballed.

1931   Major Grant dies and is succeeded by his grandson Major Douglas Mackessack.

1953   J. & J. Grant merges with George & J. G. Smith who runs Glenlivet distillery, forming The Glenlivet & Glen Grant Distillers Ltd.

1961   Armando Giovinetti and Douglas Mackessak found a friendship that leads to Glen Grant becoming the most sold malt whisky in Italy.

1965   Glen Grant No. 2 is back in production, but renamed Caperdonich.

1972   The Glenlivet & Glen Grant Distillers merges with Hill Thompson & Co. and Longmorn-Glenlivet Ltd to form The Glenlivet Distillers.

1973   Stills are increased from four to six.

1977   The Chivas & Glenlivet Group (Seagrams) buys Glen Grant Distillery. Stills are increased from six to ten.

2001   Pernod Ricard and Diageo buy Seagrams Spirits & Wine, with Pernod acquiring Chivas Group.

History continued:

2006 Campari buys Glen Grant for €115m.

2007 The entire range is re-packaged and re-launched and a 15 year old single cask is released. Reconstruction of the visitor centre.

2008 Two limited cask strengths - a 16 year old and a 27 year old - are released.

2009 Cellar Reserve 1992 is released.

2010 A 170th Anniversary bottling is released.

2011 A 25 year old is released.

2012 A 19 year old Distillery Edition is released.

2013 Five Decades is released and a bottling hall is built.

2014 A 50 year old and the Rothes Edition 10 years old is released.

2015 Glen Grant Fiodh is launched.

2016 A 12 year old and an 18 year old are launched and a 12 year old non chill-filtered is released for travel retail.

2017 A 15 year old is released for the American market.

2018 A 15 year old is released for the duty free market.

Tasting notes Glen Grant 12 year old:

GS – A blast of fresh fruit – oranges, pears and lemons – on the initial nose, before vanilla and fudge notes develop. The fruit carries over on to the palate, with honey, caramel and sweet spices. Medium in length, with cinnamon and soft oak in the finish.

12 years old   12 years old
non chill-filtered   18 years old

10 years old   The Major´s Reserve

Glengyle

[glen•gajl]

Owner: Mitchell´s Glengyle Ltd

Region/district: Campbeltown

Founded: 2004

Status: Active

Capacity: 750 000 litres

Address: Glengyle Road, Campbeltown, Argyll PA28 6LR

Website: kilkerran.com

Tel: 01586 551710

Since the start in 2004, the owners decided to make Glengyle a place for various experimental trials they did not want to run in their other distillery in Campbeltown – Springbank.

After all, the latter produces three well-known brands of malt and it would be risky to insert trial distillations into a tight production schedule. At Glengyle, on the other hand, they have run quadruple distillations over the years as well as distillations with extremely heavily peated malt. This has now resulted in the first peated release with an unusually high phenol specification (84ppm) for the barley that was used. The first releases will be labeled "peat in progress" in the same manner they launched Kilkerran before offering a core product.

The distillery was dormant for 79 years until Hedley Wright, owner of Springbank, brought it back to life. One of the reasons for Wright´s interest in reviving the old distillery, was the re-introduction of Campbeltown as a recognised whisky region in accordance with the SWA regulations. Glengyle together with already existing Springbank and Glen Scotia made up the three distilleries apparently required to be named a region.

The distillery is equipped with a 4.5 ton semi-lauter mash tun, two washbacks made of boat skin larch and two made of Douglas fir. The fermentation varies between 72 and 110 hours. There is also one set of stills. Malt is obtained from the neighbouring Springbank and operations are managed by the same staff. The capacity is 750,000 litres, but considerably smaller amounts have been produced over the years. However, production has increased in later years and the plan for 2019 is to distill 96,000 litres, 80% made up of "regular" Kilkerran and the rest of heavily peated spirit.

After many years of work in progress, the first core **12 year old** was launched in 2016. It was a vatting of bourbon- (70%) and sherry-matured (30%) whisky. In spring 2017, an **8 year old** cask strength was launched with a new batch planned for late autumn 2019. The first batch of **Kilkerran Heavily Peated** was released in spring 2019 followed by a second batch in autumn.

History:

1872 The original Glengyle Distillery is built by William Mitchell.

1919 The distillery is bought by West Highland Malt Distilleries Ltd.

1925 The distillery is closed.

1929 The warehouses (but no stock) are purchased by the Craig Brothers and rebuilt into a petrol station and garage.

1941 The distillery is acquired by the Bloch Brothers.

1957 Campbell Henderson applies for planning permission with the intention of reopening the distillery.

2000 Hedley Wright, owner of Springbank Distillery and related to founder William Mitchell, acquires the distillery.

2004 The first distillation after reconstruction takes place in March.

2007 The first limited release - a 3 year old.

2009 Kilkerran "Work in progress" is released.

2010 "Work in progress 2" is released.

2011 "Work in progress 3" is released.

2012 "Work in progress 4" is released.

2013 "Work in progress 5" is released and this time in two versions - bourbon and sherry.

2014 "Work in progress 6" is released in two versions - bourbon and sherry.

2015 "Work in progress 7" is released in two versions - bourbon and sherry.

2016 Kilkerran 12 years old is released.

2017 Kilkerran 8 year old cask strength is released.

2019 Kilkerran Heavily Peated is released.

Tasting notes Kilkerran 12 year old:

GS – Initially, quite reticent on the nose, then peaty fruit notes develop. Oily and full on the palate, with peaches and more overt smoke, plus an earthy quality. Castor oil and liquorice sticks. Slick in the medium-length finish, with slightly drying oak and enduring liquorice.

12 years old

Pioneers of Whisky

Lila and Nestor Serenelli
Founders
La Alazana Distillery, Argentina

Argentina is a country known for its excellent wines and has a long history of winemaking that goes more than five centuries back. Today it´s the fifth biggest producer of wine in the world.

It certainly takes courage to build a whisky distillery in such an environment and when La Alazana became the first single malt whisky producer in Argentina in 2011, one or two eyebrows were raised in the whisky world. The man behind the idea was Nestor Serenelli who grew up in the countryside in the province of Buenos Aires where his family owned a dairy farm and grew different kinds of grain. When he was old enough to leave home he travelled south to Patagonia where he started a hardware shop. The business grew and is now managed by his two older sons.

Living in a small town, he had the dream of moving to the countryside and when he met his wife Lila more than twenty years ago, they decided to buy a farm at the foot of the Andes in the province of Chubut. It was the perfect place to have their horses and finally they could also fulfil their dream of starting up a distillery.

The idea wasn't far fetched as Nestor´s granddad used to distil grappa and whisky had been a spirit that was often enjoyed in his ancestral home. Funding the distillery all by themselves, Nestor and Lila started off by distilling all kinds of fruit brandies but it didn't take long until the first malt whisky was produced.

"Our first whiskies were incredible and most probably the best in the world, but that was surely due to the fact that we were getting all our reviews from friends who were happy to share a free dram!"

Nestor and Lila realised that if they wanted to make an excellent whisky for the future, they needed to go to the "source" to learn. That was when they set off on their first trip to Scotland. It was the start of a network of friends that has inspired them ever since.

Nestor remembers; "One of my biggest inspirations was Jim McEwan, who was then still at Bruichladdich. I learned a lot from him about the technical aspects, and above all, I learned the most important thing needed to make whisky - passion."

And while Nestor recognised passion as an important factor, he was also aware of the fact that the understanding of the technique behind whisky-making was essential to make a good whisky. Lila responded to the challenge and travelled to Scotland where she completed a Masters in Science of Brewing and Distilling at Heriot-Watt University. The ambitious couple also contracted a laboratory in Scotland for analysing their spirit.

Scotland and Scotch are the themes running through their business. Even though Argentina is a large producer of barley, all the varieties are adapted for beer production so the couple had to import barley from the UK, and still do for the greater volume of the production. However, wanting to make an all Argentinian whisky they have for the last two years been growing their own barley on site and the 100% Patagonia single malt is now maturing in their warehouses.

When asked if whisky couldn't be produced anywhere in Argentina, Nestor says "We chose Patagonia for setting up the distillery because this region has got water of an excellent quality and an exceptional climate for achieving elegantly balanced single malt." In a way, Nestor´s reply echoes that which a wine grower anywhere in the world would say about the terroir influencing the produce.

For obvious reasons, since Argentina has no history of producing whisky, the rules and regulations surrounding the production are not as strict as those in Scotland or, indeed, the EU. Whisky for example has only to mature for two years (and not three) to be called whisky.

Regardless of that, Nestor and Lila, wanting to make a great "Scottish style" whisky, early on decided to "abide" by the Scottish rules which, amongst other things, means maturing the whisky for a minimum of three years. At the moment they are working with authorities and institutions to change the regulations so that a single malt whisky category for Argentina could be established.

When asked if there were any surprises or change of plans during the building of the distillery, Nestor says "Our distillery at this point is as it was planned initially, which was to concentrate on quality and the positioning of the brand. Our second stage, which is now underway, is expanding in size whilst still maintaining our quality."

The same hard work that the couple put into creating a good quality spirit which has consistency over time is applied to when it comes to sourcing the casks for maturation. Most of the newmake is put into ex-bourbon casks from Kentucky. For special releases they buy casks made of new American oak and send them to various Argentinian vineyards to be seasoned with chardonnay or sherry.

Currently, the distillery is equipped with a lauter mash tun and four stainless steel washbacks with a rather long fermentation – 4-6 days. Together with two stills the distillery has a capacity of producing 8,000 litres per year.

The first, limited release of La Alazana single malt was made in November 2014 and three years later it was time for the first peated version. All releases have been in small batches so far but the time for the distillery to take its place in a wider context is approaching. The goal for Lila and Nestor is to have a core expression that is at least 8-10 years old.

"As our objective is to have a standard that is no younger than eight years old, our bottlings have been very limited so far. The local market for single malt whisky is growing steadily and we are receiving numerous visitors from our country and from abroad, which is why we are selling our limited editions exclusively at the distillery. We have a lot of interest from Europe, North America and Asia; so we plan to reserve part of our production for the export market in the future."

Apart from establishing a distillery of their own, the couple has also worked as mentors for another distillery project in the Buenos Aires area run by Santiago and Carlos Mignone.

Lila´s and Nestor´s final advice
to someone wanting to start a distillery of their own;

"Work hard, do things well every day, be constant and be passionate. Whisky has been made this way for over 500 years."

Glen Keith

[glen keeth]

Owner:
Chivas Brothers
(Pernod Ricard)

Region/district:
Speyside

Founded: 1957
Status: Active
Capacity: 6 000 000 litres

Address: Station Road, Keith, Banffshire AB55 3BU

Website:
-

Tel:
01542 783042

When Seagram's (owners at the time of Chivas Brothers) founded Glen Keith in 1957, it was in order to resolve a number of challenges that the company was facing.

They had already taken ownership of Chivas Regal, a well-known deluxe blend, as well as Strathisla, a distillery dating back to the 1700s. Sam Bronfman, the owner of Seagram's, now wanted to add a cheaper, and easy-drinking whisky to the repertoire. He had seen the success that J&B and Cutty Sark had in the USA with its light flavour, but Strathisla single malt had a stronger character and, furthermore, was needed for Chivas Regal. At first he tried to acquire Cutty Sark from Berry Brothers & Rudd but the chairman, Hugh Rudd, rejected the offer. At the beginning of the 1960s, he created the 100 Pipers blend instead. In order to make it really light, he used triple distillation at Glen Keith for a period of time. During certain times in the 1970s, in the absence of their own distillery on Islay, they also distilled peated whisky for use in other whiskies. This smoky variety has on a few occasions been released by independent bottlers as single malts under the names of Craigduff and Glenisla.

Reopened in 2013, following 13 years of silence, Glen Keith is equipped with a Briggs 8 ton full lauter mash tun and six stainless steel washbacks. In the old building there are nine washbacks made of Oregon pine and six, old but refurbished stills. The distillery has the capacity to make 6 million litres with the possibility of producing 40 mashes per week.

In 2017, **Distillery Edition** became the first widely available official bottling of Glen Keith. But it didn't stop there. In July 2019, Chivas launched a new series with 15 bottlings from four distilleries (Glen Keith, Longmorn, Braeval and Caperdonich). Named The Secret Speyside Collection it will be available in the duty-free segment before being rolled out in 2020 to domestic markets. The three Glen Keith expressions, **21, 25** and **28 years old**, were all matured in American oak. Finally, there is a **22 year old cask strength** bottling in the Distillery Reserve Collection, available at all Chivas' visitor centres.

History:

1957 The Distillery is founded by Chivas Brothers (Seagrams).

1958 Production starts.

1970 The first gas-fuelled still in Scotland is installed, the number of stills increases from three to five.

1976 Own maltings (Saladin box) ceases.

1983 A sixth still is installed.

1994 The first official bottling, a 10 year old, is released as part of Seagram's Heritage Selection.

1999 The distillery is mothballed.

2001 Pernod Ricard takes over Chivas Brothers from Seagrams.

2012 The reconstruction and refurbishing of the distillery begins.

2013 Production starts again.

2017 A Distillery Edition is launched.

2019 Three bottlings in The Secret Speyside Collection are launched.

Secret Speyside
21 years old

Tasting notes Glen Keith Distillery Edition:

IR – Sweet and fruity on the nose with notes of toffee and apples. Smooth on the palate, vanilla, tropical fruits, marzipan, sponge cake, honey, pears and a hint of dry oak in the finish.

Glenkinchie

[glen•kin•chee]

Owner:
Diageo

Region/district:
Lowlands

Founded: 1837
Status: Active (vc)
Capacity: 2 500 000 litres

Address: Pencaitland, Tranent, East Lothian EH34 5ET

Website: malts.com
Tel: 01875 342004

During the first decade of the 20th century, the Scotch whisky industry survived one of its toughest crises ever.

The general economy was bad and when Lloyd George and his government in 1909 decided to raise the duty on spirits (and not on wine and beer) to help pay for social services to the citizens, the industry was forced down on its knees. This called for organisation and structure to avoid overproduction and falling prices. James Gray, the owner of Glenkinchie at that time, persuaded four other Lowland distilleries that they should merge to form Scottish Malt Distillers Ltd (SMD) so that they could weather the storm together. A similar cooperation had already been created 30 years earlier when six grain distilleries had formed Distiller's Company Limited (DCL). DCL had grown from strength to strength over the years and in 1925 they acquired SMD.

In close proximity to Edinburgh, Glenkinchie has always been a popular destination and, tellingly, 40,000 people come to the distillery every year. The visitor centre was upgraded just a couple of years ago and now it's time again. Planning approval for a huge make-over was granted in December 2018 and this is part of a greater investment from Diageo which includes a Johnnie Walker Experience in Edinburgh as well as an upgrade of visitor centres also at Clynelish, Cardhu and Caol Ila. The new centre at Glenkinchie is expected to be completed in 2020.

The distillery is equipped with a full lauter mash tun (nine tons) and six wooden washbacks with a fermentation time of 66-110 hours. There are two stills where the wash still has the biggest charge in Scotland – 21,000 litres. Worm tubs are used to cool the spirit vapours. In 2019, the distillery will be working a five-day week with 10 mashes, producing around 2 million litres of alcohol.

The core range consists of a **12 year old** and a **Distiller's Edition** with a finish in amontillado sherry casks. There is also a **distillery exclusive** with an 18 months finish in sherry casks. In 2016, a **24 year old** was launched as part of the Special Releases and in 2019 a limited version was launched together with The Royal Edinburgh Military Tattoo.

History:
- **1825** A distillery known as Milton is founded by John and George Rate.
- **1837** The Rate brothers are registered as licensees of a distillery named Glenkinchie.
- **1853** John Rate sells the distillery to a farmer by the name of Christie who converts it to a sawmill.
- **1881** The buildings are bought by a consortium from Edinburgh.
- **1890** Glenkinchie Distillery Company is founded. Reconstruction and refurbishment is on-going for the next few years.
- **1914** Glenkinchie forms Scottish Malt Distillers (SMD) with four other Lowland distilleries.
- **1939-1945** Glenkinchie is one of few distilleries allowed to maintain production during the war.
- **1968** Floor maltings is decommissioned.
- **1969** The maltings is converted into a museum.
- **1988** Glenkinchie 10 years becomes one of selected six in the Classic Malt series.
- **1998** A Distiller's Edition with Amontillado finish is launched.
- **2007** A 12 year old and a 20 year old cask strength are released.
- **2010** A cask strength exclusive for the visitor centre, a 1992 single cask and a 20 year old are released.
- **2016** A 24 year old and a distillery exclusive without age statement are released.

12 years old

Tasting notes Glenkinchie 12 years old:

GS – The nose is fresh and floral, with spices and citrus fruits, plus a hint of marshmallow. Notably elegant. Water releases cut grass and lemon notes. Medium-bodied, smooth, sweet and fruity, with malt, butter and cheesecake. The finish is comparatively long and drying, initially rather herbal.

Glenlivet

[glen•liv•it]

Owner:
Chivas Brothers
(Pernod Ricard)

Region/district:
Speyside

Founded: 1824

Status: Active (vc)

Capacity: 21 000 000 litres

Address: Ballindalloch, Banffshire AB37 9DB

Website:
theglenlivet.com

Tel:
01340 821720 (vc)

For the best part of the new millennium, Glenfiddich and Glenlivet have been the number one and two on the single malt sales list with only one year (2014) with Glenlivet in the lead. The "battle" between the two has often had the air of a friendly banter.

But behind the polished facade, rather typical for whisky brands, there is of course a struggle for awareness and shelf space. Glenlivet has on numerous occasions called themselves "the single malt that started it all" due to their rich provenance, being one of the very first distilleries to get a license in 1824. On the other hand, Glenfiddich was the first distillery to actively promote a single malt brand on a global scale in the early 1960s. In 2018 Glenfiddich sold 16,8 million bottles and Glenlivet 14,5 million but Glenlivet still has a firm grip of the most important single malt market in the world, the USA, where they have been number one for several decades.

With a second distillery commissioned in summer 2018 the total capacity of Glenlivet is 21 million litres of alcohol – by far the largest malt distillery in Scotland. With the new addition, the distillery is now equipped with two Briggs full lauter mash tuns, each with a 13.5 ton charge. The 16 wooden washbacks have been augmented by another 16 made of stainless steel. Fourteen pairs of stills are divided with four pairs in the oldest still room, three in the beautiful room that was built in 2010 and another seven in the third and latest still room.

The core range of Glenlivet is made up of **Founder's Reserve, Captain's Reserve** (new since May 2018 and with a finish in cognac casks), **12 year old, 12 year old Excellence, 15 year old French oak Reserve, 18 year old, 21 year old Archive** and **Glenlivet XXV**. As an exclusive to the American market, a **14 year old cognac finish** was released in September 2019. A special range of non-chill filtered whiskies called Nàdurra include: **Nàdurra Oloroso Cask Strength, Nàdurra first Fill Selection Cask Strength** and **Nàdurra Peated Whisky Cask Finish**. All three are available at cask strength but also bottled at 48% for duty free. The smoky notes in the latter come from a finish in casks that had previously held peated Scotch whisky. The travel retail range also includes **Triple Cask Distiller's Reserve, Triple Cask White Oak Reserve** and **Triple Cask Rare Cask**. All three are basically the same whiskies as the three in the previous Master Distiller's Reserve range, but with new names. In 2014 a 50 year old became the first bottling in a new range, The Winchester Collection (named after The Glenlivet's master distiller Alan Winchester) with a third edition, **Vintage 1967**, released in September 2018. Over the years, the owners have released mystery bottlings where very little has been revealed about the content. The fourth and latest bottling was **The Glenlivet Enigma** which was released in spring 2019. Finally there are four cask strength bottlings in the Distillery Reserve Collection, available at all Chivas' visitor centres – from **12** to **18 years old**.

History:

1817 George Smith inherits the farm distillery Upper Drummin from his father Andrew Smith who has been distilling on the site since 1774.

1840 George Smith buys Delnabo farm near Tomintoul and leases Cairngorm Distillery.

1845 George Smith leases three other farms, one of which is situated on the river Livet and is called Minmore.

1846 William Smith develops tuberculosis and his brother John Gordon moves back home to assist his father.

1858 George Smith buys Minmore farm and obtains permission to build a distillery.

1859 Upper Drummin and Cairngorm close and all equipment is brought to Minmore which is renamed The Glenlivet Distillery.

1864 George Smith cooperates with the whisky agent Andrew P. Usher and exports the whisky with great success.

1871 George Smith dies and his son John Gordon takes over.

1880 John Gordon Smith applies for and is granted sole rights to the name The Glenlivet.

1890 A fire breaks out and some of the buildings are replaced.

1896 Another two stills are installed.

1901 John Gordon Smith dies.

1904 John Gordon's nephew George Smith Grant takes over.

1921 Captain Bill Smith Grant, son of George Smith Grant, takes over.

1953 George & J. G. Smith Ltd merges with J. & J. Grant of Glen Grant Distillery and forms the company Glenlivet & Glen Grant Distillers.

1966 Floor maltings closes.

1970 Glenlivet & Glen Grant Distillers Ltd merges with Longmorn-Glenlivet Distilleries Ltd and Hill Thomson & Co. Ltd to form The Glenlivet Distillers Ltd.

History continued:

1978 Seagrams buys The Glenlivet Distillers Ltd. A visitor centre opens.

2000 French Oak 12 years and American Oak 12 years are launched

2001 Pernod Ricard and Diageo buy Seagram Spirits & Wine. Pernod Ricard thereby gains control of the Chivas group.

2004 This year sees a lavish relaunch of Glenlivet. French Oak 15 years replaces the previous 12 year old

2005 Two new duty-free versions are introduced – The Glenlivet 12 year old First Fill and Nadurra. The 1972 Cellar Collection (2,015 bottles) is launched.

2006 Nadurra 16 year old cask strength and 1969 Cellar Collection are released.

2007 Glenlivet XXV is released.

2009 Four more stills are installed and Nadurra Triumph 1991 is released.

2010 Another two stills are commissioned and capacity increases to 10.5 million litres. Glenlivet Founder´s Reserve is released.

2011 Glenlivet Master Distiller´s Reserve is released for the duty free market.

2012 1980 Cellar Collection is released.

2013 The 18 year old Batch Reserve and Glenlivet Alpha are released.

2014 Nadurra Oloroso, Nadurra First Fill Selection, The Glenlivet Guardian´s Chapter and a 50 year old are released.

2015 Founder´s Reserve is released as well as two new expressions for duty free; Solera Vatted and Small Batch.

2016 The Glenlivet Cipher and the second edition of the 50 year old are launched.

2018 Captain´s Reserve and Code are released. A new distillery is commissioned.

2019 Enigma and a 14 year old cognac finish are released.

Tasting notes Glenlivet 12 year old:

GS – A lovely, honeyed, floral, fragrant nose. Medium-bodied, smooth and malty on the palate, with vanilla sweetness. Not as sweet, however, as the nose might suggest. The finish is pleasantly lengthy and sophisticated.

Tasting notes Glenlivet Founder´s Reserve:

GS – The nose is fresh and floral, with ripe pears, pineapple, tangerines, honey and vanilla. Medium-bodied, with ginger nuts, soft toffee and tropical fruit on the smooth palate. Soft spices and lingering fruitiness in the finish.

Distiller´s Reserve White Oak Reserve The Glenlivet Enigma

21 years old Archive

Nàdurra Peated

Founder´s Reserve Captain´s Reserve 12 years old

Glenlossie

[glen•<u>loss</u>•ee]

Owner: **Region/district:**
Diageo   Speyside

Founded: **Status:**   **Capacity:**
1876   Active   3 700 000 litres

Address: Birnie, Elgin, Morayshire IV30 8SS

Website: **Tel:**
malts.com   01343 862000

Usually, any maintenance or exchange of equipment at a distillery is carried out during the silent season. This is a period of 3-5 weeks with no production so as not to interfere with the construction work.

However, every now and then, there comes a time in a distillery's life when so much needs to be done that you just have to close down the production for a longer period of time. This was the case for Glenlossie which stopped production in January 2018. The first distillation after the closure was 11th June 2019.

During the 18 months of closure, huge improvements were made. To mention but a few; new yeast house, new compressor room and new CIP (cleaning in place) room. On top of that all electrical equipment, cables, pumps and pipework were replaced and a general upgrade of buildings also took place. Finally, two new, external washbacks were installed and the old cast iron low wines receiver and feints receiver were replaced with stainless steel vessels

Usually, more washbacks means increased capacity but not in this case. The 3.7 million litre capacity is based on working a seven-day week with short fermentations (65 hours). However, the green/oily character of the newmake is dependent on a mix of short and long fermentations (mainly long at 104 hours) so the additional washbacks will enable the distillery to achieve the character even when they will be working 7 days a week.

After the upgrade, the distillery will be equipped with an eight ton stainless steel full lauter mash tun, eight washbacks made of larch and two made of stainless steel. There are three pairs of stills with the spirit stills equipped with purifiers between the lyne arms and the condensers, thus increasing the reflux. The initial production plans following the upgrade will be 12 mashes in a five-day week, looking for 2 million litres of alcohol on a yearly basis.

Glenlossie is one of the major contributors to the Haig Gold Label blend and the only official bottling of Glenlossie single malt available today is a **10 year old Flora & Fauna**.

History:

1876 John Duff, former manager at Glendronach Distillery, founds the distillery. Alexander Grigor Allan (to become part-owner of Talisker Distillery), the whisky trader George Thomson and Charles Shirres (both will co-found Longmorn Distillery some 20 years later with John Duff) and H. Mackay are also involved in the company.

1895 The company Glenlossie-Glenlivet Distillery Co. is formed. Alexander Grigor Allan passes away.

1896 John Duff becomes more involved in Longmorn and Mackay takes over management of Glenlossie.

1919 Distillers Company Limited (DCL) takes over the company.

1929 A fire breaks out and causes considerable damage.

1930 DCL transfers operations to Scottish Malt Distillers (SMD).

1962 Stills are increased from four to six.

1971 Another distillery, Mannochmore, is constructed by SMD on the premises. A dark grains plant is installed.

1990 A 10 year old is launched in the Flora & Fauna series.

2010 A Manager´s Choice single cask from 1999 is released.

10 years old

Tasting notes Glenlossie 10 years old:

GS – Cereal, silage and vanilla notes on the relatively light nose, with a voluptuous, sweet palate, offering plums, ginger and barley sugar, plus a hint of oak. The finish is medium in length, with grist and slightly peppery oak.

Pioneers of Whisky

Pär Caldenby
Founder
Smögen Distillery, Sweden

Working in a monopolised market has its difficulties, especially when your product is one that the government finds a threat to public health. Add to that each step of sales and marketing being controlled and monitored by the State and the difficulties increase. Being a lawyer, one would have thought that Pär Caldenby would be well equipped handling such issues but in Sweden there are no shortcuts. You abide by the regulations or you're out of business.

Pär still remembers when he first got interested in malt whisky "On the day that I turned 20 and thus became old enough to be allowed to purchase alcohol in Sweden, I got myself a bottle of single malt, Auchentoshan 10 year old, which at that time was one of a mere half dozen available single malts in Sweden. It was love at first nosing."

Two years later he founded a whisky club together with friends and as whisky availability increased when Sweden joined the EU in 1995, so did Pär's interest in whisky.

"I joined the Dutch branch of the SMWS, whom I helped with regards to the Swedish market, and began buying in from excellent retailers such as Richard Joynson's shop, Loch Fyne Whiskies. And thanks to eventually being able to turn my studies into salaried work in the court system, I got the means to start visiting Scotland, from 1996 and onwards."

In 2005, he wrote his first book on whisky, Enjoying Malt Whisky, and by this time Pär's interest had turned from hobbyish to what he calls a semi-professional approach. But from there to starting a distillery is a huge step. What prompted him to go into production? In a very self-conscious way his answer is simply "Basically, it was the feeling that I could do this even better."

At this time he had had enough of practising law and on resigning from the firm, he went to Scotland and spent a week at Bruichladdich Academy learning about distilling. Back in Sweden he started planning for the distillery. He willingly admits that plenty of good advice from friends made in Scotland (at Bruichladdich, Springbank, Bladnoch and Daftmill) were inspiring and helpful.

The distillery was designed by himself and using as much second hand equipment as possible, without compromising usability and quality. The stills and condensers were brand new and were supplied by Forsyths in Rothes.

Pär was lucky to be able to fund the distillery without any loans. His own funds, a number of friends buying shares in the company and sales of small casks to private individuals was sufficient. It also helped that the land where the distillery was built was his mother's. But funding is one thing – the actual planning and construction another. Pär remembers it as hard and plentiful work but it didn't come as a surprise. He was also assisted by a group of engineers from one of the rescue services that he had met in whisky circles. While the actual construction work turned out to be an intensive but rewarding task, the red tape surrounding the entire project could at times be frustrating.

"The amount of licenses, permits, approvals, controls and whathaveyous that are required is simply mind-boggling. Inspections and approvals by public servants who had no knowledge whatsoever about whisky is a bit of a test of the stiff upper lip every time they demanded access. I would be a happier man without such unhelpful nuisances, while I do also recognise that there is a general need to safeguard certain aspects of the production process."

The distillery has been running smoothly since the first distillation back in 2010. And Pär has only made one major change in the equipment set-up. The worm tub condensers that he couldn't afford in the beginning were finally installed in summer 2018, replacing the initial shell and tube condensers.

Aside from producing a high-quality whisky, character is also important to Pär.

"I want to create a substantial, clear and distinctive character in my whisky, through the use of selected malted barley, a very gentle mashing process, long fermentation times, low to medium-height stills and slow distillation. And when I got the characterful, intense but still clean spirit, this was only ever going to go into the best casks I could find. This comes at a significant cost, but good spirit also requires, and deserves, good quality casks, if there is to be excellent whisky in the end".

When asked about the character in his matured stock Pär describes it as "Substantial body, intensity of character, peated and with a very malty backdrop that offers a fruity potential, come a slightly higher age."

Smögen distillery has been running for nine years now which means that Pär a few years ago came to the stage where he needed to find customers for his brand, a word that Pär doesn't like by the way.

"I do not like to think of my whisky as generic and considered "merely" a brand – it is whisky and it is individual and very good, which is why you ought to get it."

Awareness about Smögen whisky is very much created by word of mouth. Swedish whisky enthusiasts are social and communicative when they like a whisky and make sure that their social network gets to know. Most is sold in Sweden even though the whisky can be found in a few export markets. The only drawback is that the country is a monopoly market where the government-owned retailer Systembolaget is the only place where you can buy alcohol.

As Pär puts it, "They are not making matters easier for a small but highly specialised and character-led whisky distillery such as ourselves."

Pär´s final advice
to someone wanting to start a distillery of their own:

"Be passionate and make sure you've got true knowledge of whisky itself and of the whisky making process – and realise that this is a long term project. Oh, and it helps not giving up your day job too soon, as there are a number of years where you are unlikely to be able to live off of the business as it will gulp up all the money you throw at it, including what could've been your salary."

Glenmorangie

[glen•<u>mor</u>•run•jee]

Owner:
The Glenmorangie Co
(Moët Hennessy)

Region/district:
Northern Highlands

Founded: 1843

Status:
Active (vc)

Capacity:
6 000 000 litres

Address: Tain, Ross-shire IV19 1PZ

Website:
glenmorangie.com

Tel:
01862 892477 (vc)

The routine answer to the question "what gives the whisky its taste" is that 70% is acquired from the cask, and the rest from the newmake.

It is a rare claim that the yeast can influence taste considerably. At least when talking about Scotch. In Japan, on the other hand, different strains of yeast are used to give different characters and in the USA, the yeast used is a closely guarded secret. For many years, Bourbon makers have known that the whisky one gets in the end is highly dependent on the yeast.

Bill Lumsden, Director of Distilling, Whisky Creation & Whisky Stocks at Glenmorangie, is also adamant that the yeast has a role to play. Perhaps this is not totally unexpected as he has a PhD in Microbial Physiology and Fermentation Science. Ten years ago he decided to start trials with yeasts other than the commercial ones that every distillery used. And not only that, he used wild yeast foraged from the barley fields surrounding the distillery. Long story short, the first commercial release where *Saccharomyces diaemath* had been used was Allta, released without age statement (but matured for 8,5 years) in February 2019.

The distillery is currently equipped with a full lauter mash tun with a charge of ten tons, 12 stainless steel washbacks with a fermentation time of 52 hours and six pairs of stills – the tallest in Scotland. Production for 2019 will land at 6,2 million litres of pure alcohol. Currently, a new still house with space for two additional stills is being built together with a building which will house two more washbacks. Due to be commissioned in December 2019, this will bring the capacity up to 7.2 million litres.

The core range consists of **Original** (10 year old) and **18 year old**. The 25 year old has now been discontinued. There are three wood finishes: **Quinta Ruban**, which used to be a 12 year old but is now a 14 year old has been finished in a combination of 225 litre ruby barriques and 670 litre ruby pipes. **Lasanta** is 12 years old with a finish in a combination of oloroso casks (75%) and PX sherry casks (25%). Finally, there is **Nectar D'Or** with no age statement that has been finished in Sauternes casks. Added to the core range is **Signet**, an unusual piece of work with 20% of the whisky having been made using chocolate malt. A series of bottlings, called **Private Edition**, started in 2009 with the release of Sonnalta PX. This has been followed up once a year where **Allta**, made from wild yeast, matured in ex-bourbon barrels and bottled at 51.2%, was the latest. A range called Glenmorangie's Legends is available for duty-free, including **Duthac**, a vatting of whiskies matured in bourbon casks, PX sherry casks and charred, virgin oak casks, **Tayne**, aged in amontillado sherry casks and **Cadboll**. The latter was matured in ex-bourbon casks and then finished in barriques that had been used to hold Muscat and Sémillon wines. Recent limited releases include a **16 year old** single cask in 2018 and, in spring 2019, **Grand Vintage Malt 1991** with a finish in both ex-oloroso sherry casks and ex-Burgundy casks. In July 2019, the 16 year old **Cask 1784** was released as the first of four distillery exclusives.

History:

1843 William Mathesen applies for a license for a farm distillery called Morangie, which is rebuilt by them. Production took place here in 1738, and possibly since 1703.

1849 Production starts in November.

1880 Exports to foreign destinations such as Rome and San Francisco commence.

1887 The distillery is rebuilt and Glenmorangie Distillery Company Ltd is formed.

1918 40% of the distillery is sold to Macdonald & Muir Ltd and 60 % to the whisky dealer Durham. Macdonald & Muir takes over Durham's share by the late thirties.

1931 The distillery closes.

1936 Production restarts in November.

1980 Number of stills increases from two to four and own maltings ceases.

1990 The number of stills is doubled to eight.

1994 A visitor centre opens. September sees the launch of Glenmorangie Port Wood Finish which marks the start of a number of different wood finishes.

1995 Glenmorangie´s Tain l´Hermitage is launched.

1996 Two different wood finishes are launched, Madeira and Sherry. Glenmorangie plc is formed.

2001 A limited edition of a cask strength port wood finish is released in July, Cote de Beaune Wood Finish is launched in September and Three Cask (ex-Bourbon, charred oak and ex-Rioja) is launched in October for Sainsbury's.

2002 A 20 year old Sauternes finish is launched.

2003 Burgundy Wood Finish is launched in July and a limited edition of cask strength Madeira-matured (i. e. not just finished) in August.

History continued:

2004 Glenmorangie buys the Scotch Malt Whisky Society. The Macdonald family decides to sell Glenmorangie plc (including the distilleries Glenmorangie, Glen Moray and Ardbeg) to Moët Hennessy at £300 million. A new version of Glenmorangie Tain I´Hermitage (28 years) is released and Glenmorangie Artisan Cask is launched in November.

2005 A 30 year old is launched.

2007 The entire range gets a complete makeover with 15 and 30 year olds being discontinued and the rest given new names as well as new packaging.

2008 An expansion of production capacity is started. Astar and Signet are launched.

2009 The expansion is finished and Sonnalta PX is released for duty free.

2010 Glenmorangie Finealta is released.

2011 28 year old Glenmorangie Pride is released.

2012 Glenmorangie Artein is released.

2013 Glenmorangie Ealanta is released.

2014 Companta, Taghta and Dornoch are released.

2015 Túsail and Duthac are released.

2016 Milsean, Tayne and Tarlogan are released.

2017 Bacalta, Astar and Pride 1974 are released.

2018 Spios, Cadboll and Grand Vintage Malt 1989 and 1993 are released.

2019 Allta, Cask 1784 and Grand Vintage Malt 1991 are launched.

Tasting notes Glenmorangie Original 10 year old:

GS – The nose offers fresh fruits, butterscotch and toffee. Silky smooth in the mouth, mild spice, vanilla, and well-defined toffee. The fruity finish has a final flourish of ginger.

Allta

Lasanta

Cadboll

Original 10 years old

Astar

Quinta Ruban

Glen Moray

[glen mur•ree]

Owner:
La Martiniquaise (COFEPP)

Region/district:
Speyside

Founded: 1897
Status: Active (vc)
Capacity: 5 700 000 litres

Address: Bruceland Road, Elgin, Morayshire IV30 1YE

Website:
glenmoray.com

Tel:
01343 542577

The last ten year's transformation of Glen Moray is remarkable. While still an important part of the owners' Label 5 blended Scotch, the single malt has carved out a niche where it is both respected by the connoisseurs and achieves huge sales figures.

Part of the credit goes to the owners, La Martiniquaise, but distillery manager and blender Graham Coull undoubtedly deserves his bit. Being a part of the industry since 1994, he became the manager of Glen Moray in 2005. In October 2019, however, Coull decided it was time for a new challenge and joined Dingle Distillery in Ireland as their distillery manager and master blender.

Since 2016, Glen Moray is equipped with an 11 ton full lauter mash tun. There are 14 stainless steel washbacks placed outside with a fermentation time of 50-60 hours and nine stills (3 wash and 6 spirit). The current capacity is 5.7 million litres of alcohol, but the owners have the option of reintroducing the old mash tun, adding a few more washbacks and two more wash stills which would increase the capacity to 8.9 million litres. In 2019, the distillery will be making 28 mashes per week, producing 5.7 million litres, of which 100,000 litres will be heavily peated (50ppm) spirit. In 2019, two smaller warehouses were demolished to make room for two bigger and palletised ones.

The core range consists of **Classic, Classic Port Finish, Classic Chardonnay Finish, Classic Sherry Finish, Classic Cabernet Sauvignon Finish** and **Classic Peated** as well as a **12, 15** and **18 year old**. A new expression was released in 2018 – **10 year old Fired Oak** – matured for 10 years in ex-bourbon casks and then finished for 10 months in heavily charred virgin American oak. Recent limited releases include a **21 year old portwood finish**, a **30 year old sherry finish** and a version with a two year **finish in cider casks**. A follow-up to the cider finish was a bottling in autumn 2019 with a two year **Rhum Agricole finish** in casks from Martinique. Finally, there are three **15 year olds** to be found at the distillery; all fully matured in **chenin blanc, chardonnay** and **burgundy** respectively.

History:

1897 Elgin West Brewery, dated 1830, is reconstructed as Glen Moray Distillery.

1910 The distillery closes.

1920 Financial troubles force the distillery to be put up for sale. Buyer is Macdonald & Muir.

1923 Production restarts.

1958 A reconstruction takes place and the floor maltings are replaced by a Saladin box.

1978 Own maltings are terminated.

1979 Number of stills is increased to four.

1996 Macdonald & Muir Ltd changes name to Glenmorangie plc.

1999 Three wood finishes are introduced - Chardonnay (no age) and Chenin Blanc (12 and 16 years respectively).

2004 Louis Vuitton Moët Hennessy buys Glenmorangie plc and a 1986 cask strength, a 20 and a 30 year old are released.

2006 Two vintages, 1963 and 1964, and a new Manager's Choice are released.

2007 New edition of Mountain Oak is released.

2008 The distillery is sold to La Martiniquaise.

2009 A 14 year old Port finish and an 8 year old matured in red wines casks are released.

2011 Two cask finishes and a 10 year old Chardonnay maturation are released.

2012 A 2003 Chenin Blanc is released.

2013 A 25 year old port finish is released.

2014 Glen Moray Classic Port Finish is released.

2015 Glen Moray Classic Peated is released.

2016 Classic Chardonnay Finish and Classic Sherry Finish are released as well as a 15 and an 18 year old.

2017 Glen Moray Mastery is launched.

2018 10 year old Fired Oak is released.

2019 Glen Moray Rhum Agricole is released.

GLEN MORAY
SPEYSIDE
SINGLE MALT SCOTCH WHISKY
· ELGIN HERITAGE ·
AGED 12 YEARS

12 years old

Tasting notes Glen Moray 12 years old:

GS — Mellow on the nose, with vanilla, pear drops and some oak. Smooth in the mouth, with spicy malt, vanilla and summer fruits. The finish is relatively short, with spicy fruit.

Glen Ord

[glen ord]

Owner:
Diageo

Region/district:
Northern Highlands

Founded: | **Status:** | **Capacity:**
1838 | Active (vc) | 11 000 000 litres

Address: Muir of Ord, Ross-shire IV6 7UJ

Website: | **Tel:**
malts.com | 01463 872004 (vc)

One of the greatest success stories in the Scotch whisky industry in recent years is the launch of The Singleton. Divided into three sub-brands – Glen Ord, Dufftown and Glendullan – it sells more than six million bottles per year.

One person has played a significant role for this whisky since the first release in 2004. Maureen Robinson has been with Diageo for 42 years now and became a master blender in 1986. Over the years she has been involved in the creation of many famous releases including Talisker, Lagavulin, the Classic Malts and several of the company's well-known blends. She has also been responsible for many unforgettable bottlings in the Special Releases series. In terms of volumes however, it is doubtful if any project can beat The Singleton. With Glen Ord as the best seller of the three distilleries, the owners have made it clear that their goal is that The Singleton will be a 1 million case (12 million bottles) brand in the near future. Only three other brands have achieved that – Glenfiddich, Glenlivet and Macallan.

Since 2011, Glen Ord distillery has been expanded rapidly in several stages and with the latest expansion in 2015, the distillery now has a capacity of 11 million litres. The complete set of equipment comprises of two stainless steel mashtuns, each with a 12.5 ton mash. There are 22 wooden washbacks with a fermentation time of 75 hours and no less than 14 stills. There are also drum maltings on-site providing a number of Diageo distilleries with malted barley.

The core range is the **Singleton of Glen Ord 12, 15** and **18 year old**. A sub-range, The Singleton Reserve Collection, is exclusive to duty free and consists of **Signature, Trinité, Liberté** and **Artisan**. The Forgotten Drops Series was created in 2017 for both Glen Ord and Glendullan with the aim to present old and limited releases. The first for Glen Ord was a **41 year old**, the oldest whisky ever released from the distillery and it was followed in autumn 2018 by a **42 year old** finished in amontillado casks. An **18 year old** also appeared in the 2019 Special Releases.

History:

1838 Thomas Mackenzie founds the distillery.

1855 Alexander MacLennan and Thomas McGregor buy the distillery.

1870 Alexander MacLennan dies and the distillery is taken over by his widow who marries the banker Alexander Mackenzie.

1877 Alexander Mackenzie leases the distillery.

1878 Alexander Mackenzie builds a new still house and barely manages to start production before a fire destroys it.

1896 Alexander Mackenzie dies and the distillery is sold to James Watson & Co. for £15,800.

1923 John Jabez Watson, James Watson's son, dies and the distillery is sold to John Dewar & Sons. The name changes from Glen Oran to Glen Ord.

1961 A Saladin box is installed.

1966 The two stills are increased to six.

1968 Drum maltings is built.

1983 Malting in the Saladin box ceases.

1988 A visitor centre is opened.

2002 A 12 year old is launched.

2003 A 28 year old cask strength is released.

2004 A 25 year old is launched.

2005 A 30 year old is launched as a Special Release from Diageo.

2006 A 12 year old Singleton of Glen Ord is launched.

2010 A Singleton of Glen Ord 15 year old is released in Taiwan.

2011 Two more washbacks are installed, increasing the capacity by 25%.

2012 Singleton of Glen Ord cask strength is released.

2013 Singleton of Glen Ord Signature, Trinité, Liberté and Artisan are launched.

2015 The Master´s Casks 40 years old is released.

2017 A 41 year old reserved for Asia is released.

2018 A 14 year old triple-matured is launched as part of the Special Releases.

2019 An 18 year old appears in the Special Releases.

Tasting notes Glen Ord 12 years old:

GS – Honeyed malt and milk chocolate on the nose, with a hint of orange. These characteristics carry over onto the sweet, easy-drinking palate, along with a biscuity note. Subtly drying, with a medium-length, spicy finish.

12 years old

Glenrothes

[glen•roth•iss]

Owner:
The Edrington Group

Region/district:
Speyside

Founded: 1878　　**Status:** Active

Capacity:
5 600 000 litres

Address: Rothes, Morayshire AB38 7AA

Website:
theglenrothes.com

Tel:
01340 872300

It has now been a year since The Glenrothes abandoned their strategy of releasing vintages – something that started in the mid 90s. The only other distillery in Scotland building their range on the same principle, Balblair, has recently followed suit.

The Glenrothes is a brand which relies heavily on sherry casks for the maturation. Almost 90% of the production are filled into casks that have held sherry and the vast majority of these are first fill – i. e. no whisky has been in the casks before. Sherry casks are considerably more expensive than ex-bourbon.

Glenrothes distillery is equipped with a 5.5 ton stainless steel full lauter mash tun. Twelve washbacks made of Oregon pine are in one room, whilst an adjacent tun room houses eight stainless steel washbacks – all of them with a 58 hour fermentation time. The magnificent still house has five pairs of stills performing a very slow distillation. In 2019, the distillery will be doing 44 mashes per week, producing just over 4 million litres of alcohol. In recent years more of the production is laid down for single malts rather than for the blended market.

All the old expressions in the core range have disappeared and the same goes for the vintages. They have been replaced by Soleo Collection which is based on whiskies that have matured 100% in sherry casks. The range consists of **10 year old, 12 year old, Whisky Maker's Cut, 18 year old** and **25 year old**. There is also a second range called Aqua Collection including **Whisky Maker's Dram** and a **12 year old** which will be sold on-line. The current duty free range will continue at least until September 2019; **Robur Reserve, Manse Reserve, Elder's Reserve, Minister's Reserve** and the **25 year old Ancestor's Reserve**. Recent limited releases include a **13 year old Halloween Edition** from 2018 with a new version coming out in October 2019. Summer 2019 saw the release of a **40 year old** and a **50 year old** is planned for a December 2019 release.

History:

1878 James Stuart & Co. begins planning the new distillery with Robert Dick, William Grant and John Cruickshank as partners.

1879 Production starts in December.

1884 The distillery changes name to Glenrothes-Glenlivet.

1887 William Grant & Co. joins forces with Islay Distillery Co. and forms Highland Distillers Company.

1897 A fire ravages the distillery.

1903 An explosion causes substantial damage.

1963 Expansion from four to six stills.

1980 Expansion from six to eight stills.

1989 Expansion from eight to ten stills.

1999 Edrington and William Grant & Sons buy Highland Distillers.

2002 Four single casks from 1966/1967 are launched.

2005 A 30 year old is launched together with Select Reserve and Vintage 1985.

2008 1978 Vintage and Robur Reserve are launched.

2009 The Glenrothes John Ramsay, Alba Reserve and Three Decades are released.

2010 Berry Brothers takes over the brand.

2011 Editor's Casks are released.

2013 2001 Vintage and the Manse Brae range are released.

2014 Sherry Cask Reserve and 1969 Extraordinary Cask are released.

2015 Glenrothes Vintage Single Malt is released.

2016 Peated Cask Reserve and Ancestor´s Reserve are released.

2017 The brand returns to Edrington and The Glenrothes Wine Merchant´s Collection is introduced.

2018 The entire range is revamped and four new bottlings with age statements are introduced.

2019 A 40 year old and a 50 year old are released.

12 years old

Tasting notes Glenrothes Soleo 12 year old:

IR – Fresh and fruity on the nose with notes of strawberries/raspberries and a hint of cinnamon. The taste if fruity and spicy with notes of pear, cinnamon, nutmeg, lemon zest and, in the finish, brown sugar and a little ginger.

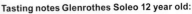

Pioneers of Whisky

Desiree Whitaker
Founder
Cardrona Distillery, New Zealand

While the Australian whisky wonder took place with more than 20 whisky distilleries being built in two decades, there were no signs of anything similar happening in New Zealand. A few distilleries tried their hand at whisky but the first purpose built whisky distillery in modern times, Cardrona, wasn't opened until 2015.

Behind Cardrona Distillery near Wanaka on the South Island, was Desiree Whitaker together with her husband Ash and it opened its doors just before Christmas 2015. Desiree came from a successful career as a farmer during which she also became the youngest ever to be elected to the Fonterra Shareholders Council. Fonterra is a co-op owned by 10,000 farmers and is also New Zealand's largest company. But this time in Desiree's life was not just about success. She and her husband at the time separated and eventually divorced and Desiree was left to run the farm alone with the support of her team. Desiree remembers;

" It was a crisis. You don't get married to get divorced. It caused me to stop and take stock and think about what I really wanted to spend my life doing. I started making lists of ideas – hundreds of ideas. Then short lists. I would research them, exhaust the list. Start over – again ideas, short lists. Whisky made my list in early 2011 and I started researching how to properly craft whisky and other spirits."

Desiree's first introduction to whisky was as a 20 year-old during a sabbatical two-year break from university. She went to London and worked in a backstreet pub in Notting Hill, the Ladbroke Arms. Single malt whisky hadn't really taken off at this time or as Desiree herself puts it, "… it was drunk really only by septuagenarians and octogenarians" but Ladbroke Arms had a good collection of malts and she became fascinated by the colours, bottles and flavours.

In May 2013, she sold her farm, moved to Wanaka and started the search for a site at Cardrona. She secured the site in late 2013, worked on getting all the consents and turned the first sod in January 2015. All she owned was invested into the distillery and, by this time, she had met her current husband Ash who sold off his business to help build the distillery. Still being short, Desiree's parents chipped in the rest of the money and a true family distillery had been borne. The first spirit ran from the stills on the 23rd October 2015 – Desiree's 37th birthday.

Most people starting a whisky distillery find inspiration and guidance from other people in the industry and Desiree is no different. When attending an American Distillers Institute conference she met the legendary (and now sadly missed) Dave Pickerell.

" He was leading a seminar and his clarity and ability to turn complex ideas into simple ones was enviable. I scoured the programme and attended every workshop he presented. A while later I took a week long Advanced Distillation Methods course with Dave. Dave mentored from afar throughout my journey and came out to New Zealand for American Thanksgiving 2016."

The next people that were instrumental to the journey of building the Cardrona Distillery were the Forsyth family, owners of

Forsyth's Coppersmiths supplying distilleries all over the world with first class copper pot stills and other distillery equipment.

"I met Richard Senior very briefly, also at an ADI conference. From there I was travelling to Scotland. I remember being in the Craigellachie Hotel. I pulled out the brochure Richard Sr had given me, and looked on Google maps to where they were based. Of course, Rothes is just only a few minutes drive from Craigellachie. I wrote Richard an email requesting a visit, and he replied quickly inviting me to visit with him and his son Richard Jr that Friday at 8am."

The result of that visit was that Forsyths became the main supplier of equipment for the new distillery. They built the stills and the mash tun, and sent two engineers from Scotland to commission the distillery. They were there for 3 weeks helping Desiree and Ash tune the distillery equipment and Richard Sr and his wife Heather flew out to formally open the distillery a short while later.

The last, but by no means the least, person that Desiree mentions is the architect Sarah Scott. She has resided in Wanaka for more than 30 years and has had a huge impact upon building design in the entire region.

"Sarah captured what was in my mind's eye and made it come to life on paper, and then in reality. Sarah guided me through the resource consent process to build in the Cardrona Outstanding Natural Landscape, and then the building consent process, for what was an unprecedented build in New Zealand Whisky history."

When asked to describe the style of the future Cardrona whisky, Desiree starts off pragmatically and ends on a more poetic note.

" I have a particular penchant for unpeated richly matured whiskies, and that has driven the style we are producing. The Cardrona is rich in honey with a thick weighted spirit on the tongue. I wanted to create the foundation for a well-made spirit, where every element was properly grounded. I have heard that a distillery has its own soul. That the distillery itself, every quirk, every nuance, every peculiarity adds to the final character of the spirit. It is the distillery itself that creates the distillery character. "

New Zealand have few standards surrounding whisky but Desiree, being something of a purist, decided early on to apply the traditions she had learned from Scotland. The spirit is matured in a range of casks, predominantly ex-bourbon and ex-sherry but also locally sourced ex-Pinot noir casks. The depth and richness imparted by the wood together with Cardrona's sweet honey new-make is what Desiree is looking for.

For every start-up distillery, distribution and sales become a reality after a couple of years. Desiree's opinion is firm:

"We would rather grow slowly making deep relationships with our customers, than quickly, where we might not be able to keep up with demand. We have just started selling in the United Kingdom, which is a big step for us, and we are really pleased with the relationships we have built there so far."

**Desiree's final advice
to someone wanting to start a distillery of their own;**

"Enjoy the journey and the wonderful friendships you will make along the way."

Glen Scotia

[glen sko•sha]

Owner: **Region/district:**
Loch Lomond Group   Campbeltown
(Hillhouse Capital Management)

Founded: **Status:**   **Capacity:**
1832   Active (vc)   800 000 litres

Address: High Street, Campbeltown, Argyll PA28 6DS

Website: **Tel:**
glenscotia.com   01586 552288

With a checkered history of closures and regular changes of ownership Glen Scotia found themselves financially secure in 2014 when Exponent Equity took over.

Colin Matthews, with a prior track record of 16 years working for Imperial Tobacco, together with former Diageo CFO Nick Rose, led a £210 m buyout of the company, financially backed by the investor Exponent Private Equity. In the first 18 months, £25m was invested in Glen Scotia and its sister distillery Loch Lomond including refurbishing of both distilleries and the creation of new product ranges. After five years, it was yet again time for new owners when the Chinese investment firm Hillhouse Capital Management acquired the Loch Lomond Group in 2019. This could mean an excellent way into a burgeoning Asian market.

Glen Scotia is equipped with a traditional 2.8 ton cast iron mash tun, nine washbacks made of stainless steel with an average fermentation time of 120 hours and one pair of stills. The shortest fermentation time is 70 hours but can reach up to 140 hours. The cut points for the middle cut are 73%-63% and slightly lower for the peated version. The production in 2019 will be ten mashes per week resulting in 520,000 litres of pure alcohol. The peated share of the production has increased since last year and is now three weeks each of heavily peated (55ppm) and medium peated (22ppm). The distillery recently went from heating the boiler with heavy fuel oil to LPG (Liquified Petroleum Gas), thus reducing the carbon footprint.

The core range consists of **Double Cask, 15, 18** and **25 year old** and the gently peated **Victoriana** which has been bottled at cask strength. Since 2017, there is also **Glen Scotia Harbour**, a 100% first fill bourbon, which is an exclusive to Waitrose and Marks & Spencer. The owners also released the first bottlings for duty free in 2017; the **Glen Scotia Campbeltown 1832** finished in PX sherry casks and a **16 year old**. A limited **2003 Vintage** finished in rum casks was launched in July 2019 and a **45 year old** was announced for a release in late 2019.

History:

1832 The families of Stewart and Galbraith start Scotia Distillery.

1895 The distillery is sold to Duncan McCallum.

1919 Sold to West Highland Malt Distillers.

1924 West Highland Malt Distillers goes bankrupt and Duncan MacCallum buys back the distillery.

1928 The distillery closes.

1930 Duncan MacCallum commits suicide and the Bloch brothers take over.

1933 Production restarts.

1954 Hiram Walker takes over.

1955 A. Gillies & Co. becomes new owner.

1970 A. Gillies & Co. becomes part of Amalgated Distillers Products.

1979 Reconstruction takes place.

1984 The distillery closes.

1989 Amalgated Distillers Products is taken over by Gibson International and production restarts.

1994 Glen Catrine Bonded Warehouse Ltd takes over and the distillery is mothballed.

1999 The distillery re-starts under Loch Lomond Distillery supervision using labour from Springbank.

2000 Loch Lomond Distillers runs operations with its own staff from May onwards.

2005 A 12 year old is released.

2006 A peated version is released.

2012 A new range (10, 12, 16, 18 and 21 year old) is launched.

2014 A 10 year old and one without age statement are released - both heavily peated.

2015 A new range is released; Double Cask, 15 year old and Victoriana.

2017 A 25 year old and an 18 year old as well as two bottlings for duty-free are released.

2019 The distillery is sold to Hillhouse Capital Management. A 2003 Vintage and a 45 year old are released.

Tasting notes Glen Scotia Double Cask:

GS – The nose is sweet, with bramble and redcurrant aromas, plus caramel and vanilla. Smooth mouth-feel, with ginger, sherry and more vanilla. The finish is quite long, with spicy sherry and a final hint of brine.

Double Cask

Glen Spey

[glen spey]

Owner: Diageo	**Region/district:** Speyside
Founded: 1878	**Status:** Active	**Capacity:** 1 400 000 litres

Address: Rothes, Morayshire AB38 7AU

| **Website:** malts.com | **Tel:** 01340 831215 |

One of Scotland's most anonymous distilleries in terms of single malt offerings is at the same time an important producer of whisky for the world famous J&B blended Scotch.

In 2008, J&B was the third biggest blend in the world selling 64 million bottles. Since then figures have been decreasing and the brand has lost more than 40% of the volume in the last decade. The main reason for this is that key markets such as Spain, Portugal and Greece have been declining for many years now due to a weakened economy and a stronger interest in other spirits.

The distillery is one of four in Rothes and if you arrive from the North on the A941, you travel through the roundabout where Glen Grant lies, heading on straight passing Station Hotel on the left and crossing the Spey river. A hundred metres further on you will see the gates of the distillery on your right side. Unfortunately there is no point entering through the gates as the distillery isn't open to visitors. Together with Knockando, Glen Spey is the third smallest of the 28 Diageo malt distilleries with Lochnagar and Oban behind.

The distillery is equipped with a 4.4 ton semi-lauter mash tun, eight stainless steel washbacks with both short (46 hours) and long (100 hours) fermentations and two pairs of stills. Usually, heating the stills is done by using internal coils or pans but the wash stills at Glen Spey have radiators. The two spirit stills are equipped with purifiers which add reflux and also help eliminate the heavier esters. Due to a cloudy wort, the Glen Spey new make is nutty and slightly oily. Even though a new control room was installed in 2017, Glen Spey is still run largely as a manual distillery. In 2019, the distillery will be doing 18 mashes per week (ten short and eight long) and 1.5 million litres of pure alcohol in the year.

The single malt from Glen Spey is important in the blend J&B, where it is one of the signature malts, and the only official single malt is the **12 year old Flora & Fauna** bottling. In 2010, two limited releases were made – a **1996 single cask** from new American oak and a **21 year old** with maturation in ex-sherry American oak.

History:

1878 James Stuart & Co. founds the distillery which becomes known by the name Mill of Rothes.

1886 James Stuart buys Macallan.

1887 W. & A. Gilbey buys the distillery for £11,000 thus becoming the first English company to buy a Scottish malt distillery.

1920 A fire breaks out and the main part of the distillery is re-built.

1962 W. & A. Gilbey combines forces with United Wine Traders and forms International Distillers & Vintners (IDV).

1970 The stills are increased from two to four.

1972 IDV is bought by Watney Mann which is then acquired by Grand Metropolitan.

1997 Guiness and Grand Metropolitan merge to form Diageo.

2001 A 12 year old is launched in the Flora & Fauna series.

2010 A 21 year old is released as part of the Special Releases and a 1996 Manager´s Choice single cask is launched.

12 years old

Tasting notes Glen Spey 12 years old:

GS – Tropical fruits and malt on the comparatively delicate nose. Medium-bodied with fresh fruits and vanilla toffee on the palate, becoming steadily nuttier and drier in a gently oaky, mildly smoky finish.

Glentauchers

[glen•tock•ers]

Owner:	**Region/district:**
Chivas Brothers	Speyside
(Pernod Ricard)	

Founded:	**Status:**	**Capacity:**
1897	Active	4 200 000 litres

Address: Mulben, Keith, Banffshire AB55 6YL

Website:	**Tel:**
-	01542 860272

The release of official bottlings of Glenburgie, Miltonduff and Glentauchers in 2017 was widely noticed. The story told was that this was the first time that a blended Scotch brand, Ballantines, entered the single malt category.

It is true that these three single malts constitute the backbone of the second largest blend in the world. On the other hand, the key malts that make up various versions of Johnnie Walker (the number one blend), for example Cardhu, Caol Ila and Clynelish, have been available as official bottlings for decades. So, if one takes the cynical approach, it was about time that Chivas Brothers released something from three of their more unknown distilleries. Regardless, this is an excellent opportunity to try three malts that used to be bottled by the owners as single casks and only available at the Chivas visitor centres. According to Chivas Brothers, Miltonduff is the foundation of Ballantine's bringing warmth and power, Glenburgie is the heart contributing fruitiness while Glentauchers brings on a smooth and delicate finish.

The distillery is equipped with a 12.2 ton stainless steel full lauter mash tun. There are six washbacks made of Oregon pine and three pairs of stills. The distillery is now doing 18 mashes per week and a total of 4 million litres per year. Most of the process at Glentauchers is done mechanically using traditional methods. The thought behind this is that new employees and trainees from Chivas Brothers will be able to work here for a while to learn the basic techniques of whisky production.

The role of Glentauchers has always been to produce malt whisky for blends – Buchanans Black & White, Teachers and today it is an integral part of Ballantines. Official bottlings have been more or less non-existent but in 2017 a **15 year old** was launched as a part of the Ballantine's Single Malt Series (together with Glenburgie and Miltonduff). There are also two cask strength bottlings in the Distillery Reserve Collection, available at all Chivas´ visitor centres – **10 and 12 years old**.

History:

1897 James Buchanan and W. P. Lowrie, a whisky merchant from Glasgow, found the distillery.

1898 Production starts.

1906 James Buchanan & Co. takes over the whole distillery and acquires an 80% share in W. P. Lowrie & Co.

1915 James Buchanan & Co. merges with Dewars.

1923 Mashing house and maltings are rebuilt.

1925 Buchanan-Dewars joins Distillers Company Limited (DCL).

1930 Glentauchers is transferred to Scottish Malt Distillers (SMD).

1965 The number of stills is increased from two to six.

1969 Floor maltings is decommissioned.

1985 DCL mothballs the distillery.

1989 United Distillers (formerly DCL) sells the distillery to Caledonian Malt Whisky Distillers, a subsidiary of Allied Distillers.

1992 Production recommences in August.

2000 A 15 year old Glentauchers is released.

2005 Chivas Brothers (Pernod Ricard) become the new owner through the acquisition of Allied Domecq.

2017 A 15 year old is released in the Ballantine´s Single Malt Series.

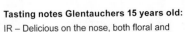
15 years old

Tasting notes Glentauchers 15 years old:

IR – Delicious on the nose, both floral and fruity, vanilla, pastry, heather and honey. Still fruity on the palate with additional notes of roasted nuts, toffee and milk chocolate..

Glenturret

[glen•turr•et]

Owner: **Region/district:**
Lalique Group/Hansjörg Wyss   Southern Highlands

Founded: **Status:**   **Capacity:**
1775   Active (vc)   340 000 litres

Address: The Hosh, Crieff, Perthshire PH7 4HA

Website: **Tel:**
theglenturret.com   01764 656565

In June 2018 Edrington announced plans to sell Glenturret distillery but no information about the new owner was revealed. Six months later it became apparent that the distillery was taken over by luxury goods company Lalique together with the Swiss billionaire Hansjörg Wyss.

Lalique specialises in luxury products such as crystal glassware, jewellery and furniture. They also have a subsidiary, Art & Terroire, in the wine business. Between 2006 and 2016, Lalique and Macallan had a collaboration where the Swiss company produced crystal decanters for some of Macallans rarest malts.

Glenturret and the blend Famous Grouse have had strong ties throughout the years, not least through a well-visited Visitor's Centre. But as Edrington keeps Famous Grouse this connection ceases. The new owners will focus on rebranding the Glenturret single malt. The deal, in which Lalique paid £15.5m for 50% of the distillery, included one million litres of maturing stock.

The distillery is equipped with a 1.05 ton stainless steel, open mash tun, the only one left in Scotland where the mash is stirred by hand. There are eight Douglas fir washbacks with a fermentation time of up to 120 hours and one pair of stills. The production target for 2019 is 190,000 litres of alcohol. Usually, a small part of the production is made up of the heavily peated (80ppm) Ruadh Maor but for 2019 all the spirit will be unpeated. The new owners plan to increase production, without adding any new equipment, to 500,000 litres within the next couple of years.

New products will be introduced in 2020 under supervision of the former Macallan master blender Bob Dalgarno. The current core range consists of **10 year old, Glenturret Sherry, Glenturret Triple Wood** and **Glenturret Peated**. Recent limited releases include two 29 year olds - **Cameron's Cut** and **Jamieson's Jigger Edition** - as well as the **Peated Drummond Edition**, a 100% peated Glenturret, bottled at cask strength and available only at the distillery.

History:

1775 Whisky smugglers establish a small illicit farm distillery named Hosh Distillery.

1818 John Drummond is licensee until 1837.

1826 A distillery in the vicinity is named Glenturret, but is decommissioned before 1852.

1852 John McCallum is licensee until 1874.

1875 Hosh Distillery takes over the name Glenturret Distillery and is managed by Thomas Stewart.

1903 Mitchell Bros Ltd takes over.

1921 Production ceases and the buildings are used for whisky storage only.

1929 Mitchell Bros Ltd is liquidated, the distillery dismantled and the facilities are used as storage for agricultural needs.

1957 James Fairlie buys the distillery and re-equips it.

1959 Production restarts.

1981 Remy-Cointreau buys the distillery and invests in a visitor centre.

1990 Highland Distillers takes over.

1999 Edrington and William Grant & Sons buy Highland Distillers for £601 million. The purchasing company, 1887 Company, is a joint venture between Edrington (70%) and William Grant (30%).

2002 The Famous Grouse Experience, a visitor centre costing £2.5 million, is inaugurated.

2003 A 10 year old Glenturret replaces the 12 year old as the distillery's standard release.

2007 Three new single casks are released.

2013 An 18 year old bottled at cask strength is released as a distillery exclusive.

2014 A 1986 single cask is released.

2015 Sherry, Triple Wood and Peated are released.

2016 Fly's 16 Masters is released.

2017 Cameron's Cut, Jamieson's Jigger Edition and Peated Drummond Edition are launched.

2019 Lalique Group and Hansjörg Wyss buy the distillery.

10 years old

Tasting notes Glenturret 10 years old:

GS – Nutty and slightly oily on the nose, with barley and citrus fruits. Sweet and honeyed on the full, fruity palate, with a balancing note of oak. Medium length in the sweet finish.

Highland Park

[hi•land park]

Owner:
The Edrington Group

Region/district:
Highlands (Orkney)

Founded: 1798
Status: Active (vc)
Capacity: 2 500 000 litres

Address: Holm Road, Kirkwall, Orkney KW15 1SU

Website: highlandparkwhisky.com
Tel: 01856 874619

The number of visitors to Orkney last year was close to 200,000 not counting in the 130,000 that visited the islands from one of the 140 visiting cruise liners. The importance for the local economy is huge but questions have been raised whether it is sustainable.

The impact on the environment and the infrastructure risk reaching those of Isle of Skye in recent years. Still, the total value of tourism to the area, well over £50m, cannot be ignored and Highland Park are determined to receive its share. 20,000 visitors come to the visitor centre annually, but in the beginning of May 2019 the owners also opened a shop right in the centre of Kirkwall. Apart from selling Highland Park whisky, the shop will house a gallery and an education area for the distillery's community training programme.

The distillery is equipped with a semi-lauter mash tun, twelve Oregon Pine washbacks with a fermentation time between 50 and 80 hours, and two pairs of stills. The mash tun has a capacity of 12 tons but is only filled to 50%. The plan for 2019 is to make 22 mashes per week which means a total of 2.5 million litres of alcohol. Highland Park is malting 30% of its malt themselves and there are five malting floors with a capacity of almost 36 tons of barley. The phenol content is 30-40 ppm in its own malt and the malt which has been bought from Simpson's is unpeated. There are also 19 dunnage warehouses and four racked on site.

The core range of Highland Park consists of **10 year old Viking Scars, 12 year old Viking Honour, 18 year old Viking Pride** as well as **25, 30** and **40** year olds. A new addition the range was made in 2019 – a **21 year old**. Included are also **Dragon Legend** and **Viking Tribe**, the last one released in late 2018 as an exclusive to Amazon UK. The duty free range, called the Warrior Series, has been around for several years now and the more expensive ones, **Ragnvald** and **Thorfinn**, will be available for another year or so. The rest, however, have been replaced by four new expressions that were launched in autumn 2018; **Spirit of the Bear** (40% and matured mainly in American Oak ex-sherry), **Loyalty of the Wolf** (14 years old, bottled at 42.3% and matured in a combination of American Oak ex-sherry and ex-bourbon), **Wings of the Eagle** (16 years old, bottled at 44.5% and predominantly from European Oak ex-sherry) and a duty free version of the **18 year old Viking Pride** bottled at the higher strength of 46%. **Voyage of the Raven** is a limited edition also available in travel retail.

Recent limited expressions include a new edition of the **50 year old**, the 16 year old **Twisted Tattoo** which was partly matured in ex-Rioja casks, the third instalment in the Viking Legend series, **Valfather** using a higher proportion of their own malt, a **26 year old** matured in ex-sherry and first fill bourbon honouring the photographer Soren Solkjaer, **Ness of Brodgar's Legacy**, available only at the distillery and **Triskelion**. The latter was created by the distillery's three master whisky makers – Gordon Motion, John Ramsay and Max McFarlane.

History:

1798 David Robertson founds the distillery. The local smuggler and businessman Magnus Eunson previously operated an illicit whisky production on the site.

1816 John Robertson, an Excise Officer who arrested Magnus Eunson, takes over production.

1826 Highland Park obtains a license and the distillery is taken over by Robert Borwick.

1840 Robert's son George Borwick takes over but the distillery deteriorates.

1869 The younger brother James Borwick inherits Highland Park and attempts to sell it as he does not consider the distillation of spirits as compatible with his priesthood.

1895 James Grant (of Glenlivet Distillery) buys Highland Park.

1898 The distillery is expanded from two to four stills.

1937 Highland Distilleries buys Highland Park.

1979 Highland Distilleries invests considerably in marketing Highland Park as single malt which increases sales markedly.

1986 A visitor centre, considered one of Scotland's finest, is opened.

1997 Two new Highland Park are launched, an 18 year old and a 25 year old.

1999 Highland Distillers are acquired by Edrington Group and William Grant & Sons.

2000 Visit Scotland awards Highland Park "Five Star Visitor Attraction".

2005 Highland Park 30 years old is released. A 16 year old for the Duty Free market and Ambassador's Cask 1984 are released.

2006 The second edition of Ambassador's Cask, a 10 year old from 1996, is released.

Dragon Legend   Valfather   Twisted Tattoo

History continued:

2007 The Rebus 20, a 21 year old duty free exclusive, a 38 year old and a 39 year old are released.

2008 A 40 year old and the third and fourth editions of Ambassador´s Cask are released.

2009 Two vintages and Earl Magnus 15 year are released.

2010 A 50 year old, Saint Magnus 12 year old, Orcadian Vintage 1970 and four duty free vintages are released.

2011 Vintage 1978, Leif Eriksson and 18 year old Earl Haakon are released.

2012 Thor and a 21 year old are released.

2013 Loki and a new range for duty free, The Warriors, are released.

2014 Freya and Dark Origins are released.

2015 Odin is released.

2016 Hobbister, Ice Edition, Ingvar and King Christian I are released.

2017 Valkyrie, Dragon Legend, Voyage of the Raven, Shiel, Full Volume, The Dark and The Light are released.

2018 New bottlings in the duty free range include Spirit of the Bear, Loyalty of the Wolf and Wings of the Eagle. The limited Valknut is also released.

2019 Twisted Tattoo, Valfather and Triskelion are released.

Tasting notes Highland Park 12 year old:

GS – The nose is fragrant and floral, with hints of heather and some spice. Smooth and honeyed on the palate, with citric fruits, malt and distinctive tones of wood smoke in the warm, lengthy, slightly peaty finish.

10 years old

12 years old

Loyalty of the Wolf

Inchgower

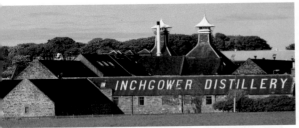

INCHGOWER DISTILLERY

[inch•gow•er]

Owner: | **Region/district:**
Diageo | Speyside

Founded: | **Status:** | **Capacity:**
1871 | Active | 3 200 000 litres

Address: Buckie, Banffshire AB56 5AB

Website: | **Tel:**
malts.com | 01542 836700

Inchgower single malt is virtually unknown amongst the general whisky drinkers. The aficionados, on the other hand, seek out the occasional releases provided by independent bottlers.

With a nutty and waxy newmake it makes for a spicy and rather robust whisky after a decade or more in ex-bourbon barrels. The owners wish to avoid the fruity esters and so the middle cut starts as low as at 70% abv. On the other hand, they are eager to catch the heavier compounds at the end of the spirit distillation and come off spirit as low as 55%.

This makes it an ideal choice as one of the signature malts for Bells blended Scotch (together with Blair Athol). The famous blend is still one the best sellers in the UK and on a global scale it comes in at 11th place with a very consistent 25 million bottles sold yearly during the last decade.

Inchgower is situated on the South side of Moray Firth and is difficult to miss as it is situated just at the A98 near the small fishing port of Buckie. If one is driving from Elgin towards Banff, it is even easier to spot the distillery as the name is visible on the roof.

The distillery is equipped with an 8.4 ton stainless steel semilauter mash tun where the cloudy wort adds to the spirit character. Six washbacks made from Oregon pine with an average fermentation time of 48-53 hours are complemented by two pairs of stills. In 2017, the distillery was closed from June-October to replace the five existing malt bins with three larger ones. The production plan for 2019 is a five-day operation which means short (40-45 hours) and long (90-92 hours) fermentations and a total production of 2 million litres of pure alcohol.

Besides the official **Flora & Fauna 14 year old**, there have also been a few limited bottlings of Inchgower single malt. The most recent was a **27 year old** in autumn 2018 which was part of the yearly Special Releases.

History:

1871 Alexander Wilson & Co. founds the distillery. Equipment from the disused Tochineal Distillery, also owned by Alexander Wilson, is installed.

1936 Alexander Wilson & Co. becomes bankrupt and Buckie Town Council buys the distillery and the family's home for £1,600.

1938 The distillery is sold on to Arthur Bell & Sons for £3,000.

1966 Capacity doubles to four stills.

1985 Guinness acquires Arthur Bell & Sons.

1987 United Distillers is formed by a merger between Arthur Bell & Sons and DCL.

1997 Inchgower 1974 (22 years) is released as a Rare Malt.

2004 Inchgower 1976 (27 years) is released as a Rare Malt.

2010 A single cask from 1993 is released.

2018 A 27 year old is launched as part of the Special Releases.

14 years old

Tasting notes Inchgower 14 years old:

GS – Ripe pears and a hint of brine on the light nose. Grassy and gingery in the mouth, with some acidity. The finish is spicy, dry and relatively short.

Jura

[joo•rah]

Owner:
Whyte & Mackay
(Emperador Inc)

Region/district:
Highlands (Jura)

Founded: 1810

Status: Active (vc)

Capacity: 2 400 000 litres

Address: Craighouse, Isle of Jura PA60 7XT

Website:
isleofjura.com

Tel:
01496 820240

In terms of sales, Jura single malt has been on an incredible journey with an increase of 150% during the last decade. The brand now sells around 1.7 million bottles worldwide.

It is an amazing development for a distillery that was closed and destined for oblivion only 60 years ago. Thanks to a group of Jura landowners, the distillery, which had been dormant since 1901, was rebuilt and re-equipped.

Jura distillery has a 5 ton semi-lauter mash tun, six stainless steel washbacks with a fermentation time of 54 hours and two pairs of stills – the second tallest in Scotland. Working a seven-day week since 2011, they will make 28 mashes per week and 2.3 million litres of alcohol during 2019, which will include four weeks of peated production (at 45ppm). The owners, Emperador Inc., have recently upgraded the Dalmore visitor centre and a similar investment will be destined for Jura in the next year or so.

The entire core range was discontinued in 2018 and replaced by **Journey** (matured in American oak), **10 year old** (finished in oloroso sherry casks), **12 year old** (also an oloroso finish), **Seven Wood** (a vatting of whiskies matured in seven types of French oak as well as ex-bourbon barrels) and **18 year old** (finished in red wine casks). All the expressions have an amount of peated Jura in the recipe. Four new expressions available for duty-free include **The Sound, The Road, The Loch** and **The Paps**. All of them have been finished in PX casks that have held sherry for varying amounts of time. A fifth version, exclusive to Asia, is the 12 year old **The Bay**. One more expression was added to the duty free range in summer 2019 when the 21 year old **Jura Time**, finished in ex-peated malt casks, was launched. Recent limited releases include a **Vintage 1988** finished in port pipes, a **Vintage 1989,** the 21 year old **Jura Tide** with a finish in virgin American oak casks, **French Oak** finished in French oak casks and the 13 year old **Two-One-Two** (referring to the roughly 212 people living on the island) which was finished in casks made of Chinkapin oak.

12 years old

History:

1810 Archibald Campbell founds a distillery named Small Isles Distillery.

1853 Richard Campbell leases the distillery to Norman Buchanan from Glasgow.

1867 Buchanan files for bankruptcy and J. & K. Orr takes over the distillery.

1876 Licence transferred to James Ferguson & Sons.

1901 Ferguson dismantles the distillery.

1960 Charles Mackinlay & Co. extends the distillery. Newly formed Scottish & Newcastle Breweries acquires Charles Mackinlay & Co.

1963 The first distilling takes place.

1985 Invergordon Distilleries acquires Charles Mackinlay & Co., Isle of Jura and Glenallachie from Scottish & Newcastle Breweries.

1993 Whyte & Mackay (Fortune Brands) buys Invergordon Distillers.

1996 Whyte & Mackay changes name to JBB (Greater Europe).

2001 The management buys out the company and changes the name to Kyndal.

2002 Isle of Jura Superstition is launched.

2003 Kyndal reverts back to its old name, Whyte & Mackay. Isle of Jura 1984 is launched.

2006 The 40 year old Jura is released.

2007 United Spirits buys Whyte & Mackay. The 18 year old Delmé-Evans and an 8 year old heavily peated expression are released.

2008 A series of four different vintages, called Elements, is released.

2009 The peated Prophecy and three new vintages called Paps of Jura are released.

2012 The 12 year old Jura Elixir is released.

2013 Camas an Staca, 1977 Juar and Turas-Mara are released.

2014 Whyte & Mackay is sold to Emperador Inc.

2016 The 22 year old "One For The Road" is released.

2017 The limited One and All is released.

2018 A new core range is released; 10, 12 and 18 year old as well as Journey and Seven Wood.

2019 A new range for duty-free is released.

Tasting notes Jura 10 years old:

GS – Resin, oil and pine notes on the delicate nose. Light-bodied in the mouth, with malt and drying saltiness. The finish is malty, nutty, with more salt, plus just a wisp of smoke.

Kilchoman

[kil•ho•man]

Owner:
Kilchoman Distillery Co.

Region/district:
Islay

Founded: **Status:** **Capacity:**
2005 Active (vc) 480 000 litres

Address: Rockside farm, Bruichladdich,
Islay PA49 7UT

Website: **Tel:**
kilchomandistillery.com 01496 850011

Since 2005 when production started, Kilchoman has been referred to as the new kid on the block. Not anymore. The latest addition to the Islay whisky scene, Ardnahoe, opened up in late 2018.

When Anthony Wills started production at Kilchoman it was the first distillery to be built on Islay for 124 years. Admittedly Malt Mill started in 1908 making use of two old buildings on the site of Lagavulin but as they were also using Lagavulin's mash tun they could not be considered a completely new distillery.

The success for Kilchoman single malt has now forced the owners to increase production. A second distillery, mirroring the existing one, has been built leading to a doubled capacity.

The new still house was commissioned in April 2019 and with the recent expansion the equipment now consists of two 1.2 ton stainless steel semi-lauter mash tuns, 12 stainless steel, 6,000 litre washbacks with an average fermentation time of 90 hours and two pairs of stills. In 2018, the distillery produced 220,000 litres of pure alcohol and the forecast for 2019 is 300,000 litres.

Meanwhile, a new malting floor and kiln have been opened which means that the distillery is now able to produce 30% of their malt requirement themselves, typically with a phenol content of 20ppm. The rest (50ppm) is bought from Port Ellen.

The core range consists of **Machir Bay** and **Sanaig**. The latter was released in 2016 and has been matured in a combination of ex-bourbon and ex-oloroso sherry casks. Limited, but regular releases are **Loch Gorm**, the only expression in the range to be fully matured in sherry casks and **100% Islay** made from 100% barley grown and malted on the island. Other limited releases include **STR Cask Matured** from May 2019. Made from 50ppm barley and bottled at 50% it has matured in shaved, toasted and re-charred red wine casks. There is also a 9 year old **Vintage 2009**. The special **Feis Ile 2019 bottling** was a vatting of a 2007 ex-bourbon cask and a 2008 ex-oloroso sherry butt. For the UK duty free market there is **Coull Point** and for global duty free, **Saligo Bay** is available.

History:

2002 Plans are formed for a new distillery at Rockside Farm on western Islay.

2005 Production starts in June.

2006 A fire breaks out in the kiln causing a few weeks´ production stop but malting has to cease for the rest of the year.

2007 The distillery is expanded with two new washbacks.

2009 The first single malt, a 3 year old, is released on 9th September followed by a second release.

2010 Three new releases and an introduction to the US market. John Maclellan from Bunnahabhain joins the team as General Manager.

2011 Kilchoman 100% Islay is released as well as a 4 year old and a 5 year old.

2012 Machir Bay, the first core expression, is released together with Kilchoman Sherry Cask Release and the second edition of 100% Islay.

2013 Loch Gorm and Vintage 2007 are released.

2014 A 3 year old port cask matured and the first duty free exclusive, Coull Point, are released.

2015 A Madeira cask maturation is released and the distillery celebrates its 10th anniversary.

2016 Sanaig and a Sauternes cask maturation are released.

2017 A Portugese red wine maturation and Vintage 2009 are released.

2018 Original Cask Strength and 2009 Vintage are released.

2019 Capacity is doubled with two more stills. A limited STR Cask Matured is released.

BARLEY GROWING FOR KILCHOMAN 100% ISLAY

KILCHOMAN
ISLAY SINGLE MALT SCOTCH WHISKY
MACHIR BAY

Machir Bay

Tasting notes Kilchoman Machir Bay:

GS – A nose of sweet peat and vanilla, undercut by brine, kelp and black pepper. Filled ashtrays in time. A smooth mouth-feel, with lots of nicely-balanced citrus fruit, peat smoke and Germolene on the palate. The finish is relatively long and sweet, with building spice, chili and a final nuttiness.

Kininvie

[kin•in•vee]

Owner: | **Region/district:**
William Grant & Sons | Speyside

Founded: | **Status:** | **Capacity:**
1990 | Active | 4 800 000 litres

Address: Dufftown, Keith, Banffshire AB55 4DH

Website: | **Tel:**
- | 01340 820373

Kininvie malt is a vital part of a hugely successful brand, namely Monkey Shoulder. It was first introduced in 2005 and while being defined as a blended malt today, it was actually called a vatted malt according to the current terminology back then.

This was at a time when some of the big producers had taken an interest in a category that could serve as an entry level to single malts. Edrington had a special segment of their Famous Grouse brand dedicated to blended malts and Diageo launched Johnnie Walker Green Label. William Grant & Sons realised they should be on this train as well and launched Monkey Shoulder which was made up of their own three single malts at the time; Glenfiddich, Balvenie and, not least, Kininvie. With Ailsa Bay single malt having reached a considerable age, the company has indicated that this may also in the future be a part of the recipe as well as other Speyside single malts. Monkey Shoulder is nothing but a success story. It is especially appreciated by bartenders across the world not just for the quality but also for its versatility when it comes to cocktails. Drink International publish a yearly survey called the Brands Report where well over 100 of the world's best bars reveal their best selling Scotch whiskies. In 2018, Monkey Shoulder managed to claim first place and forced the mega brand Johnnie Walker down to second place.

Kininvie distillery consists of one still house with three wash stills and six spirit stills, tucked away behind Balvenie. There is a 9.6 ton stainless steel full lauter mash tun which is placed next to Balvenie's in the Balvenie distillery and ten Douglas fir washbacks with a minimum fermentation time of 75 hours can be found in two separate rooms next to the Balvenie washbacks. Production in 2019 will be just under four million litres of pure alcohol.

It wasn't until 2013 that Kininvie single malt was launched as a limited bottling in select markets. The end of 2015 saw the release of a **23 year old** core bottling, matured in a combination of ex-bourbon and ex-sherry. A **17 year old** is available in the duty free market.

History:

1990 Kininvie distillery is inaugurated and the first distillation takes place on 25th June.

1994 Another three stills are installed.

2006 The first expression of a Kininvie single malt is released as a 15 year old under the name Hazelwood.

2008 In February a 17 year old Hazelwood Reserve is launched at Heathrow's Terminal 5.

2013 A 23 year old Kininvie is launched in Taiwan.

2014 A 17 year old and batch 2 of the 23 year old are released.

2015 Batch 3 of the 23 year old is released and later in the year, the batches are replaced by a 23 year old signature bottling. Three 25 year old single casks are launched.

23 years old

Tasting notes Kininvie 17 years old:

GS – The nose offers tropical fruits, coconut and vanilla custard, with a hint of milk chocolate. Pineapple and mango on the palate, accompanied by linseed oil, ginger, and developing nuttiness. The finish dries slowly, with more linseed, plenty of spice, and soft oak.

Knockando

[nock•an•doo]

Owner: | **Region/district:**
Diageo. | Speyside

Founded: | **Status:** | **Capacity:**
1898 | Active | 1 400 000 litres

Address: Knockando, Morayshire AB38 7RT

Website: | **Tel:**
malts.com | 01340 882000

John Thompson founded Knockando in 1898 but was already forced to close it two years later, in the aftermath of the disastrous Pattison crash.

When the famous English gin producer W&A Gilbey bought the distillery in 1903 it was a real bargain at £3,500 (which is the equivalent of £370,000 in today's value). This was their third acquisition of a whisky distillery in Scotland. The first was Glen Spey in 1887 (at £11,000) followed by Strathmill in 1895 (at £9,500). All three distilleries were of approximately the same size, with just the one pair of stills at the time, the price tag probably showing just how desperate John Thompson was to get rid of the distillery. Knockando single malt soon became a vital part of Gilbey's blended Scotch Spey Royal and, after Gilbey's and Justerini & Brooks joined forces in 1962, it became one of the signature malts of J&B. It was also one of the first to be bottled as a single malt in the 1970s and, by the early 1990s, it was one of the top ten single malt brands of the world. Today it is found around place 30 on that list. Knockando is beautifully situated on the Spey river at the end of the road with Tamdhu as its nearest distillery neighbour.

The distillery is equipped with a small (4.4 ton), semi-lauter mash tun, eight Douglas fir washbacks and two pairs of stills. Knockando has always worked a five-day week with 16 mashes per week, 8 short fermentations (50 hours) and 8 long (100 hours). The distillery has been closed for the better part of two years due to a major refurbishment. Knockando's nutty character, a result of the cloudy worts coming from the mash tun, has given it its fame. However, in order to balance the taste, the distillers also wish to create the typical Speyside floral notes by using boiling balls on the spirit stills to increase reflux.

Knockando is Diageo's 8th most sold single malt (around 600,000 bottles) and has for many years been especially popular in France, Spain and Greece. The core range consists of **12 year old, 15 year old Richly Matured, 18 year old Slow Matured** and the **21 year old Master Reserve**. In 2011 a **25 year old** matured in first fill European oak was released as part of the Special Releases.

History:

1898 John Thompson founds the distillery. The architect is Charles Doig.

1899 Production starts in May.

1900 The distillery closes in March and J. Thompson & Co. takes over administration.

1903 W. & A. Gilbey purchases the distillery for £3,500 and production restarts in October.

1962 W. & A. Gilbey merges with United Wine Traders (including Justerini & Brooks) and forms International Distillers & Vintners (IDV).

1968 Floor maltings is decommissioned.

1969 The number of stills is increased to four.

1972 IDV is acquired by Watney Mann who, in its turn, is taken over by Grand Metropolitan.

1978 Justerini & Brooks launches a 12 year old Knockando.

1997 Grand Metropolitan and Guinness merge and form Diageo; simultaneously IDV and United Distillers merge to United Distillers & Vintners.

2010 A Manager's Choice 1996 is released.

2011 A 25 year old is released.

12 years old

Tasting notes Knockando 12 years old:

GS – Delicate and fragrant on the nose, with hints of malt, worn leather, and hay. Quite full in the mouth, smooth and honeyed, with gingery malt and a suggestion of white rum. Medium length in the finish, with cereal and more ginger.

Knockdhu

[nock•doo]

Owner:
Inver House Distillers
(Thai Beverages plc)

Region/district:
Highland

Founded: 1893
Status: Active (vc)
Capacity: 2 000 000 litres

Address: Knock, By Huntly, Aberdeenshire AB54 7LJ

Website: ancnoc.com
Tel: 01466 771223

Knockdhu distillery lies in q fertile part of the Highlands, a few miles east of Keith. The beautiful distillery can be seen from the A95 to Banff, but the best distance view is from the B9022 heading to Portsoy.

While still looking to build a more up-scale visitor centre, the distillery opened up for visitors a couple of years ago. Two daily tours (Mon-Fri) are offered and they are not the traditional 45 minute sharp visits. On the contrary, the staff themselves say "a tour can take between one and two hours depending on the questions being asked". An unusual and relaxed way of greeting the fans!

Knockdhu distillery is equipped with a 5.1 ton stainless steel lauter mash tun, eight washbacks made of Oregon pine (two of them used as intermediate vats), with fermentation time now increased to 65 hours and one pair of stills with worm tubs. For 2019, they've moved from a five-day week to seven-days which means 20 mashes per week and a total of 1.7 million litres of pure alcohol in the year. Around 400,000 litres of that will be heavily peated (45ppm). The spirit is filled mainly into bourbon casks with an additional 15% of sherry butts. There are also three dunnage and one racked warehouse on site.

The core range consists of **12, 18, 24** and **35 years old**. In addition to that there is the peated range where almost ten different expressions have replaced each other over the past few years. The latest addition from 2017, **Peatheart**, however has now become a part of the core range. Up until now, the phenol content on the label has referred to the ppm in the matured whisky. With Peatheart (40ppm), the owners have decided to state the ppm of the malted barley. Every year a new vintage is released and in spring 2017 it was a **2002** which replaced the 2000. More recent limited bottlings, launched in September 2019 to commemorate the 125th anniversary of the distillery, were a **16 year old cask strength** and a peaty expression with an **extra maturation in Spanish oak**. In 2015, two new expressions were released for duty-free; **Black Hill Reserve** and the peated (13.5ppm) **Barrow**. Both have matured in bourbon casks and they were complemented by **Rùdhan** in autumn 2016.

History:

1893 Distillers Company Limited (DCL) starts construction of the distillery.

1894 Production starts in October.

1930 Scottish Malt Distillers (SMD) takes over production.

1983 The distillery closes in March.

1988 Inver House buys the distillery from United Distillers.

1989 Production restarts on 6th February.

1990 First official bottling of Knockdhu.

1993 First official bottling of anCnoc.

2001 Pacific Spirits purchases Inver House Distillers at a price of $85 million.

2003 Reintroduction of anCnoc 12 years.

2004 A 14 year old from 1990 is launched.

2005 A 30 year old from 1975 and a 14 year old from 1991 are launched.

2006 International Beverage Holdings acquires Pacific Spirits UK.

2007 anCnoc 1993 is released.

2008 anCnoc 16 year old is released.

2011 A Vintage 1996 is released.

2012 A 35 year old is launched.

2013 A 22 year old and Vintage 1999 are released.

2014 A peated range with Rutter, Flaughter, Tushkar and Cutter is introduced.

2015 A 24 year old, Vintage 1975 and Peatlands are released as well as Black Hill Reserve and Barrow for duty free.

2016 Vintage 2001, Blas and Rùdhan are released.

2017 Vintage 2002 and Peatheart are released.

2019 A 16 year old cask strength is released.

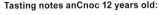

12 years old

Tasting notes anCnoc 12 years old:

GS – A pretty, sweet, floral nose, with barley notes. Medium bodied, with a whiff of delicate smoke, spices and boiled sweets on the palate. Drier in the mouth than the nose suggests. The finish is quite short and drying.

Lagavulin

LAGAVULIN

[lah•gah•**voo**•lin]

Owner: Diageo
Region/district: Islay

Founded: 1816
Status: Active (vc)
Capacity: 2 530 000 litres

Address: Port Ellen, Islay, Argyll PA42 7DZ

Website: malts.com
Tel: 01496 302749 (vc)

One surname, Mackie, is prominent throughout Lagavulin's history. The first person bearing this name turns up as James Logan Mackie who became part-owner in 1861, but his nephew, Peter, has had a stronger impact.

He took over the distillery in 1889 and the year after he launched White Horse, a blended whisky with Lagavulin single malt as a major contributor. Mackie was agent for neighbouring Laphroaig at the same time and desired more than anything else to acquire that distillery as well. The owners refused, however, and were also of the opinion that Peter Mackie marginalised Laphroaig in favour of his own whisky. The agreement was torn up and Mackie became furious. In 1908 he decided to build another distillery on Lagavulin's premises which would make an exact copy of Laphroaig. Such attempts are usually effectless and this was also the case with Malt Mill. Despite that, production was continued until 1962 when it was closed for good. Malt Mill has become the contemporary Holy Grail of the whisky world and few believe there is still whisky available from the distillery. There is, however, a bottle of Malt Mill newmake exhibited in Lagavulin's visitor centre.

The distillery is equipped with a 4.4 ton stainless steel full lauter mash tun with a 4 hour mash cycle and ten washbacks made of larch, filled with 22,000 litres and with a 55 hour fermentation cycle. There are two pairs of stills where the spirit stills are actually larger than the wash stills and are filled almost to the brim. This diminishes the copper contact and that, together with a slow distillation, creates the rich and pungent character of Lagavulin single malt. The newmake is, almost without exception, filled into ex-bourbon hogsheads. There are only 5,000 casks maturing at the distillery and all of the new production is now shipped to the mainland for maturation. During 2019, the distillery will be working 24/7 with 28 mashes per week and the volume is between 2.5 and 2.6 million litres of alcohol.

The core range of Lagavulin is unusually limited and only consists of a **12 year old cask strength** (which actually forms part of the Special Releases but new bottlings appear every year), a **16 year old** and the **Distiller's Edition**, a Pedro Ximenez sherry finish. A fourth addition to the permanent range appeared in autumn 2017 in the way of an **8 year old** bottled at 48%. Recent limited bottlings include three bottlings in 2016 to celebrate the distillery's bicentenary – an **8 year old** (later to become part of the core range), a **25 year old** bottled at cask strength and matured in sherry casks and a single cask **Lagavulin 1991**. The Islay Festival special release for 2019, bottled at 53.8%, was a **19 year old** matured in three different casks; refill American Oak, ex-sherry American Oak and an ex-sherry European Oak puncheon. In spring 2019, a **9 year old House Lannister** was released as part of the game of Thrones series and in August 2019 a **10 year old** was released exclusively for duty-free.

History:

1816 John Johnston founds the distillery.

1825 John Johnston takes over the adjacent distillery Ardmore.

1836 John Johnston dies and the two distilleries are merged and operated under the name Lagavulin. Alexander Graham, a wine and spirits dealer from Glasgow, buys the distillery.

1861 James Logan Mackie becomes a partner.

1867 The distillery is acquired by James Logan Mackie & Co. and refurbishment starts.

1878 Peter Mackie is employed.

1889 James Logan Mackie passes away and nephew Peter Mackie inherits the distillery.

1890 J. L. Mackie & Co. changes name to Mackie & Co. Peter Mackie launches White Horse onto the export market with Lagavulin included in the blend.

1908 Peter Mackie uses the old distillery buildings to build a new distillery, Malt Mill, on the site.

1924 Peter Mackie passes away and Mackie & Co. changes name to White Horse Distillers.

1927 White Horse Distillers becomes part of Distillers Company Limited (DCL).

1930 The distillery is administered under Scottish Malt Distillers (SMD).

1952 An explosive fire breaks out and causes considerable damage.

1962 Malt Mills distillery closes and today it houses Lagavulin's visitor centre.

1974 Floor maltings are decommissioned and malt is bought from Port Ellen instead.

1988 Lagavulin 16 years becomes one of six Classic Malts.

1998 A Pedro Ximenez sherry finish is launched as a Distillers Edition.

History continued:

2002 Two cask strengths (12 years and 25 years) are launched.

2006 A 30 year old is released.

2007 A 21 year old from 1985 and the sixth edition of the 12 year old are released.

2009 A new 12 year old appears as a Special Release.

2010 A new edition of the 12 year old, a single cask exclusive for the distillery and a Manager´s Choice single cask are released.

2011 The 10th edition of the 12 year old cask strength is released.

2012 The 11th edition of the 12 year old cask strength and a 21 year old are released.

2013 A 37 year old and the 12th edition of the 12 year old cask strength are released.

2014 A triple matured for Friends of the Classic Malts and the 13th edition of the 12 year old cask strength are released.

2015 The 14th edition of the 12 year old cask strength is released.

2016 An 8 year old and a 25 year old are launched.

2017 A new edition of the 12 year old cask strength is released.

2018 An 18 year old is released for Feis Ile.

2019 A 19 year old is released for Feis Ile, a 9 year old House Lannister in the Game of Thrones series and a 10 year old for duty free.

Tasting notes Lagavulin 16 year old:

GS – Peat, iodine, sherry and vanilla merge on the rich nose. The peat and iodine continue on to the expansive, spicy, sherried palate, with brine, prunes and raisins. Peat embers feature in the lengthy, spicy finish.

House Lannister

19 years old
Feis Ile 2019

12 years old
cask strength

16 years old

8 years old

Distiller´s Edition

Laphroaig

[lah•froyg]

Owner: **Region/district:**
Beam Suntory   Islay

Founded: **Status:**   **Capacity:**
1815   Active (vc)   3 300 000 litres

Address: Port Ellen, Islay, Argyll PA42 7DU

Website: **Tel:**
laphroaig.com   01496 302418

Laphroaig was founded in 1815 by Alexander and Donald Johnston and their family remained owners until 1954. This was unusual after the 1920s, when most distilleries were acquired by larger companies.

One of the family members, Ian Hunter arrived on Islay in 1908 and took over as owner in 1927. Hunter would be the person who took Laphroaig into the twentieth century. He doubled the capacity and travelled the USA to develop new markets. It was there that he discovered how used bourbon casks could influence his whisky and he became, in this respect, a forerunner for the whole Scotch whisky industry. As a tribute to him, a new range of whiskies was introduced in 2019 - The Ian Hunter Story.

The distillery is equipped with a 5.5 ton stainless steel full lauter mash tun and six stainless steel washbacks with an average fermentation time of 50-55 hours. The distillery uses an unusual combination of three wash stills and four spirit stills, all fitted with ascending lyne arms and with unusually long foreshots of 45 minutes. The middle cut start at 72% and goes down to 60%. It is one of very few distilleries with its own maltings which, using two malting floors, produces 20% of its requirements. The malt is dried for 12-15 hours using peat and then for another 10 hours on hot air. The in-house malt has a phenol specification of 40-60ppm, while the remaining malt from Port Ellen or the mainland lies between 35 and 45ppm. The distillery is running at full capacity which means 34 mashes per week and 3.3 million litres in the year. Due to increased sales volumes, discussions about an increase of production at the distillery have been ongoing for the past three years but so far no decision has been taken whether or not to increase the capacity. Around 70% of the production is destined to be bottled as single malt and the rest is used for blends.

Laphroaig has one of the best visitor centres in the industry with a huge shop and an excellent tasting bar. There is a wide variety of tours, including one where you can try your hand at cutting peat.

The core range consists of **Select** without age statement, **10 year old, 10 year old cask strength** (11th batch released in 2019), **Quarter Cask, Triple Wood, Lore** and a **25 year old**. The travel retail range consists of **Four Oak** and the **1815 Edition**. Four Oak, is a vatting from four different casks while The 1815 Edition is a mix of first-fill, heavily charred bourbon barrels and new European oak hogsheads. **PX Cask** is also a part of the duty free range. Recent limited releases include a **27 year old** with a maturation in first-fill ex-bourbon barrels and refill quarter casks and a **28 year old**, matured in a combination of quarter casks, ex-bourbon barrels and oloroso sherry butts. The **Cairdeas** range include bottlings exclusively for the Friends of Laphroaig; **Quarter Cask, Fino** and a **15 year old**. The Feis Ile bottling for 2019 was a **Cairdeas Triple Wood** bottled at 51,4% and without age statement. The first instalment in the new Ian Hunter series was a **30 year old** released in autumn 2019.

History:

1815 Brothers Alexander and Donald Johnston found Laphroaig.

1836 Donald buys out Alexander and takes over operations.

1837 James and Andrew Gairdner found Ardenistiel a stone's throw from Laphroaig.

1847 Donald Johnston is killed in an accident in the distillery. The Manager of neigh-bouring Lagavulin, Walter Graham, takes over.

1857 Operation is back in the hands of the Johnston family when Donald's son Dougald takes over.

1877 Dougald, being without heirs, passes away and his sister Isabella, married to their cousin Alexander takes over.

1907 Alexander Johnston dies and the distillery is inherited by his two sisters Catherine Johnston and Mrs. William Hunter (Isabella Johnston).

1908 Ian Hunter arrives in Islay to assist his mother and aunt with the distillery.

1924 The two stills are increased to four.

1927 Catherine Johnston dies and Ian Hunter takes over.

1928 Isabella Johnston dies and Ian Hunter becomes sole owner.

1950 Ian Hunter forms D. Johnston & Company

1954 Ian Hunter passes away and management of the distillery is taken over by Elisabeth "Bessie" Williamson.

1967 Seager Evans & Company buys the distillery through Long John Distillery, having already acquired part of Laphroaig in 1962. The number of stills is increased from four to five.

1972 Bessie Williamson retires. Another two stills are installed bringing the total to seven.

1975 Whitbread & Co. buys Seager Evans (now renamed Long John International) from Schenley International.

History continued:

1989 The spirits division of Whitbread is sold to Allied Distillers.

1991 Allied Distillers launches Caledonian Malts. Laphroaig is one of the four malts included.

1994 The Friends of Laphroaig is founded.

1995 A 10 year old cask strength is launched.

2001 A 40 year old is released.

2004 Quarter Cask is launched.

2005 Fortune Brands becomes new owner.

2007 A vintage 1980 (27 years old) and a 25 year old are released.

2008 Cairdeas, Cairdeas 30 year old and Triple Wood are released.

2009 An 18 year old is released.

2010 A 20 year old for French Duty Free and Cairdeas Master Edition are launched.

2011 Laphroaig PX and Cairdeas - The Ileach Edition are released.

2012 Brodir and Cairdeas Origin are launched.

2013 QA Cask, An Cuan Mor, 25 year old cask strength and Cairdeas Port Wood Edition are released.

2014 Laphroaig Select and a new version of Cairdeas are released.

2015 A 21 year old, a 32 year old sherry cask and a new Cairdeas are released and the 15 year old is re-launched.

2016 Lore, Cairdeas 2016 and a 30 year old are released.

2017 Four Oak, The 1815 Edition and a 27 year old are released.

2018 A 28 year old and Cairdeas Fino are released.

2019 A 30 year old is the first release in a new series named The Ian Hunter Story.

Tasting notes Laphroaig Select:

GS – The nose offers chocolate and malt notes set against peat, citrus fruit and iodine. Citrus fruit is most apparent on the relatively light palate, along with ginger, cinnamon and dried fruits. The peat is muted. The finish offers bright spices, new oak and medicinal notes.

Tasting notes Laphroaig 10 year old:

GS – Old-fashioned sticking plaster, peat smoke and seaweed leap off the nose, followed by something a little sweeter and fruitier. Massive on the palate, with fish oil, salt and plankton, though the finish is quite tight and increasingly drying

Select   Quarter Cask   Lore

Cairdeas Triple Wood   The 1815 Edition

10 years old

Triple Wood

Cairdeas Quarter Cask

Linkwood

[link•wood]

Owner: Diageo

Region/district: Speyside

Founded: 1821

Status: Active

Capacity: 5 600 000 litres

Address: Elgin, Morayshire IV30 8RD

Website: malts.com

Tel: 01343 862000

Linkwood has received a lot of attention from the owners, Diageo, in recent times. It all started in 1971 when a second distillery was built on the site.

This was followed up in 2011 when the distillery was closed for four months for a major upgrade including a new mash tun and automation by way of a new control system for the stills. Only two years later, the distillery closed temporarily again when two more stills and an additional six washbacks were installed. The capacity increased by 60% to 5,6 million litres.

Every distillery is subjected to refurbishment now and then but Linkwood is particularly important to Diageo. For more than one hundred years, single malt from the distillery has played a vital part in whisky from many different producers, not just Diageo's. where for instance Johnnie Walker and White Horse both depend on Linkwood for their character. Blenders have all agreed that Linkwood contributes to both flavour and body to a blend and the malt is essential for many of Diageo's top blends.

The old part of the distillery, which worked in tandem with the new site but stopped producing in 1996, was equipped with worm tubs and had a slightly different character than the Linkwood of today. In connection with the second upgrade in 2013, the old distillery buildings facing Linkwood Road were demolished and an extension of the current still house, which houses two of the stills and the tunroom, was conducted. The only original buildings from 1872 left standing are No. 6 warehouse and the redundant, old kiln with the pagoda roof. The set up of equipment now is one 12.5 ton full lauter mash tun, 11 wooden washbacks and three pairs of stills. The fermentation time during five-day week production varies between 65 and 105 hours. Production during the last couple of years has varied between 3.6 and 5.6 million litres of alcohol, depending on having a five or seven-day production week.

The only official core bottling is a **12 year old Flora & Fauna**. In October 2016, a **37 year old** distilled in 1978 and bottled at 50.3%, was launched as part of the Special Releases.

History:

1821 Peter Brown founds the distillery.

1868 Peter Brown passes away and his son William inherits the distillery.

1872 William demolishes the distillery and builds a new one.

1897 Linkwood Glenlivet Distillery Company Ltd takes over operations.

1902 Innes Cameron, a whisky trader from Elgin, joins the Board and eventually becomes the major shareholder and Director.

1932 Innes Cameron dies and Scottish Malt Distillers takes over in 1933.

1962 Major refurbishment takes place.

1971 The two stills are increased by four. Technically, the four new stills belong to a new distillery referred to as Linkwood B.

1985 Linkwood A (the two original stills) closes.

1990 Linkwood A is in production again for a few months each year until 1996.

2002 A 26 year old from 1975 is launched as a Rare Malt.

2005 A 30 year old from 1974 is launched as a Rare Malt.

2008 Three different wood finishes (all 26 year old) are released.

2009 A Manager´s Choice 1996 is released.

2013 Expansion of the distillery including two more stills.

2016 A 37 year old is released.

12 years old

Tasting notes Linkwood 12 years old:

GS – Floral, grassy and fragrant on the nutty nose, while the slightly oily palate becomes increasingly sweet, ending up at marzipan and almonds. The relatively lengthy finish is quite dry and citric.

Pioneers of Whisky

David Zibell
Founder
Golani Distillery, Israel

David Zibell is definitely one of the whisky pioneers in Israel. Almost at the same time as Milk & Honey distillery started production in Tel Aviv, so did David at his Golani distillery in the Golan Heights in the very north of Israel. Born in Paris and raised in Canada, Israel and the USA, David had a deli and bakery in San José, Costa Rica and after that he was running a real estate brokerage in Canada.

It was in Costa Rica that he fell in love with whisky. Even though he had enjoyed it for as long as he could remember, a Caol Ila 12 year old offered to him by his restaurant manager was (in his own words) his first wow moment with whisky.

In spring 2014 he went on a trip to Israel with his family and ended up in the small town of Katzrin. He saw the abundance of natural springs and fell in love with the place. With a strong support from his wife, who had just started enjoying whisky herself, he decided this was the place to stay and build a distillery of his own. As he himself puts it, "I'd like to make this gold myself, even if it's a few barrels in my basement."

Funding the first part of the project meant selling off some of their assets, but at that time David didn't want to bring in any external investment. The income was enough to finance their first mash tun, a fermentor, a copper pot still, the grain needed and the first 40 casks.

The next step was a crowdfunding campaign which gave him the means to cover the first year of rent as well as more raw material. After that he managed to get a business loan which gave him more fermenters and stills. Since then the growth has been financed by regular bank loans. He also sold 5% of the shares in the company to finance the purchase of a dunam (1,000 m²) of land to build yet another distillery and a visitor centre. The buyer of the shares wasn't just any investor but the master welder responsible for all the copper and metal work crucial to the business.

For the maturation of his whisky, David first decided on ex-bourbon barrels like so many new distilleries do. Worried about the availability and the cost, David changed his mind.

"I decided to recycle local wine casks. We use some that have been freshly dumped, but most get shaved down to the new oak, toasted and charred giving us a new, charred cask that gets its first fill for our Golani Black version of our single grain."

Some of his casks are obtained from a Golan Heights winery in the same town, a producer that David considers to be one of the best in Israel.

For startup Scottish distilleries, the matter of where you order your stills is easily settled. Forsyths in Rothes are the main suppliers or you can order from Frilli in Italy. David wanted to acquire locally but realised that there were no copper smiths in his part of the country. He simply decided to open a copper still factory together with master welder Yishai Socher to provide stills for not only Golani but also for his second distillery, focused on peated whisky, and a third which will be producing corn whisky. Golani Copper Works will also supply external distilleries with stills and other equipment.

When asked if anything took him by surprise when he started production, David says "I think the greatest element of surprise came in the barrel room. Ageing in the hot Israeli climate is just insane! During the summer months daytime temperatures are above 40°C, then night comes and they drop down to the 20's and so much happens in the casks. You can see the weekly evolution and it's tremendous."

David currently sells 15,000 bottles a year and the first ten or so expressions show tremendous maturity in spite of the young age. In many ways this is thanks to the temperature differences which absorb the whisky into the wood when it's hot and extract it again when it gets colder.

Creating a house style is often a challenge to a new producer. Single malt was a given for David but he also wanted something different and local.

"After trial and error, I decided on a single grain based on a 50/50 wheat and malted barley. Wheat and barley are two of Israel's seven fruits of the land going back to the bible and as Israelis are relatively new to whisky, a softer spirit would suit them. Israel is a hot country which a lighter whisky also could match."

Until recently, David had to rely on imported barley for his production but in spring 2019, he harvested the first locally grown barley in Israel intended for whisky production - a variety that was developed especially for Israeli conditions.

When asking David the question if regulations and laws surrounding whisky production have been an issue when starting the company, he replies with a simple "No" which probably makes whisky entrepreneurs in other parts of the world quite envious.

Apart from opening up new distilleries, David is now focused on selling his matured Golani single malt and the idea is to do it without a distributor. It will be marketed and distributed directly without "selling their souls to any distributor."

"We don't need to be just another product in a catalogue. We can choose our customers and, more importantly, know our customers. In today's digital world it is much easier to listen to and serve your customers and I think that in such a competitive market having the best product based on the demand of the customer gives you a real chance to succeed."

Except for having released a number of young whiskies that have aged tremendously well in the hot and humid climate of Israel, David has also produced a number of other spirits; gin, absinth, arrak and brandy,

David's final advice
to someone wanting to start a distillery of their own;

"Follow your dreams, be creative, make something that makes you happy. Don't be afraid to do something different (obviously listen to your customers if nobody likes it besides you). Don't just try to do what's being done and don't listen to the naysayers. There will be plenty of the latter. Get as many answers and advise from other distillers. Learning from others' mistakes can save you so much time and money."

Loch Lomond

[lock low•mund]

Owner:
Loch Lomond Group
(Hillhouse Capital Management)

Region/district:
Western Highlands

Founded: 1965
Status: Active
Capacity: 5 000 000 litres

Address: Lomond Estate, Alexandria G83 0TL

Website: lochlomondwhiskies.com
Tel: 01389 752781

In 2017, the owners of Loch Lomond secured a distribution deal for their whiskies in the expanding Chinese market and in June 2019, the Chinese connection was taken a step further.

Exponent Equity, who took over Loch Lomond in 2014, announced that they had sold the company to the Chinese investment firm Hillhouse Capital Management and a price of $500-$550m was mentioned.

Loch Lomond has an extremely unusual equipment setup. Founded in 1966, one pair of straight neck pot stills was installed. Yet another pair were installed in 1990 and four years later a grain distillery was opened. One pair of traditional swan neck pot stills was installed in 1998 and in 2007 a single grain coffey still was added. Complemented with a third pair of straight neck stills the distillery now has 13 stills of four different kinds!

Of the 5 million litres on the malt side, 3.5 million is distilled in the Coffey still with Loch Lomond Single Grain (actually a single malt but not accepted as such by the SWA) the big seller. Loch Lomond is distilled in the traditional pot stills but mixed before bottling with whisky from the straight neck stills. Inchmurrin comes from the straight neck stills. So does the heavily peated Inchmoan and other styles not being promoted as official bottlings (Inchfad, Glen Douglas, Craiglodge, and Croftengea).

The rest of the equipment consists of a 9.5 ton full lauter mash tun, ten stainless steel washbacks (with a fermentation time of 92 to 160 hours) for the malt side of the production and another 18 for the grain side.

The core range is divided between three brands; **Loch Lomond** with **Classic, Original,** a **10, 12** and an **18 year old; Inchmurrin,** with a **12** and an **18 year old** and a **Madeira wood finish** as well and **Inchmoan 12 year old** and a **1992 vintage.** For duty-free there is **Loch Lomond Single Grain, Signature blended Scotch** and a **12 year old single malt** together with **Inchmurrin Madeira finish,** a **10 year old single cask** and a **10 year old Inchmoan.** Limited expressions include **Loch Lomond The Open Special Edition** as well as a **50 year old.**

History:

1965 The distillery is built by Littlemill Distillery Company Ltd owned by Duncan Thomas and American Barton Brands.

1966 Production commences.

1971 Duncan Thomas is bought out.

1984 The distillery closes.

1985 Glen Catrine Bonded Warehouse Ltd buys Loch Lomond Distillery.

1987 The distillery resumes production.

1993 Grain spirits are also distilled.

1997 A fire destroys 300,000 litres of maturing whisky.

1999 Two more stills are installed.

2005 Inchmoan and Craiglodge as well as Inchmurrin 12 years are launched.

2006 Inchmurrin 4 years, Croftengea 1996 (9 years), Glen Douglas 2001 (4 years) and Inchfad 2002 (5 years) are launched.

2010 A peated Loch Lomond with no age statement is released as well as a Vintage 1966.

2012 New range for Inchmurrin released – 12, 15, 18 and 21 years.

2014 The distillery is sold to Exponent Private Equity. Organic versions of 12 year old single malt and single blend are released.

2015 Loch Lomond Original Single Malt is released together with a single grain and two blends, Reserve and Signature.

2016 A 12 year old and an 18 year old are launched.

2017 This year's releases include Inchmoan 12 year old and Inchmurrin 12 and 18 year old.

2018 A 50 year old Loch Lomond is released.

2019 The distillery is sold to Hillhouse Capital Management and a 50 year old is released.

Loch Lomond Original

Tasting notes Loch Lomond Original:

GS – Initially earthy on the nose, with malt and subtle oak. The palate is rounded, with allspice, orange, lime, toffee, and a little smokiness. Barley, citrus fruits and substantial spiciness in the finish.

Longmorn

[long•morn]

Owner:
Chivas Brothers
(Pernod Ricard)

Region/district:
Speyside

Founded: 1894   **Status:** Active   **Capacity:** 4 500 000 litres

Address: Longmorn, Morayshire IV30 8SJ

Website: -

Tel: 01343 554139

The vast number of distilleries being planned, built or expanded in Scotland at the moment, is of concern to some observers who argue that the industry is at risk of overcapacity due to its optimism.

But these concerns are not new. This is what the Northern Scot newspaper wrote in connection to Longmorn being established in 1893: "Still another distillery! I wonder how many new distilleries have been erected, or are in the course of construction, or have been arranged for within the last three years... to this must be joined the fact that at the same time, the old and existing distilleries have been lengthening their stakes and widening their stills to an unprecedented extent... When is all this going to end?"

Longmorn distillery is equipped with a modern 8.5 ton Briggs full lauter mash tun which replaced the old, traditional tun in 2012. At the same time, seven of the eight, old stainless steel washbacks were moved to the new tun room and an additional three were installed. The eight, onion-shaped stills with declining lyne arms are big and fitted with sub-coolers and the wash stills have external heat exchangers. The production capacity was also increased in 2012 by 30% to 4.5 million litres. Currently, the production runs for five days per week with 18 mashes. This means roughly 3 million litres of alcohol over the year. The style of the newmake is fruity yet robust.

The core range consists of **The Distiller's Choice** (with no age statement), a **16 year old** and a **23 year old**. In July 2019, another three bottlings appeared in the new Chivas series named The Secret Speyside Collection. Two of them, **18** and **23 year old**, had been matured in American oak barrels and hogsheads and were bottled at 48%. The third, a **25 year old**, had been filled into a combination of ex-bourbon and ex-sherry and was bottled at 52.2%. All three are available for the first year in duty-free and will then be made available to domestic markets in 2020. Finally, there are three cask strength bottlings in the Distillery Reserve Collection, available at all Chivas' visitor centres – from **12 to 20 years old**.

History:

1893 John Duff & Company, which founded Glenlossie already in 1876, starts construction. John Duff, George Thomson and Charles Shirres are involved in the company. The total cost amounts to £20,000.

1894 First production in December.

1897 John Duff buys out the other partners.

1898 John Duff builds another distillery next to Longmorn which is called Benriach (at times aka Longmorn no. 2). Duff declares bankruptcy and the shares are sold by the bank to James R. Grant.

1970 The distillery company is merged with The Glenlivet & Glen Grant Distilleries and Hill Thomson & Co. Ltd. Own floor maltings ceases.

1972 The number of stills is increased from four to six. Spirit stills are converted to steam firing.

1974 Another two stills are added.

1978 Seagrams takes over through The Chivas & Glenlivet Group.

1994 Wash stills are converted to steam firing.

2001 Pernod Ricard buys Seagram Spirits & Wine together with Diageo and Pernod Ricard takes over the Chivas group.

2004 A 17 year old cask strength is released.

2007 A 16 year old is released replacing the 15 year old.

2012 Production capacity is expanded.

2015 The Distiller's Choice is released.

2016 A 16 year old and a 23 year old are released.

2019 Three expressions in the new The Secret Speyside Collection are released.

Longmorn 25 year old

Tasting notes Longmorn Distiller's Choice:

GS – Barley sugar, ginger, toffee and malt on the sweet nose. The palate reveals caramel and milk chocolate, with peppery Jaffa orange. Toffee, barley and a hint of spicy oak in the medium-length finish.

Macallan

[mack•al•un]

Owner:	**Region/district:**
Edrington Group	Speyside
Founded: **Status:**	**Capacity:**
1824 Active (vc)	15 000 000 litres

Address: Easter Elchies, Craigellachie, Morayshire AB38 9RX

Website:	**Tel:**
themacallan.com	01340 871471

Given Macallan's current position of being one of the three great malts that sell more than one million cases per year, it is hard to imagine that only forty years ago it was virtually unknown as a brand. The vast majority of the production went into blends.

The person credited with this remarkable transition to a mega-brand is Willie Phillips, the Macallan managing director from 1978 to 1996. But by his side was also a man with exceptional marketing skills – Hugh Mitcalfe. After having helped Glen Grant establish themselves in the Italian market, Mitcalfe joined the Macallan team in 1978. The first thing he realised was that Macallan had an unrivalled stock of old whiskies. He introduced the Anniversary expressions, whiskies of considerable age, and together with a London ad agency he introduced a number of small ads featuring drawings that were placed next to The Times crossword. When he left the company in 1996, Macallan was the sixth biggest malt selling two million bottles. Today the brand sells 12 million bottles. Hugh Mitcalfe died in January 2019 aged 84.

The new distillery is equipped with a full lauter mash tun with a 17 ton mash and 21 washbacks made of stainless steel with a fermentation time of 60 hours. There are 12 wash stills, each paired to two spirit stills – a total of 36 stills. The capacity is 15 million litres of pure alcohol per year and the plan for 2019 is to do 35-40 mashes per week resulting in 11-11.5 million litres of alcohol. The old distillery has been mothballed and with a capacity of 11 million litres, it is equipped with two mash tuns, 22 stainless steel washbacks and six made of wood, seven wash stills and 14 spirit stills. On site there are 16 dunnage and 31 racked warehouses.

The core range of Macallan consists of three styles; **Sherry Oak** (100% maturation in sherry casks) was the only available version until 2004 when Fine Oak was introduced. Today it is represented by **12, 18, 25** and **30 years old**. **Triple Cask** (formerly known as Fine Oak) is a combination of whisky matured in ex-bourbon and ex-sherry casks and is represented by **12, 15** and **18 year old**. **Double Cask**, finally, is the youngest addition to the range and is currently made up of a **12 year old** and **Gold** without age statement. In this case, Double cask means a mix of sherry casks from both American and European Oak. The Macallan Quest Collection is reserved for duty-free including **Quest, Lumina, Terra** and **Enigma**. A fifth member is **Aurora**, recenly released exclusively for Taiwan. Another duty-free bottling appeared in early 2019 when **Concept No. 1**, the first in a new range, was released. A range of prestige bottlings called The Macallan Masters Decanter Series includes; **Reflexion, No 6, M Rare Cask** and **M Black**. Recent limited releases include **Edition No. 5, Classic Cut** and **Genesis**. Macallan holds an impressive stock of old whiskies and in 2018 a **72 year old** and a **52 year old** were released. In 2019 a new core release, **Macallan Estate** made from barley grown on the Macallan estate, was released. Finally, there is the **Fine and Rare Collection** with single casks from 1926 to 1990.

History:

1824 The distillery is licensed to Alexander Reid under the name Elchies Distillery.

1847 Alexander Reid passes away and James Shearer Priest and James Davidson take over.

1868 James Stuart takes over the licence. He founds Glen Spey distillery a decade later.

1886 James Stuart buys the distillery.

1892 Stuart sells the distillery to Roderick Kemp from Elgin. Kemp expands the distillery and names it Macallan-Glenlivet.

1909 Roderick Kemp passes away and the Roderick Kemp Trust is established to secure the family's future ownership.

1965 The number of stills is increased from six to twelve.

1966 The trust is reformed as a private limited company.

1968 The company is introduced on the London Stock Exchange.

1974 The number of stills is increased to 18.

1975 Another three stills are added, now making the total 21.

1984 The first official 18 year old single malt is launched.

1986 Japanese Suntory buys 25% of Macallan-Glenlivet plc stocks.

1996 Highland Distilleries buys the remaining stocks. 1874 Replica is launched.

1999 Edrington and William Grant & Sons buys Highland Distilleries for £601 million through The 1887 Company with 70% held by Edrington and 30% by William Grant & Sons. Suntory still holds 25% in Macallan.

2000 The first single cask from Macallan (1981) is named Exceptional 1.

History continued:

2001 A new visitor centre is opened.

2002 Elegancia replaces 12 year old in the duty-free range. 1841 Replica, Exceptional II and Exceptional III are also launched.

2003 1876 Replica and Exceptional IV, single cask from 1990 are released.

2004 Exceptional V, single cask from 1989 is released as well as Exceptional VI, single cask from 1990. The Fine Oak series is launched.

2005 New expressions are Macallan Woodland Estate, Winter Edition and the 50 year old.

2006 Fine Oak 17 years old and Vintage 1975 are launched.

2007 1851 Inspiration and Whisky Maker´s Selection are released as a part of the Travel Retail range.

2008 Estate Oak and 55 year old Lalique are released.

2009 The mothballed No. 2 stillhouse is re-opened. The Macallan 1824 Collection and a 57 year old Lalique bottling is released.

2010 Oscuro is released for Duty Free.

2011 Macallan MMXI is released for duty free.

2012 Macallan Gold, the first in the new 1824 series, is launched.

2013 Amber, Sienna and Ruby are released.

2014 1824 Masters Series (with Rare Cask, Reflexion and No. 6) is released.

2015 Rare Cask Black is released.

2016 Edition No. 1 and 12 year old Double Cask are released.

2017 Folio 2 is released. The new distillery is commissioned.

2018 Fine Oak changes name to Triple Cask, The Quest Collection is released for duty free and Macallan M Black and Genesis are launched. Concept No. 1 and a 72 and a 52 year old are released.

2019 Macallan Estate and Edition No. 5 are released.

Tasting notes Macallan 12 year old Sherry oak:

GS – The nose is luscious, with buttery sherry and Christmas cake characteristics. Rich and firm on the palate, with sherry, elegant oak and Jaffa oranges. The finish is long and malty, with slightly smoky spice.

Tasting notes Macallan 12 year old Triple Cask:

GS – The nose is perfumed and quite complex, with marzipan and malty toffee. Expansive on the palate, with oranges, marmalade, milk chocolate and oak. Meidum in length, balanced and comparatively sweet.

Gold

12 yo Sherry Oak

12 yo Triple Cask

Enigma

52 years old

Concept No. 1

Double Cask 12 yo

Macallan Estate

Rare Cask Black

Rare Cask Batch 2

Macduff

[mack•duff]

Owner:
John Dewar & Sons Ltd
(Bacardi)

Region/district:
Highlands

Founded: 1960
Status: Active
Capacity: 3 400 000 litres

Address: Banff, Aberdeenshire AB45 3JT

Website: lastgreatmalts.com
Tel: 01261 812612

History:

1960 The distillery is founded by Marty Dyke, George Crawford, James Stirrat and Brodie Hepburn (who is also involved in Tullibardine and Deanston). Macduff Distillers Ltd is the name of the company.

1964 The number of stills is increased from two to three.

1967 Stills now total four.

1972 William Lawson Distillers, part of General Beverage Corporation which is owned by Martini & Rossi, buys the distillery from Glendeveron Distilleries.

1990 A fifth still is installed.

1993 Bacardi buys Martini Rossi (including William Lawson) and eventually transfered Macduff to the subsidiary John Dewar & Sons.

2013 The Royal Burgh Collection (16, 20 and 30 years old) is launched for duty free.

2015 A new range is launched - 10, 12 and 18 years old.

With the same number of working weeks and the same number of mashes per week as last year, the distillery expects to lose around 5% in production during 2019. This is, however, not isolated to Macduff but affects all distilleries.

The reason is that the predominating barley variety used for whisky production in the last decade, Concerto, had a poor harvest in 2018. What was particularly alarming was that the nitrogen level , which whisky producers wish to keep low, had increased. This ultimately affects yields. Usually, a distillery will get 410-415 litres of alcohol per ton of malted barley. After a poor harvest with higher nitrogen levels, this could easily drop by 5-10 litres per ton.

While yield isn't everything, it is still an important factor for the larger producers. Maltsters, as well as farmers and whisky producers, are always on the hunt for a better barley variety. For almost two decades Optic was the variety preferred and was then succeeded by Concerto. It seems nowadays every new variety is having a shorter "shelf life" and the next variety, Laureate, is already on the horizon.

Macduff is equipped with a 6.75 ton stainless steel semi-lauter mash tun and nine washbacks (29,800 litres) made of stainless steel with a fermentation time of 55 hours. There is also a rather unusual set-up of five stills – two wash stills and three spirit stills. In order to fit the stills into the still room, the lyne arms on four of the stills are bent in a peculiar way and on one of the wash stills it is U-shaped. In 2019 the distillery will be making 26 mashes per week for 48 weeks, producing 3.26 million litres of alcohol.

Since 2015, the core range from the distillery is known under the name The Deveron and consists of a **10 year old**, exclusive to France, as well as a **12 and 18 year old**. For duty free, a range first launched in 2013, is available named Glen Deveron encompassing a **16**, a **20** and a **30 year old**.

12 years old

Tasting notes The Deveron 12 years old:

GS – Soft, sweet and fruity on the nose, with vanilla, ginger, and apple blossom. Medium-bodied, gently spicy, with butterscotch and Brazil nuts. Caramel contrasts with quite dry spicy oak in the finish.

Mannochmore

[man•och•moor]

Owner: Diageo

Region/district: Speyside

Founded: 1971

Status: Active

Capacity: 6 000 000 litres

Address: Elgin, Morayshire IV30 8SS

Website: malts.com

Tel: 01343 862000

Speyside is full of distilleries – actually more than fifty – and most of them are easy to find. A few can even be seen from miles around. Mannochmore does, however, not belong to that category.

If you travel along the A941 towards Rothes and stop to take pictures of Longmorn on your left side, in the distance on your right hand side you see the rooftops of warehouses and a huge, white building with smoke coming out of the chimney. Mannochmore you think but you're only half right. The building is a dark grains plant sitting on the same site as the distillery.

Even if you drive right up to the distillery, you'll have difficulties spotting it. What you see is Glenlossie's old kiln house with the pagoda roof. To be able to detect Mannochmore, you have to enter the site (which you shouldn't as visitors are not allowed) where you will find a recently expanded still house with large windows and another building for the mash tun and eight of the washbacks.

Mannochmore was built in the early 1970s when optimism was high in the industry and its main purpose was, and still is, to produce malt for blends. For many years Mannochmore and Glenlossie were run as "sisters distilleries" with the same workforce alternating between the two distilleries with each distillery being in production for half a year at the time

Since 2013 the distillery is equipped with an 11.1 ton Briggs full lauter mash tun, eight wooden washbacks and another eight external made of stainless steel and four pairs of stills. Clear wort and long fermentations (up to 100 hours) creates a new-make spirit with a fruity character and the distillery is currently running a seven-day production.

Mannocmore is the signature malt for Haig, a brand first introduced in the late 1880s. The Haig family belonged to the elite of Scottish whisky aristocracy and had been involved in distilling since the middle of the 17th century.

The only current official bottling is a **12 year old Flora & Fauna**. In autumn 2016, a **25 year old** distilled in 1990 and bottled at 53.4%, was launched as part of the Special Releases.

History:

1971 Distillers Company Limited (DCL) founds the distillery on the site of their sister distillery Glenlossie. It is managed by John Haig & Co. Ltd.

1985 The distillery is mothballed.

1989 In production again.

1992 A Flora & Fauna series 12 years old becomes the first official bottling.

2009 An 18 year old is released.

2010 A Manager´s Choice 1998 is released.

2013 The number of stills is increased to eight.

2016 A 25 year old cask strength is released.

12 years old

Tasting notes Mannochmore 12 years old:

GS – Perfumed and fresh on the light, citric nose, with a sweet, floral, fragrant palate, featuring vanilla, ginger and even a hint of mint. Medium length in the finish, with a note of lingering almonds.

Miltonduff

[mill•ton•<u>duff</u>]

Owner:
Chivas Brothers
(Pernod Ricard)

Region/district:
Speyside

Founded: 1824   **Status:** Active   **Capacity:** 5 800 000 litres

Address: Miltonduff, Elgin, Morayshire IV30 8TQ

Website: -   **Tel:** 01343 547433

In 1935, the Canadian distiller Hiram Walker, Gooderham & Worts took over the blended Scotch Ballantine's and one year later they acquired two distilleries, Miltonduff and Glenburgie, that have since then been the backbone of this famous blend

Ballantine's is the second most sold Scotch in the world – way behind the unbeatable number one, Johnnie Walker, but far ahead of its other competitors. And what's even more important, the trend is in favour of Ballantine's. Since 2014, sales have increased by 25% to reach 89 million bottles in 2018. There's only one other Scotch whisky brand amongst the Top 30 with a better record – Black & White but on the other hand their increase of 93% in the same period of time came from substantially lower volumes.

Official bottlings of Miltonduff, Glenburgie and Glentauchers, the third signature malt in Ballantine's, were virtually non existent a couple of years ago. In 2017, however, this changed when all three appeared on the shelves in the shape of 15 year olds. To emphasise the important connection between the three malts and the blend, the series is called Ballantine's Single Malt Series.

Miltonduff distillery is equipped with an eight ton full lauter mash tun with a copper dome, 16 stainless steel washbacks with a fermentation time of 56 hours and six, large stills. The lyne arms are all sharply descending which allows for very little reflux. This makes for a rather robust and oily new make in contrast to the lighter and more floral Glenburgie. In 1964, two Lomond stills were installed at Miltonduff. They were equipped with columns with adjustable plates with the intention of distilling different styles of whisky from the same still. They were dismantled in 1981 but the special whisky produced in the still, Mosstowie, can still, with a bit of luck, be found.

The new official bottling is a **15 year old** matured in ex-bourbon casks. There is also a **19 year old** exclusive to the Taiwanese market and a **12 year old cask strength** bottling in the Distillery Reserve Collection, available at all Chivas' visitor centres.

History:

1824 Andrew Peary and Robert Bain obtain a licence for Miltonduff Distillery. It has previously operated as an illicit farm distillery called Milton Distillery but changes name when the Duff family buys the site it is operating on.

1866 William Stuart buys the distillery.

1895 Thomas Yool & Co. becomes new part-owner.

1936 Thomas Yool & Co. sells the distillery to Hiram Walker Gooderham & Worts. The latter transfers administration to the newly acquired subsidiary George Ballantine & Son.

1964 A pair of Lomond stills is installed to produce the rare Mosstowie.

1974 Major reconstruction of the distillery.

1981 The Lomond stills are decommissioned and replaced by two ordinary pot stills, the number of stills now totalling six.

1986 Allied Lyons buys 51% of Hiram Walker.

1987 Allied Lyons acquires the rest of Hiram Walker.

1991 Allied Distillers follow United Distillers´ example of Classic Malts and introduce Caledonian Malts in which Tormore, Glendro-nach and Laphroaig are included in addition to Miltonduff. Tormore is later replaced by Scapa.

2005 Chivas Brothers (Pernod Ricard) becomes the new owner through the acquisition of Allied Domecq.

2017 A 15 year old is released.

15 years old

Tasting notes Miltonduff 15 years old:

IR – Fresh citrus and honey on the nose together with heather, ginger and peaches. More spicy on the palate with cinnamon and clove, vanilla, honey, red berries and liquorice.

Mortlach

[mort•lack]

Owner:
Diageo

Region/district:
Speyside

Founded: 1823   **Status:** Active   **Capacity:** 3 800 000 litres

Address: Dufftown, Keith, Banffshire AB55 4AQ

Website:
mortlach.com, malts.com

Tel:
01340 822100

Mortlach was the first distillery to be built in Dufftown which, thanks to its six distilleries, is now known as the whisky capital of the world. At least among whisky aficionados

Every year in May, it is also the epicentre of the Spirit of Speyside festival and this is also the only time when Mortlach distillery is opened to visitors. In 2019, tickets for the ten tours sold out instantly. While Dufftown is a fairly young village, founded in 1817 to provide accommodation and employment after the Napoleonic wars, there was a much older village preceding it. Parts of the current Mortlach church from the 19th century may date to the 8th century or possibly even earlier.

The distillery is equipped with a 12 ton full lauter mash tun and six washbacks made of Douglas fir, currently with six short fermentations (55 hours) and six long (110 hours). There are three wash stills and three spirit stills and the No. 3 pair acts as a traditional double distillation. The low wines from wash stills No. 1 and 2 are directed to the remaining two spirit stills according to a certain distribution. In one of the spirit stills, called Wee Witchie, the charge is redistilled twice and, with all the various distillations taken into account, Mortlach is distilled 2.81 times. All the stills are attached to worm tubs for cooling the spirit vapours. During 2019, the distillery will be working a five-day week with 12 mashes per week with a target of making 2.6 million litres.

The previous range relied heavily on maturation in ex-bourbon barrels. With the new expressions it's the other way around – the two older ones have been matured in ex-sherry casks while the youngest is a vatting of ex-sherry and ex-bourbon. The range now consists of **12 year old Wee Witchie, 16 year old Distiller's Dram, 20 year old Cowie's Blue Seal** and, for duty-free, the **14 year old Alexander´s Way**. In April 2019, the first in Mortlach's Singing Stills series appeared when a **47 year old** was released and later in the year a **26 year old** appeared in the Special Releases.

History:

- 1823 The distillery is founded by James Findlater.
- 1824 Donald Macintosh and Alexander Gordon become part-owners.
- 1831 The distillery is sold to John Robertson for £270.
- 1832 A. & T. Gregory buys Mortlach.
- 1837 James and John Grant of Aberlour become part-owners. No production takes place.
- 1842 The distillery is now owned by John Alexander Gordon and the Grant brothers.
- 1851 Mortlach is producing again after having been used as a church and a brewery for some years.
- 1853 George Cowie joins and becomes part-owner.
- 1867 John Alexander Gordon dies and Cowie becomes sole owner.
- 1896 Alexander Cowie joins the company.
- 1897 The number of stills is increased from three to six.
- 1923 Alexander Cowie sells the distillery to John Walker & Sons.
- 1925 John Walker becomes part of Distillers Company Limited (DCL).
- 1964 Major refurbishment.
- 1968 Floor maltings ceases.
- 1996 Mortlach 1972 is released as a Rare Malt.
- 1998 Mortlach 1978 is released as a Rare Malt.
- 2004 Mortlach 1971, a 32 year old cask strength is released.
- 2014 Four new bottlings are released - Rare Old, Special Strength, 18 year old and 25 year old.
- 2018 A new range is presented; 12 year old Wee Witchie, 16 year old Distiller´s Dram and 20 year old Cowie´s Blue Seal.
- 2019 The oldest official Mortlach bottling ever, 47 years, is released and a 26 year old appears in the Special Releases.

12 years old

Tasting notes Mortlach 12 years old:

IR – Fresh and intense on the nose with notes of sherry, apple cider, dark plums, tobacco and toffee. The palate is robust with orange marmalade, dark chocolate, espresso and chili pepper.

Oban

[oa•bun]

Owner:		Region/district:
Diageo		Western Highlands

Founded:	Status:	Capacity:
1794	Active (vc)	870 000 litres

Address: Stafford Street, Oban, Argyll PA34 5NH

Website:	Tel:
malts.com	01631 572004 (vc)

These days, urban distilleries are rare in Scotland. Most of the distilleries are located miles away from any town. Bowmore and Jura are exceptions, both very much a part of their respective villages. Glen Garioch and Springbank even more so.

There is no doubt however that Oban is the most urban distillery to be found. Completely surrounded by shops and restaurants in the busiest part of Oban, the distillery stands no chance of expanding even if they wanted to. The distillery was there first though and the village, officially founded in 1820, was built around it. But long before that, more than 6,000 years ago, there were people living in the very grounds of todays distillery. In 1890, a cave in the tree-covered Creag A' Bharrain cliff, rising just behind the distillery, was discovered with the remains of four adults and four children.

The equipment consists of a seven ton traditional stainless steel mash tun with rakes, four washbacks made of European larch and one pair of stills. The washbacks are now being replaced by new ones, one each year starting 2018. Attached to the stills is a rectangular, stainless steel, double worm tub to condense the spirit vapours. One washback will fill the wash still twice. However, the character of Oban single malt is dependent on long fermentations (110 hours), hence they can only manage six mashes per week, giving it five long fermentations and one short (65 hours). The production for 2019 will be around 840,000 litres.

At least to my knowledge, Oban is the only Diageo distillery where the entire production is destined to be bottled as single malt. At both Lagavulin and Talisker, some of the whisky still seem to be intended for blends.

The core range consists of **Little Bay**, a **14 year old**, an **18 year old** exclusive for USA and a **Distiller's Edition** with a montilla fino sherry finish. In spring 2019, the **Night's Watch - Oban Bay Reserve** was released as part of the Game of Thrones series and in summer the same year, **Oban Old Teddy**, a distillery excluisve bottled at 51,7% matured in ex-bodega sherry casks was released.

History:
1793 John and Hugh Stevenson found the distillery.
1820 Hugh Stevenson dies.
1821 Hugh Stevenson's son Thomas takes over.
1829 Bad investments force Thomas Stevenson into bankruptcy. His eldest son John takes over.
1830 John buys the distillery from his father's creditors for £1,500.
1866 Peter Cumstie buys the distillery.
1883 Cumstie sells Oban to James Walter Higgins who refurbishes and modernizes it.
1898 The Oban & Aultmore-Glenlivet Co. takes over with Alexander Edwards at the helm.
1923 The Oban Distillery Co. owned by Buchanan-Dewar takes over.
1925 Buchanan-Dewar becomes part of Distillers Company Limited (DCL).
1931 Production ceases.
1937 In production again.
1968 Floor maltings ceases and the distillery closes for reconstruction.
1972 Reopening of the distillery.
1979 Oban 12 years is on sale.
1988 United Distillers launches Classic Malts and Oban 14 year old is included.
1998 A Distillers' Edition is launched.
2002 The oldest Oban (32 years) so far is launched.
2004 A 1984 cask strength is released.
2009 Oban 2000, a single cask, is released.
2010 A no age distillery exclusive is released.
2013 A limited 21 year old is released.
2015 Oban Little Bay is released.
2016 A distillery exclusive without age statement is released.
2018 A 21 year old is launched as a part of the Special Releases.
2019 Night's Watch - Oban Bay Reserve and Oban Old Teddy are released.

Tasting notes Oban 14 years old:
GS – Lightly smoky on the honeyed, floral nose. Toffee, cereal and a hint of peat. The palate offers initial cooked fruits, becoming spicier. Complex, bittersweet, oak and more gentle smoke. The finish is quite lengthy, with spicy oak, toffee and new leather.

Night's Watch
- Oban Bay Reserve

Pulteney

[poolt•ni]

Owner:
Inver House Distillers
(Thai Beverages plc)

Region/district:
Northern Highlands

Founded:
1826

Status:
Active (vc)

Capacity:
1 800 000 litres

Address: Huddart St, Wick, Caithness KW1 5BA

Website:
oldpulteney.com

Tel:
01955 602371

Despite being located on the northern tip of the Scottish mainland, the distillery gets its fair share of visitors – around 14,000 last year. Thanks to a new cooperation with other distilleries in the northern Highlands, the owners hope to see these figures rise.

The inaugural Highland Whisky Festival, with eight distilleries from Tomatin in the South to Wolfburn in the North, took place in May 2019. Each of the eight distilleries hosted an open day with tastings and other special events. The decision to create the Highland Whisky Festival was in part inspired by the fact that Scotland's Highland and Islands were listed as number five on Lonely Planet's "Best in Travel" list in 2018.

With the new festival, May/early June has turned into a hectic season for the whisky enthusiasts with the Spirit of Speyside festival at the beginning of the month and Campbeltown Malts Festival and Feis Ile towards the end.

The distillery is equipped with a stainless steel semi-lauter mash tun with a copper canopy. There are seven washbacks made of stainless steel with a fermentation time between 50 and 110 hours. The wash still, equipped with a huge boil ball and a very thick lye pipe, is quaintly chopped off at the top. Both stills use stainless steel worm tubs for condensing the spirit. Around 1.6 million litres of alcohol are produced yearly.

In August 2018, the entire core range of Old Pulteney was revamped and is now made up of **12 years old**, matured in ex-bourbon casks and bottled at 40%, the smoky **Huddart** without age statement and **15** and **18 years old**, both matured in a combination of ex-bourbon and ex-sherry casks. Two months later it was time for the duty-free range to get an overhaul. The old line with names after lighthouses was replaced by a **10 year old** matured in ex-bourbon barrels, a **16 year old** with a finish ex-oloroso casks and a **Vintage 2006** which had spent time in first fill ex-bourbon barrels. As usual, a number of limited single casks have also been released

History:

1826 James Henderson founds the distillery.

1920 The distillery is bought by James Watson.

1923 Buchanan-Dewar takes over.

1930 Production ceases.

1951 In production again after being acquired by the solicitor Robert Cumming.

1955 Cumming sells to James & George Stodart, a subsidiary to Hiram Walker & Sons.

1958 The distillery is rebuilt.

1959 The floor maltings close.

1961 Allied Breweries buys James & George Stodart Ltd.

1981 Allied Breweries changes name to Allied Lyons.

1995 Allied Domecq sells Pulteney to Inver House Distillers.

1997 Old Pulteney 12 years is launched.

2001 Pacific Spirits (Great Oriole Group) buys Inver House at a price of $85 million.

2004 A 17 year old is launched.

2005 A 21 year old is launched.

2006 International Beverage Holdings acquires Pacific Spirits UK.

2010 WK499 Isabella Fortuna is released.

2012 A 40 year old and WK217 Spectrum are released.

2013 Old Pulteney Navigator, The Lighthouse range (3 expressions) and Vintage 1990 are released.

2014 A 35 year old is released.

2015 Dunnet Head and Vintage 1989 are released.

2017 Three vintages (1983, 1990 and 2006) are released together with a 25 year old.

2018 A completely new core range is launched; 12 years old, Huddart, 15 years old and 18 years old.

12 years old

Tasting notes Old Pulteney 12 years old:

GS – The nose presents pleasingly fresh malt and floral notes, with a touch of pine. The palate is comparatively sweet, with malt, spices, fresh fruit and a suggestion of salt. The finish is medium in length, drying and decidedly nutty.

Royal Brackla

Owner:
John Dewar & Sons
(Bacardi)

Region/district:
Highlands

Founded: 1812
Status: Active
Capacity: 4 100 000 litres

Address: Cawdor, Nairn, Nairnshire IV12 5QY

Website:
lastgreatmalts.com

Tel:
01667 402002

Perhaps the least known of Dewar's five distilleries, Royal Brackla has spent the last twenty years producing single malt for blends. It took a dedicated marketing manager and a skilful master blender to change that.

Stephen Marshall realised around 2010 that the distilleries owned by Bacardi all made beautiful whisky, each in its own style. The problem, in his mind, was that virtually none of it was bottled as single malt. He struggled within the company to get acceptance for a release of what he called The Last Great Malts.

He was soon supported by master blender Stephanie Macleod. She had been with the company since 1998 and took over as Master Blender in 2006. Her main duty is to maintain and develop the company's blends – Dewar's and Lawson's – but has for the past six years been instrumental in creating new expressions of the five single malts.

Royal Brackla is equipped with a 12.5 ton full lauter mash tun. There are six wooden washbacks and another two made of stainless steel which have been placed outside – all with a fermentation time of 70 hours. Finally, there are also two pairs of stills. In 2019, the distillery will be doing 17 mashes per week which translates to 3.97 million litres of alcohol which is more or less the full capacity of the distillery. With a mash tun producing a clear wort, long fermentations, long foreshots (30 minutes), a slow distillation and lots of reflux from the stills due to ascending lyne arms – the house style of Brackla is elegant and fruity. In 2015 a biomass boiler (fired with wood-chips) replaced the old, heavy fuel oil boiler. Not only will this contribute to a 5,000 ton reduction of CO_2 emissions, but it will also be 50% more energy efficient.

The new core range introduced in 2015 and replacing the previous 10 year old, consists of a **12, 16** and **21 year old**. The latest limited release was in June 2019 when a **20 year old** was released for duty free Asia. It was a double maturation with the final 11 years in casks that had previously held the Tuscan wine Sassicaia.

History:

1812 The distillery is founded by Captain William Fraser.

1833 Brackla becomes the first of three distilleries allowed to use 'Royal' in the name.

1852 Robert Fraser & Co. takes over the distillery.

1897 The distillery is rebuilt and Royal Brackla Distillery Company Limited is founded.

1919 John Mitchell and James Leict from Aberdeen purchase Royal Brackla.

1926 John Bisset & Company Ltd takes over.

1943 Scottish Malt Distillers (SMD) buys John Bisset & Company Ltd and thereby acquires Royal Brackla.

1964 The distillery closes for a big refurbishment
-1966 and the number of stills is increased to four. The maltings closes.

1970 Two stills are increased to four.

1985 The distillery is mothballed.

1991 Production resumes.

1993 A 10 year old Royal Brackla is launched in United Distillers' Flora & Fauna series.

1997 UDV spends more than £2 million on improvements and refurbishing.

1998 Bacardi–Martini buys Dewar's from Diageo.

2004 A new 10 year old is launched.

2014 A 35 year old is released for Changi airport in Singapore.

2015 A new range is released; 12, 16 and 21 year old.

2019 A limited 20 year old is released.

Tasting notes Royal Brackla 12 years old:

GS – Warm spices, malt and peaches in cream on the nose. The palate is robust, with spice and mildly smoky soft fruit. Quite lengthy in the finish, with citrus fruit, mild spice and cocoa powder.

12 years old

Royal Lochnagar

[royal loch•nah•gar]

Owner:
Diageo

Region/district:
Eastern Highlands

Founded: **Status:**
1845   Active (vc)

Capacity:
500 000 litres

Address: Crathie, Ballater, Aberdeenshire AB35 5TB

Website:
malts.com

Tel:
01339 742700

Of the 28 malt distilleries owned by Diageo, Royal Lochnagar is definitely the smallest! At the same time it is the home of Diageo's Malt Advocate courses, where employees, during a five-day course, can learn about all aspects of whisky.

The distillery is set in beautiful surroundings with Royal Deeside and the imposing Lochnagar Mountain situated to the south and Balmoral, the Queen's summer residence, just a stone's throw to the north,

The distillery is equipped with a 5.4 ton open, traditional stainless steel mash tun. There are two wooden washbacks (a third that hadn't been used for years has now been removed), with short fermentations of 70 hours and long ones of 110 hours. The two stills are quite small with a charge in the wash still of 6,100 litres and 4,000 litres in the spirit still and the spirit vapours are condensed in cast iron worm tubs. The whole production is filled on site with 1,000 casks being stored in its only warehouse, while the rest is sent to Glenlossie. Four mashes per week during 2019 will result in 450,000 litres of pure alcohol.

Royal Lochnagar is the signature malt of Windsor, the best selling Scotch blend in Korea since 2006. first launched in 1996 the brand peaked in 2014 in terms of sales when 8.4 million bottles were sold. Since then the Korean economic slowdown has affected sales and one way to mitigate this has been to launch whiskies with a lower abv (35%). This special version is called W and there are now three varieties W Ice, W Signature 12 years and W Signature 17 years.

The official core range of single malts consists of the **12 year old** and **Selected Reserve**. The latter is a vatting of casks, usually around 18-20 years of age. In autumn 2015 one of the oldest bottlings from the distillery was launched, a **36 year old** single cask. In February 2019, well before the last episodes of Game of Thrones, Diageo launched a series of eight single malts named after the Houses of Westeros. A **12 year old** Royal Lochnagar, **House Baratheon**, was one of them.

History:

1823   James Robertson founds a distillery in Glen Feardan on the north bank of River Dee.

1826   The distillery is burnt down by competitors but Robertson decides to establish a new distillery near the mountain Lochnagar.

1841   This distillery is also burnt down.

1845   A new distillery is built by John Begg, this time on the south bank of River Dee. It is named New Lochnagar.

1848   Lochnagar obtains a Royal Warrant.

1882   John Begg passes away and his son Henry Farquharson Begg inherits the distillery.

1896   Henry Farquharson Begg dies.

1906   The children of Henry Begg rebuild the distillery.

1916   The distillery is sold to John Dewar & Sons.

1925   John Dewar & Sons becomes part of Distillers Company Limited (DCL).

1963   A major reconstruction takes place.

2004   A 30 year old cask strength from 1974 is launched in the Rare Malts series (6,000 bottles).

2008   A Distiller´s Edition with a Moscatel finish is released.

2010   A Manager´s Choice 1994 is released.

2013   A triple matured expression for Friends of the Classic Malts is released.

2016   A distillery exclusive without age statement is released.

2019   House Baratheon is released as part of the Game of Thrones series.

House Baratheon
12 years old

Tasting notes Royal Lochnagar 12 years old:

GS – Light toffee on the nose, along with some green notes of freshly-sawn timber. The palate offers a pleasing and quite complex blend of caramel, dry sherry and spice, followed by a hint of liquorice before the slightly scented finish develops.

Scapa

[ska•pa]

Owner:
Chivas Brothers
(Pernod Ricard)

Region/district:
Highlands (Orkney)

Founded: **Status:** **Capacity:**
1885 Active 1 300 000 litres

Address: Scapa, St Ola, Kirkwall, Orkney KW15 1SE

Website: **Tel:**
scapawhisky.com 01856 876585

There was a chance of Scapa and Glendronach becoming stablemates with Bruichladdich 15 years ago when a huge ownership restructuring saw a number of distilleries changing hands.

Pernod Ricard had been eying Allied Domecq since 1999 and in a joint venture with Fortune Brands in 2005 they finally took over the company. Allied owned Scotch brands such as Teachers and Ballantines at that time, as well as Beefeater, Courvoisier and Canadian Club together with a number of malt distilleries including Scapa and Glendronach. The £7.4bn deal went through but not without competition regulations being enforced. A deal between the two buyers where they split the assets between them was decided but Mark Reynier, who had bought Bruichladdich from Fortune Brands five years earlier, saw an opportunity of acquiring more distilleries. He placed a bid on Scapa and Glendronach but in the end both distilleries landed with Pernod Ricard.

The equipment consists of a 2.9 ton semi-lauter mash tun with a copper dome, twelve washbacks and two stills. Until recently, there were eight washbacks with four of them made from Corten steel. With the Corten replaced by stainless steel and another four added, there are now twelve made of stainless steel. At the same time, the boiler has also been replaced. Due to increased production, fermentation time was down to 52 hours from the previous 160 but with the additional washbacks, it is now possible to increase fermentation time again. The wash still, sourced from Glenburgie distillery in 1959, is only one of two surviving Lomond stills in the industry but on the Scapa still, the adjustable plates were removed in 1979.

The previous 16 year old was replaced by **Scapa Skiren** in 2015. Matured in first fill bourbon, it doesn't carry an age statement. This was followed up in autumn 2016 with **Scapa Glansa**, matured in American oak and then finished in casks that previously held peated whisky. There are also no less than seven different cask strength bottlings in the Distillery Reserve Collection, available at all Chivas´ visitor centres from **10** to **25 years old** - two from sherry butts and the rest matured in ex-bourbon.

History:

1885 Macfarlane & Townsend founds the distillery with John Townsend at the helm.

1919 Scapa Distillery Company Ltd takes over.

1934 Scapa Distillery Company goes into voluntary liquidation and production ceases.

1936 Production resumes.

1936 Bloch Brothers Ltd (John and Sir Maurice) takes over.

1954 Hiram Walker & Sons takes over.

1959 A Lomond still is installed.

1978 The distillery is modernized.

1994 The distillery is mothballed.

1997 Production takes place a few months each year using staff from Highland Park.

2004 Extensive refurbishing takes place at a cost of £2.1 million. Scapa 14 years is launched.

2005 Production ceases in April and phase two of the refurbishment programme starts. Chivas Brothers becomes the new owner.

2006 Scapa 1992 (14 years) is launched.

2008 Scapa 16 years is launched.

2015 The distillery opens for visitors and Scapa Skiren is launched.

2016 The peated Glansa is released.

Tasting notes Scapa Skiren:

GS – Lime is apparent on the early nose, followed by musty peaches, almonds, cinnamon, and salt. More peaches on the palate, with tinned pear and honey. Tingling spices in the drying finish, which soon becomes slightly astringent.

Scapa Skiren

Speyburn

[spey•burn]

Owner:
Inver House Distillers
(Thai Beverages plc)

Region/district:
Speyside

Founded:
1897

Status:
Active

Capacity:
4 500 000 litres

Address: Rothes, Aberlour, Morayshire AB38 7AG

Website:
speyburn.com

Tel:
01340 831213

Back in 2009, Old Pulteney was the best selling brand in the Inver House range of single malts. Since then, however, Speyburn has taken over the number one spot every year.

Well over 500,000 bottles were sold in 2018 and while it still flies a bit under the radar in Europe, it's a well-known brand in the USA and has been that for many years.

An impressive expansion of the distillery was completed in 2015. The expansion cost £4m and included a new 6.25 ton stainless steel mash tun. Four of the six wooden washbacks were kept but they have also expanded with no less than 15 washbacks made of stainless steel. Finally, the existing wash still was converted to a spirit still of exactly the same shape as the other one, while a new and much larger wash still was installed. The two spirit stills are connected to a worm tub while the wash still is fitted with a shell and tube condenser. The distillery has recently changed from a five-day week to working seven-days which means the fermentation time will be a flat 72 hours. With 38 mashes per week, the owners are looking to get 4.2 million litres of alcohol in the year. In 1900, Speyburn was the first distillery to abandon floor malting in favour of a new method – drum malting. In the late sixties, the malting closed but the equipment is still there to see, protected by Historic Scotland.

The core range of Speyburn single malt is a **10 year old**, a **15 year old** and **Bradan Orach** without age statement. A new addition to the range, **18 years old**, was launched in December 2018. Although a permanent member of the core range, it is limited to 9,000 bottles per year. In 2015, **Arranta Casks** was released as a limited USA exclusive and this was followed up in 2017 by **Companion Cask**, a series of single casks matured in first fill ex Buffalo Trace bourbon casks. Recently, expressions for the duty free market have also been released. The **10 year old**, a combination of ex-bourbon and ex-sherry is bottled at 46% as opposed to the standard 40% and there is also a **Hopkins Reserve** that has been matured in casks that previously held a peated whisky. In autumn 2018, a **16 year old** aged in ex-bourbon barrels was added to the range.

History:

1897 Brothers John and Edward Hopkins and their cousin Edward Broughton found the distillery through John Hopkins & Co. They already own Tobermory. The architect is Charles Doig. Building the distillery costs £17,000 and the distillery is transferred to Speyburn-Glenlivet Distillery Company.

1916 Distillers Company Limited (DCL) acquires John Hopkins & Co. and the distillery.

1930 Production stops.

1934 Productions restarts.

1962 Speyburn is transferred to Scottish Malt Distillers (SMD).

1968 Drum maltings closes.

1991 Inver House Distillers buys Speyburn.

1992 A 10 year old is launched as a replacement for the 12 year old in the Flora & Fauna series.

2001 Pacific Spirits (Great Oriole Group) buys Inver House for $85 million.

2005 A 25 year old Solera is released.

2006 Inver House changes owner when International Beverage Holdings acquires Pacific Spirits UK.

2009 The un-aged Bradan Orach is introduced for the American market.

2012 Clan Speyburn is formed.

2014 The distillery is expanded.

2015 Arranta Casks is released.

2017 A 15 year old and Companion Casks are launched.

2018 Two expressions for duty free are released - a 10 year old and Hopkins Reserve. A core 18 year old is launched.

10 years old

Tasting notes Speyburn 10 years old:

GS – Soft and elegant on the spicy, nutty nose. Smooth in the mouth, with vanilla, spice and more nuts. The finish is medium, spicy and drying.

Speyside

[spey•side]

Owner:
Speyside Distillers Co.

Region/district:
Speyside

Founded: 1990
Status: Active
Capacity: 600 000 litres

Address: Glen Tromie, Kingussie, Inverness-shire PH21 1NS

Website:
speysidedistillery.co.uk

Tel:
01540 661060

When Harvey's of Edinburgh bought the distillery in 2012 they could instantly benefit from the new owners connections with the Asian market. Since the 1990s, John Harvey McDonough had created an excellent network in the Taiwanese spirits business.

Spey whisky soon became the third biggest single malt in Taiwan and the presence in that market now seems to have aroused interest from mainland China. In April 2019 the owners announced a major distribution deal with a Chinese distributor, Luzhou Laojiao, which will result in the distillery having to raise its production from 600,000 litres to 1 million litres per year. The agreement will make Spey single malt present not only in duty free stores across China but also in the US and other key markets. On top of that, Luzhou Laojiao will also be selling the whisky across China in its own retail stores.

Speyside distillery, set in beautiful surroundings near Kingussie, is equipped with a 4.2 ton semi-lauter mash tun, four stainless steel washbacks with a 70-120 hour fermentation time and one pair of stills. For the last couple of years they have been working a six-day week with a total production of 600,000 litres of alcohol. There are no warehouses on site. Instead, the spirit is tankered away to the company's warehouses in Glasgow.

The core range of Spey single malt is made up of **Tenné** (with a 6 months port finish), **18 year old** (sherry matured), **Chairman's Choice** and **Royal Choice**. The latter two are multi-vintage marriages from both American and European oak. Two new core bottlings were added in 2017; **Trutina** which is a 100% bourbon maturation and **Fumare**, similar to Trutina but distilled from peated barley. Also part of the core range is **Beinn Dubh** which replaced the black whisky Cu Dubh. Recent limited releases include the second batch of **cask strength versions** of Tenné, Trutina and Fumare, a **10 year old bourbon/port marriage**, a **single cask** for Speyside Whisky Festival as well as single cask releases for select markets.

History:

1956 George Christie buys a piece of land at Drumguish near Kingussie.

1957 George Christie starts a grain distillery near Alloa.

1962 George Christie (founder of Speyside Distillery Group in the fifties) commissions the drystone dyker Alex Fairlie to build a distillery in Drumguish.

1986 Scowis assumes ownership.

1987 The distillery is completed.

1990 The distillery is on stream in December.

1993 The first single malt, Drumguish, is launched.

1999 Speyside 8 years is launched.

2000 Speyside Distilleries is sold to a group of private investors including Ricky Christie, Ian Jerman and Sir James Ackroyd.

2001 Speyside 10 years is launched.

2012 Speyside Distillers is sold to Harvey´s of Edinburgh.

2014 A new range, Spey from Speyside Distillery, is launched (NAS, 12 and 18 year old).

2015 The range is revamped again. New expressions include Tenné, 12 years old and 18 years old.

2016 "Byron´s Choice - The Marriage" and Spey Cask 27 are released.

2017 Trutina and Fumare are released.

2019 Cask strength versions of Tenné, Trutina and Fumare are released.

Trutina

Tasting notes Spey Trutina:

IR – A floral nose, with lemon, granola, shortbread and dried grass. A sweet start on the palate, honey, white chocolate, sweet red apples and then ends with a dry, oaky note.

Pioneers of Whisky

Grant Stevely
Founder
Dubh Glas Distillery, Canada

Almost all of the new whisky pioneers have one thing in common. Founding a distillery was not their first experience of working life. Most of them were in the middle of a career doing something completely different when the idea to become a whisky producer surfaced. The same goes for Grant Stevely, founder of Dubh Glas Distillery in the Okanagan Valley of British Columbia in Canada.

Commonly known as Stevely, he had been working for an international ski resort in Banff, Alberta for 18 years. Over the years as part of the management team, Stevely was involved in security, firefighting, risk management and worked as an emergency medical technician. So what made him give that up, I ask him.

"The idea was prompted by a few too many whiskies amongst whisky-appreciating friends. I was looking for a career change and the idea was hatched."

Actually it wasn't Stevely's first and only alternative for a new career. He made a list of requirements that the new job should include; opportunity for growth, something that was social and that he was passionate about and a business he could enjoy into retirement. Starting a whisky distillery ticked all the boxes. Stevely made some research about the costs involved, attended a distilling class in Arizona and decided this would be his future business.

Whisky, in particular Scotch and single malts from all over the world, wasn't all that new to him. He had been to Scotland on whisky trips and hosted whisky tastings at the ski resort. He even tried his hand at whisky distillation doing small batches in the staff accommodations shower at the resort!

The first question for Stevely was where he would build the distillery and how he would fund it.

"I sold almost everything I owned in Alberta and moved to the Okanagan of British Columbia. With the help of a close friend, we purchased the distillery land and I funded the distillery project itself through an amazing financial lender (we think of them as a business partner), Community Futures, that focuses on helping small businesses in rural communities. Our whisky program itself is partially funded by a Cask Futures program that allows investors to purchase portions of barrels in advance at a discounted rate before it comes to maturity."

When I ask him if there were people in the whisky business that he could lean on for advice and help, Stevely confirms that there were many but mentions a few that inspired him in particular;

"I looked at Canadian Whisky trail blazer John Hall, the founder of Forty Creek Distillery and John Glaser of Compass Box. Innovation respecting tradition but not bound by it was reflective in what was being produced and blended by these pioneers. I received advice from Mike Nicolson; long-time retired Scotch whisky distiller, who had retired to Vancouver Island and was very helpful before and during start-up."

The construction of the distillery was a challenge, Stevely admits, but so was also the staffing of the distillery. At first his plan was to

hire people to take care of the tasting room and promotion but he soon realised that the visitors wanted to talk to the owner and distiller. They wanted to hear the story straight from the horse's mouth. To give Stevely more time with consumers, a mashman was hired to help keeping up production levels as the demand increased.

Production and sales of alcohol is a highly regulated and taxed industry in Canada. That didn't come as a surprise and while Stevely says it was a challenge he adds that it was not one that they couldn´t handle. He still would like to see some relaxation in the legislation.

"We, as a small distillery, would like to see some changes that encourages small operations like ours. A reduction or elimination of the Federal Excise Tax of the first 50,000 litres ageing if using 100% Canadian grown agricultural goods would certainly do a lot to increase the viability of our industry."

Talking about the character of the whisky, Stevely explains what he is looking for.

"I certainly wanted to make a distinctive Canadian Single Malt Whisky. Canadian defining the form and character and Single Malt defining the Scottish heritage in the use of malted barley. Our hot and dry climate was going to lead us the way in richer maturation and we have certainly embraced the form and character of our region and grain available."

They use 100% British Columbia malted barley in the making of all their spirits. He also wanted to get richer maturation from ageing whisky in a shorter period of time. The Okanagan Valley is one of the warmest places in Canada. This trade off of richer maturation has increased their Angel's Share to over 10%.

The first release from the distillery was Noteworthy Gin followed by Virgin Spirits Barley, basically a newmake spirit to get attention from customers. The very first single malt from the distillery were the 107 bottles of the peated Against All Odds that were released in June 2019.

Stevely has plans on extending the tasting room and retail area but realised that at the moment storage for glass bottles, and more importantly, barrels of maturing whisky was a higher priority. That's why building a storage warehouse is the next phase of the Dubh Glas project.

I'm curious to know how Stevely is working on getting people to know about their products. He explains;

"I think the biggest thing is to take away the risk of the consumers hard earned money being spent on your product. Allowing them to taste your products is the largest risk reducer and value added option. The more tastings we do, education we share and first hand stories we tell add value in to the cost of the product."

Stevely´s final advice
to someone wanting to start a distillery of their own;

"Other than the obvious? Money - which I'm sure everyone says. I think the most important is to surround yourself with great, helpful people that want to see you succeed. We are only successful so far because of the friends, family and business mentors we have helping us."

Lionel Trudel Photography

Springbank

[spring•bank]

Owner: **Region/district:**
Springbank Distillers   Campbeltown
(J & A Mitchell)

Founded: **Status:**   **Capacity:**
1828   Active (vc)   750 000 litres

Address: Well Close, Campbeltown, Argyll PA28 6ET

Website: **Tel:**
springbankwhisky.com   01586 551710

Whisky enthusiasts have always wondered why a renowned distillery such as Springbank with a capacity of 750,000 litres only produces a mere 15% of that yearly. It seems that they could easily sell way more than that every year.

The point is that the owner, Hedley Wright, a direct descendant of the founder, has always run the company with caution. This means producing a volume that he knows will sell regardless of any ups and downs in the whisky market and without taking any financial risks. With a staff of more than 80 people in a town where employment can be an issue, the company is an important provider of jobs and they are determined to remain so for many years to come. It thus came as a surprise when the company recently increased its production to almost 400,000 litres, split between their two distilleries – Springbank and Glengyle. Apparently, and with good reason, the owners feel secure that there is a market for more of this cult whisky.

The distillery is equipped with a 3.5 ton open cast iron mash tun, six washbacks made of Scandinavian larch with a fermentation time of up to 110 hours, one wash still and two spirit stills. The wash still is unique in Scotland, as it is fired by both an open oil-fire and internal steam coils. Ordinary condensers are used to cool the spirit vapours, except in the first of the two spirit stills, where a worm tub is used. Springbank is also the only distillery in Scotland that malts its entire need of barley using own floor maltings. Currently there are nine warehouses on site (dunnage and racked) shared between Spingbank and its sister distillery Glengyle.

Springbank produces three distinctive single malts with different phenol contents in the malted barley. Springbank is distilled two and a half times (12-15ppm), Longrow is distilled twice (50-55 ppm) and Hazelburn is distilled three times and unpeated. When Springbank is produced, the malted barley is dried using 6 hours of peat smoke and 30 hours of hot air, while Longrow requires 48 hours of peat smoke. In 2019 a total of 275,000 litres will be produced of which 10% is Longrow and 10% Hazelburn.

The core range is **Springbank 10, 15** and **18 year old**, as well as **12 year old cask strength** (latest batch in February 2019). There are also limited but yearly releases of a **21 year old** and a **25 year old**. Other recent limited releases include **Springbank Local Barley 9 years old** and a **15 year old Springbank rum wood**. Longrow is represented by **Longrow without age statement** and the **18 year old** (latest batch in April 2019). A limited yet annual release is **Longrow Red** with an **11 year old Pinot Noir finish** released in February 2019 being the latest. This will be followed by a **13 year old** finished in **Chilean red wine casks**. A limited, unusually old **Longrow (21 years)** was released in autumn 2019. For **Hazelburn**, the core expression is a **10 year old** complemented by limited annual releases. In April 2019 it was a **Hazelburn 14 year old** matured in oloroso casks.

History:

1828   The Reid family, in-laws of the Mitchells (see below), founds the distillery.

1837   The Reid family encounters financial difficulties and John and William Mitchell buy the distillery.

1897   J. & A. Mitchell Co Ltd is founded.

1926   The depression forces the distillery to close.

1933   The distillery is back in production.

1960   Own maltings ceases.

1969   J. & A. Mitchell buys the independent bottler Cadenhead.

1979   The distillery closes.

1985   A 10 year old Longrow is launched.

1987   Limited production restarts.

1989   Production restarts.

1992   Springbank takes up its maltings again.

1997   First distillation of Hazelburn.

1998   Springbank 12 years is launched.

1999   Dha Mhile (7 years), the world's first organic single malt, is launched.

2000   A 10 year old is launched.

2001   Springbank 1965 'Local barley' (36 years), 741 bottles, is launched.

2002   Number one in the series Wood Expressions is a 12 year old with five years on Demerara rum casks.

2004   Springbank 10 years 100 proof is launched as well as Longrow 14 years old, Springbank 32 years old and Springbank 14 years Port Wood.

2005   Springbank 21 years, the first version of Hazelburn (8 years) and Longrow Tokaji Wood Expression are launched.

2006   Longrow 10 years 100 proof, Springbank 25 years, Springbank 9 years Marsala finish, Springbank 11 years Madeira finish and a new Hazelburn 8 year old are released.

History continued:

2007 Springbank Vintage 1997 and a 16 year old rum wood are released.

2008 The distillery closes temporarily. Three new releases of Longrow - CV, 18 year old and 7 year old Gaja Barolo.

2009 Springbank Madeira 11 year old, Springbank 18 year old, Springbank Vintage 2001 and Hazelburn 12 year old are released.

2010 Springbank 12 year old cask strength and a 12 year old claret expression together with new editions of the CV and 18 year old are released.

2011 Longrow 18 year old and Hazelburn 8 year old Sauternes wood expression are released.

2012 Springbank Rundlets & Kilderkins, Springbank 21 year old and Longrow Red are released.

2013 Longrow Rundlets & Kilderkins, a new edition of Longrow Red and Springbank 9 year old Gaja Barolo finish are released.

2014 Hazelburn Rundlets & Kilderkins, Hazelburn 10 year old and Springbank 25 years old are launched.

2015 New releases include Springbank Green 12 years old and a new edition of the Longrow Red.

2016 Springbank Local Barley and a 9 year old Hazelburn barolo finish are released.

2017 Springbank 14 year old bourbon cask and Hazelburn 13 year old sherrywood are released.

2018 Local Barley 10 year old, 14 year old Longrow Sherry Wood and a new Longrow Red are released.

2019 Springbank 25, Hazelburn 14 and a new Longrod Red are released.

Tasting notes Springbank 10 years old:

GS – Fresh and briny on the nose, with citrus fruit, oak and barley, plus a note of damp earth. Sweet on the palate, with developing brine, nuttiness and vanilla toffee. Long and spicy in the finish, coconut oil and drying peat.

Tasting notes Longrow NAS:

GS – Initially slightly gummy on the nose, but then brine and fat peat notes develop. Vanilla and malt also emerge. The smoky palate offers lively brine and is quite dry and spicy, with some vanilla and lots of ginger. The finish is peaty with persistent, oaky ginger.

Tasting notes Hazelburn 10 years old:

GS – Pear drops, soft toffee and malt on the mildly floral nose. Oiliness develops in time, along with a green, herbal note and ultimately brine. Full-bodied and supple on the smoky palate, with barley and ripe, peppery orchard fruits. Developing cocoa and ginger in the lengthy finish.

Longrow 18 years

Springbank 18 years

Hazelburn 10 years

Springbank 10 years

Springbank 25

Longrow Red

Springbank 12 years c.s.　　Longrow NAS

Strathisla

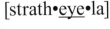

[strath•eye•la]

Owner: **Region/district:**
Chivas Bros (Pernod Ricard)   Speyside

Founded: **Status:**   **Capacity:**
1786   Active (vc)   2 450 000 litres

Address: Seafield Avenue, Keith,
Banffshire AB55 5BS

Website: **Tel:**
chivas.com   01542 783044

Strathisla must be one of the most photographed distilleries in Scotland. With the double pagodas and the water wheel in the foreground it epitomises the classic Scottish whisky distillery.

Until recently, the visitor centre was in the same classic style with dark wood panels and Chesterfield furniture. In the past year, however, the owners have spent time and money changing it into something more contemporary. There is special emphasis on the art of blending since Strathisla is the brand home of Chivas Regal as well. There is also a lab where visitors can try their skills at blending whisky and leave the distillery with a small bottle of their very own blend.

Strathisla is a key malt in the blended Scotch Chives Regal, currently the fourth biggest blend in the world. The brand has gone through some tough years but in 2018 sales were up (+7%) as they sold 54 million bottles. The owners have also launched a campaign with the message that "Success is a blend". One of the messages in the campaign is that "Malt's Good, Grain's Great, Blend's Better". Chivas Distillery Collection is a part of the campaign with a special release of Strathisla 12 year old and the grain whisky Strathclyde 12 year old – both vital components of the Chivas Regal blend.

Strathisla is equipped with a 5.12 ton traditional mash tun with a raised copper canopy, seven washbacks made of Oregon Pine and three of larch – all with a 54 hour fermentation cycle. There are two pairs of stills in a cramped, but very charming stillroom. The wash stills are of lantern type with descending lyne arms and the spirit stills have boiling balls with the lyne arms slightly ascending. Most of the spirit produced at Strathisla is piped to nearby Glen Keith distillery for filling or to be tankered away.

The core expression is the **12 year old** but there are also four cask strength bottlings in the Distillery Reserve Collection, available at all Chivas´ visitor centres – from **13** to **16 years old**.

History:

1786   Alexander Milne and George Taylor found the distillery under the name Milltown, but soon change it to Milton.

1823   MacDonald Ingram & Co. purchases the distillery.

1830   William Longmore acquires the distillery.

1870   The distillery name changes to Strathisla.

1880   William Longmore retires and hands operations to his son-in-law John Geddes-Brown. William Longmore & Co. is formed.

1890   The distillery changes name to Milton.

1942   Jay Pomeroy acquires a majority of the shares in William Longmore & Co. Pomeroy is jailed as a result of dubious business transactions and the distillery goes bankrupt in 1949.

1950   Chivas Brothers buys the run-down distillery at a compulsory auction for £71,000 and starts restoration.

1951   The name reverts to Strathisla.

1965   The number of stills is increased from two to four.

1970   A heavily peated whisky, Craigduff, is produced but production stops later.

2001   The Chivas Group is acquired by Pernod Ricard.

2019   Chivas Distillery Collection Strathisla 12 year old is released.

12 years old

Tasting notes Strathisla 12 years old:

GS – Rich on the nose, with sherry, stewed fruits, spices and lots of malt. Full-bodied and almost syrupy on the palate. Toffee, honey, nuts, a whiff of peat and a suggestion of oak. The finish is medium in length, slightly smoky and a with a final flash of ginger.

Strathmill

[strath•mill]

Owner:	**Region/district:**
Diageo	Speyside
Founded: **Status:**	**Capacity:**
1891 Active	2 600 000 litres

Address: Keith, Banffshire AB55 5DQ

Website:	**Tel:**
malts.com	01542 883000

Ever so often, when travelling around Speyside in search for a distillery, you stumble across other enterprises that are whisky related.

To find Strathmill distillery, entering Keith from the West, you take a right turn just before the A96 crosses the river Isla. On the right hand side of the road is a small cooperage and, being so close to the distillery, it's fair to assume that there is some kind of long-term relationship between the two. But this is not the case. Isla Cooperage, founded by Ricky Proctor as late as in 1989, was sold in 2009 to one of the largest cooperages in the world – Tonnellerie Francois Frères (TFF) which is based in St Romain in the Burgundy region. The year before, the French company had also acquired a much larger and more well-known cooperage in Scotland, Speyside Cooperage.

If you follow the road, you come to a dead end where you will find Strathmill distillery. Climb the tiny slope to take some decent photos but don't expect to walk around the distillery as it doesn't accept visitors. This is a true working distillery with a main purpose of producing malt whisky to be used in blends, in particular J&B which currently is in fifth place on the Top Ten list of blends with close to 40 million bottles sold annually.

The equipment at Strathmill consists of a 9.1 ton stainless steel semi-lauter mash tun and six stainless steel washbacks. Currently, the distillery is working a five-day week which means both short (65 hours) and long (120 hours) fermentations, producing around two million litres of pure alcohol in the year. There are two pairs of stills and Strathmill is one of few distilleries still using purifiers on the spirit stills. This device is mounted between the lyne arm and the condenser and acts as a mini-condenser, allowing the lighter alcohols to travel towards the condenser and forcing the heavier alcohols to go back into the still for another distillation. The result is a lighter spirit. In Strathmill's case both purifiers and condensers are fitted on the outside of the still house to optimise energy savings.

The only official bottling is the **12 year old Flora & Fauna**, but a limited **25 year old** was launched in 2014 as part of the Special Releases.

History:

1891 The distillery is founded in an old mill from 1823 and is named Glenisla-Glenlivet Distillery.

1892 The inauguration takes place in June.

1895 The gin company W. & A. Gilbey buys the distillery for £9,500 and names it Strathmill.

1962 W. & A. Gilbey merges with United Wine Traders (including Justerini & Brooks) and forms International Distillers & Vintners (IDV).

1968 The number of stills is increased from two to four and purifiers are added.

1972 IDV is bought by Watney Mann which later the same year is acquired by Grand Metropolitan.

1993 Strathmill becomes available as a single malt for the first time since 1909 as a result of a bottling (1980) from Oddbins.

1997 Guinness and Grand Metropolitan merge and form Diageo.

2001 The first official bottling is a 12 year old in the Flora & Fauna series.

2010 A Manager's Choice single cask from 1996 is released.

2014 A 25 year old is released.

Tasting notes Strathmill 12 years old:

GS – Quite reticent on the nose, with nuts, grass and a hint of ginger. Spicy vanilla and nuts dominate the palate. The finish is drying, with peppery oak.

12 years old

Talisker

[tal•iss•kur]

Owner: **Region/district:**
Diageo   Highlands (Skye)

Founded: **Status:**   **Capacity:**
1830   Active (vc)   3 300 000 litres

Address: Carbost, Isle of Skye,
Inverness-shire IV47 8SR

Website: **Tel:**
malts.com   01478 614308 (vc)

There used to be a sign in the stillhouse that said "Talisker - the only distillery on the Isle of Skye". This has now been changed from "the oldest" to "the oldest working". The reason for that is the foundation of Torabhaig distillery a couple of years ago.

The owners of Talisker do not seem too worried about the competition. On the contrary, they have, with Torabhaig and the distilleries on Raasay and Harris, created the Hebridean Whisky Trail with the purpose of luring more tourists. Talisker is of course by far the oldest and most well-known of the four. In 2018 the brand sold well over 3 million bottles worldwide and is thereby Diageo's second best selling single malt efter The Singleton.

The distillery is equipped with a stainless steel lauter mash tun with a capacity of 8 tonnes and eight washbacks made of Oregon pine. Before mashing, the malted barley is mixed to a ratio of 25% unpeated and 75% peated which has a phenol specification of 20-25ppm. There are five stills – two wash stills and three spirit stills. Two of the spirit stills were replaced in February 2019. The wash stills are equipped with a special type of purifiers, which use the colder outside air, and have a u-bend in the lyne arm. The purifiers and the peculiar bend of the lyne arms allow for more copper contact and increases the reflux during distillation. All of the stills are connected to wormtubs and Talisker has an unusual system where seawater is used to cool down the hot water in the tubs. Installed a few years ago, it was used for the first time successfully in the extremely hot summer of 2018. The fermentation time is quite long (60-65 hours) and the middle cut from the spirit still is collected between 76% and 65% which, together with the phenol specification, gives a medium peated spirit. In 2019 production will be 20 mashes per week which accounts for 3.3 million litres of alcohol.

Talisker's core range consists of **Skye** and **Storm,** both without age statement, **10, 18, 25** and **30 year old, Distiller's Edition** with an Amoroso sherry finish, **Talisker 57° North** which is released in small batches, and **Port Ruighe**, finished in ruby port casks. There is also **Dark Storm**, the peatiest Talisker so far, which is exclusive to duty free together with **Neist Point**. A new range of limited bottlings was introduced in summer 2018 – the Bodega Series which will explore the impact of different sherry cask finishes. The first installment was a **40 year old** which had been finished in casks that once held 40 year old amontillado sherry. The second release, in July 2019, was a **41 year old** finished in Manzanilla casks that were over 100 years old. This was also the oldest expression ever released from the distillery. An **8 year old** matured in first fill ex-bourbon and bottled at 59.4% was launched in autumn 2018 as part of the yearly Special Releases and **Talisker Select Reserve House Greyjoy** was released in spring 2019 as part of the Game of Thrones series. A **15 year old** appeared in autumn as part of the yearly Special Releases and there is a special bottling exclusively available at the distillery.

History:

1830 Hugh and Kenneth MacAskill found the distillery.

1848 The brothers transfer the lease to North of Scotland Bank and Jack Westland from the bank runs the operations.

1854 Kenneth MacAskill dies.

1857 North of Scotland Bank sells the distillery to Donald MacLennan for £500.

1863 MacLennan experiences difficulties in making operations viable and puts the distillery up for sale.

1865 MacLennan, still working at the distillery, nominates John Anderson as agent in Glasgow.

1867 Anderson & Co. from Glasgow takes over.

1879 John Anderson is imprisoned after having sold non-existing casks of whisky.

1880 New owners are now Alexander Grigor Allan and Roderick Kemp.

1892 Kemp sells his share and buys Macallan Distillery instead.

1894 The Talisker Distillery Ltd is founded.

1895 Allan dies and Thomas Mackenzie, who has been his partner, takes over.

1898 Talisker Distillery merges with Dailuaine-Glenlivet Distillers and Imperial Distillers to form Dailuaine-Talisker Distillers Company.

1916 Thomas Mackenzie dies and the distillery is taken over by a consortium consisting of, among others, John Walker, John Dewar, W. P. Lowrie and Distillers Company Limited (DCL).

1928 The distillery abandons triple distillation.

1960 On 22nd November the distillery catches fire and substantial damage occurs.

1962 The distillery reopens after the fire.

History continued:

1972 Own malting ceases.

1988 Classic Malts are introduced, Talisker 10 years included. A visitor centre is opened.

1998 A new stainless steel/copper mash tun and five new worm tubs are installed. Talisker is launched as a Distillers Edition with an amoroso sherry finish.

2004 Two new bottlings appear, an 18 year old and a 25 year old.

2005 To celebrate the 175th birthday of the distillery, Talisker 175th Anniversary is released. The third edition of the 25 year old cask strength is released.

2006 A 30 year old and the fourth edition of the 25 year old are released.

2007 The second edition of the 30 year old and the fifth edition of the 25 year old are released.

2008 Talisker 57° North, sixth edition of the 25 year old and third edition of the 30 year old are launched.

2009 New editions of the 25 and 30 year old are released.

2010 A 1994 Manager's Choice single cask and a new edition of the 30 year old are released.

2011 Three limited releases - 25, 30 and 34 year old.

2012 A limited 35 year old is released.

2013 Four new expressions are released – Storm, Dark Storm, Port Ruighe and a 27 year old.

2014 A bottling for the Friends of the Classic Malts is released.

2015 Skye and Neist Point are released.

2016 A distillery exclusive without age statement is released.

2018 A 40 year old, the first in the new Bodega Series, and an 8 year old Special Release are launched.

2019 A 41 year old Bodega Series and House Greyjoy in the Game of Thrones series are released.

Tasting notes Talisker 10 years old:

GS – Quite dense and smoky on the nose, with smoked fish, bladderwrack, sweet fruit and peat. Full-bodied and peaty in the mouthy; complex, with ginger, ozone, dark chocolate, black pepper and a kick of chilli in the long, smoky tail.

Tasting notes Talisker Storm:

GS – The nose offers brine, burning wood embers, vanilla, and honey. The palate is sweet and spicy, with cranberries and blackcurrants, while peat-smoke and black pepper are ever-present. The finish is spicy, with walnuts, and fruity peat.

Port Ruighe

Storm

Skye

House Greyjoy

Dark Storm

10 years old

Bodega Series 41 yo

Distiller's Edition

Tamdhu

[tam•doo]

Owner:	**Region/district:**
Ian Macleod Distillers	Speyside

Founded:	**Status:**	**Capacity:**
1897	Active	4 000 000 litres

Address: Knockando, Aberlour,
Morayshire AB38 7RP

Website:	**Tel:**
tamdhu.com	01340 872200

For remote distilleries like Tamdhu and neighbouring Knockando, a railway was essential for bringing in goods such as barley and coal as well as for sending out casks with whisky

The Strathspey Railway was opened in 1863, stretching from Boat of Garten to Dufftown. One of the stations along the way was Dalbeallie which can still be seen next to Tamdhu. The station was later renamed Knockando. The Strathspey Railway was closed in 1966 but the station has from time to time been used as a visitor centre by Tamdhu. At the moment it is opened on rare occasions, working as a distillery shop.

The distillery is equipped with an 11.8 ton semilauter mash tun, nine Oregon pine washbacks with a fermentation time of 59 hours and three pairs of stills. There are a total of 20 warehouses (a mix of dunnage, racked and palletised) and at the moment another four palletised warehouses are being built. Production for 2019 will be 16 mashes per week which translates to 3.1 million litres for the entire year. An on-site cooperage for repairing and testing casks became operational in March 2019.

The owners are committed to maturing Tamdhu single malt in sherry casks but the flavour is further enhanced by the fact that they are using both American and European oak. The core range consists of a **10 year old** matured in first and second fill sherry casks and the non-chill filtered **Batch Strength** (currently the fourth annual release). A **12 year old** was released globally in August 2018 and a limited **15 year old** core bottling appeared later that year. Limited releases include a **50 year old**, matured in a first fill European oak sherry butt released in 2017 to celebrate the 120th anniversary of the distillery. Recently, we have seen the release of part two of **Dalbeallie Dram** and also **Distillery Manager's Edition** which was selected ty Sandy McIntyre who was appointed Manager of the Year 2019 in Whisky Magazine's Icons of Whisky. In 2019 there was also the release of the first travel retail exclusives from the distillery; **Ámbar 14 year old** as well as the **Gran Reserva First Edition**.

History:

1896 The distillery is founded by Tamdhu Distillery Company, a consortium of whisky blenders with William Grant as the main promoter. Charles Doig is the architect.

1897 The first casks are filled in July.

1898 Highland Distillers Company, which has several of the 1896 consortium members in managerial positions, buys Tamdhu Distillery Company.

1911 The distillery closes.

1913 The distillery reopens.

1928 The distillery is mothballed.

1948 The distillery is in full production again in July.

1950 The floor maltings is replaced by Saladin boxes when the distillery is rebuilt.

1972 The number of stills is increased from two to four.

1975 Two stills augment the previous four.

1976 Tamdhu 8 years is launched as single malt.

2005 An 18 year old and a 25 year old are released.

2009 The distillery is motbalded.

2011 The Edrington Group sells the distillery to Ian Macleod Distillers.

2012 Production is resumed.

2013 The first official release from the new owners – a 10 year old.

2015 Tamdhu Batch Strength is released.

2017 A 50 year old is released.

2018 A 12 year old and the Dalbeallie Dram are released.

2019 Two expressions for duty-free - Ámbar and Gran Reserva First Edition.

12 years old

Tasting notes Tamdhu 12 years old:

IR – Distinct sherry notes on the nose with raisins and prunes as well as menthol and green leaves. The taste is wellbalanced with dried fruit, crème brûlée, roasted nuts, bananas and cinnamon.

Tamnavulin

[tam•na•voo•lin]

Owner:	**Region/district:**
Whyte & Mackay (Emperador)	Speyside
Founded: **Status:**	**Capacity:**
1966 Active	4 000 000 litres

Address: Tomnavoulin, Ballindalloch,
Banffshire AB3 9JA

Website:	**Tel:**
www.tamnavulinwhisky.com	01807 590285

One of the most anonymous distilleries in Speyside has recently had its fair share of attention. Not just by way of an increased range of whiskies but also by taking part in the annual Spirit of Speyside Festival.

Driving the B9008 from Bridge of Avon towards Tomintoul, the distillery is on your left hand side after you've passed Glenlivet. The signs and the distillery are there but unfortunately there is no visitor centre yet.

Tamnavulin distillery is equipped with a full lauter mash tun with an 11 ton charge, nine washbacks made of stainless steel with a fermentation time of 54-60 hours (an increase from the previous 48) and three pairs of stills. The wash stills, with horizontal lyne arms, are all equipped with sub-coolers while the spirit stills with their descending lyne arms have purifiers. On the environmental side, the distillery is since September 2018 running on LPG (liquefied petroleum gas) rather than heavy fuel oil and a new bioplant has been installed to take care of the residues from the distillation. During 2019, the owners will be doing 18 mashes per week which means a total of 3.5 million litres of alcohol. From 2010 to 2013, part of the yearly production (around 5%) was heavily peated with a phenol specification in the barley of 55ppm. There has been no peated production since. There are two racked warehouses on site, holding 40,000 casks.

The character of Tamnavulin new make is slightly grassy with a 25 minute foreshot and a middle cut running from 75% down to 60%. These days, the spirit always goes into first fill bourbon and part of it is then finished in sherry casks. However, it hasn't always been that way and recently a re-racking programme has been implemented. The result is a vast improvement of the quality of the whisky.

The core expression, introduced in 2016, is **Double Cask** with a sherry finish. In 2019 a **Sherry Cask Edition** with a finish in three types of oloroso sherry casks has been launched together with a **Tempranillo finish** for duty free. Limited bottlings include **four vintages** (from 1970 to 2000) released exclusively for the Taiwanese market.

History:

1966 Tamnavulin-Glenlivet Distillery Company, a subsidiary of Invergordon Distillers Ltd, founds Tamnavulin.

1993 Whyte & Mackay buys Invergordon Distillers.

1995 The distillery closes in May.

1996 Whyte & Mackay changes name to JBB (Greater Europe).

2000 Distillation takes place for six weeks.

2001 Company management buy out operations for £208 million and rename the company Kyndal.

2003 Kyndal changes name to Whyte & Mackay.

2007 United Spirits buys Whyte & Mackay. Tamnavulin is opened again in July after having been mothballed for 12 years.

2014 Whyte & Mackay is sold to Emperador Inc.

2016 Tamnavulin Double Cask is released.

2019 Sherry Cask Edition and Tempranillo Finish are released.

Tasting notes Tamnavulin Double Cask:

GS – The nose offers malt, soft toffee, almonds and tangerines. Finally, background earthiness. Smooth on the palate, with ginger nut biscuits, vanilla and orchard fruits, plus walnuts. The finish is medium in length, with lingering fruity spice.

Double Cask

Teaninich

[tee•<u>ni</u>•nick]

Owner:
Diageo

Region/district:
Northern Highlands

Founded: 1817

Status: Active

Capacity: 10 200 000 litres

Address: Alness, Ross-shire IV17 0XB

Website: malts.com

Tel: 01349 885001

There are two distilleries in the small town of Alness north of Inverness. As uniquely beautifully situated with its view over Cromarty firth as Dalmore is, as uniquely dull are the surroundings of Teaninich.

The distillery is found in an industrial area south of the town centre, looking very streamlined and efficient. When Alfred Barnard was here in 1896, he told a different story. While Dalmore was described as being merely "favourably situated", he noted that Teaninich was "beautifully situated". Since then, the two distilleries have diverged in their development. In the heydays of Scotch whisky in the early 1970s, Teaninich was expanded in a grand way and was at a time the largest distillery in Scotland. In the year 2000, they were also pioneers when they installed the first mash filter in the Scotch whisky industry.

The mashing technique is what sets Teaninich apart from all other Scotch distilleries save Inchdairnie. The malted barley is ground into a fine flour without husks in an Asnong hammer mill. This is the basis for a higher spirit yield but also means that one can use barley varieties that wouldn't mill as well in a traditional Porteus mill with rollers. The grist is mixed with water in a conversion vessel. Once the conversion from starch to sugar is done, the mash passes through a Meura 2001 mash filter which consists of a number of mesh bags. The filter compresses the bags and the wort is collected for the next step – fermentation.

A huge expansion of the distillery was conducted in 2015 which led to a doubled capacity. The equipment now consists of a filter with a 2 x 7 ton mash. There are also 18 wooden washbacks and two made of stainless steel – all with a fermentation time of 75 hours. Three of the original wash stills were altered into spirit stills so that the old still house now houses all six spirit stills, while a new house was been built for the six, new wash stills. At the moment, the distillery is alternating between 16 and 28 mashes per week which gives a total of four million litres of pure alcohol.

The only official core bottling is a **10 year old** in the Flora & Fauna series but a limited **17 year old** matured in refill American oak was launched in autumn 2017 as part of the Special Releases.

History:

1817 Captain Hugh Monro, owner of the estate Teaninich, founds the distillery.

1831 Captain Munro sells the estate to his younger brother John.

1850 John Munro, who spends most of his time in India, leases Teaninich to the infamous Robert Pattison from Leith.

1869 John McGilchrist Ross takes over the licence.

1895 Munro & Cameron takes over the licence.

1898 Munro & Cameron buys the distillery.

1904 Robert Innes Cameron becomes sole owner of Teaninich.

1932 Robert Innes Cameron dies.

1933 The estate of Robert Innes Cameron sells the distillery to Distillers Company Limited.

1970 A new distillation unit with six stills is commissioned and becomes known as the A side.

1975 A dark grains plant is built.

1984 The B side of the distillery is mothballed.

1985 The A side is also mothballed.

1991 The A side is in production again.

1992 United Distillers launches a 10 year old Teaninich in the Flora & Fauna series.

1999 The B side is decommissioned.

2000 A mash filter is installed.

2009 Teaninich 1996, a single cask in the new Manager´s Choice range is released.

2015 The distillery is expanded with six new stills and the capacity is doubled.

2017 A 17 year old is launched as part of the Special Releases.

10 years old

Tasting notes Teaninich 10 years old:

GS – The nose is initially fresh and grassy, quite light, with vanilla and hints of tinned pineapple. Mediumbodied, smooth, slightly oily, with cereal and spice in the mouth. Nutty and slowly drying in the finish, with pepper and a suggestion of cocoa powder notes.

Tobermory

[tow•bur•mo•ray]

Owner: | **Region/district:**
Distell International Ltd. | Highland (Mull)

Founded: | **Status:** | **Capacity:**
1798 | Active (vc) | 1 000 000 litres

Address: Tobermory, Isle of Mull, Argyllsh. PA75 6NR

Website: | **Tel:**
tobermorydistillery.com | 01688 302647

Tobermory was due for a major refurbishment when it closed in spring 2017. It would take more than two years until it opened up again. This was longer than expected and partly because the owners, Distell International, were simultaneously upgrading one of their other distilleries, Bunnahabhain.

An exciting by-product from the long closure was that the distillery could experiment with gin production. A small pilot still was used to create Tobermory Hebridean Gin which was first released in spring 2019. The gin is a mix of NGS (neutral grain spirit) and Tobermory newmake with the addition of botanicals from the Isle of Mull. A proper, larger gin still will now be installed.

The distillery has been known as Tobermory from 1798 until its closure in 1930. After that it has been alternating between being called Tobermory and Ledaig, and single malt is released under both names with Ledaig being reserved for the peated versions with a phenol content in the barley of 38-40ppm.

Tobermory used to be equipped with a 45 year old traditional five ton cast iron mash tun which was replaced like for like in 2019. There are four, new wooden washbacks with a fermentation time of 48 to 100 hours and two pairs of stills. Two of the stills were replaced in August 2014 and the other two in summer 2019. Before the temporary closure, the production was eight mashes per week and 750,000 litres of alcohol with a 55/45% split between Ledaig and Tobermory. If everything goes according to plan, the goal is to make 500,000 litres during 2019.

The core range from Tobermory distillery consists of the new **12 year old Tobermory** which has replaced the 10 year old and the **10 and 18 year old Ledaig**. Recent limited expressions include a **Tobermory 19 year old Marsala Finish**, a **port matured 13 year old Ledaig** matured in port casks and a **21 year old Ledaig Manzanilla finish**. Available at the distillery at the time of writing were a **2007 Tobermory sherry finish** and a **1997 Ledaig port finish**.

History:

1798 John Sinclair founds the distillery.

1837 The distillery closes.

1878 The distillery reopens.

1890 John Hopkins & Company buys the distillery.

1916 Distillers Company Limited (DCL) takes over John Hopkins & Company.

1930 The distillery closes.

1972 A shipping company in Liverpool and the sherrymaker Domecq buy the buildings and embark on refurbishment. When work is completed it is named Ledaig Distillery Ltd.

1975 Ledaig Distillery Ltd files for bankruptcy and the distillery closes again.

1979 The estate agent Kirkleavington Property buys the distillery, forms a new company, Tobermory Distillers Ltd and starts production.

1982 No production. Some of the buildings are converted into flats and some are rented to a dairy company for cheese storage.

1989 Production resumes.

1993 Burn Stewart Distillers buys Tobermory for £600,000.

2002 CL Financial buys Burn Stewart Distillers.

2005 A 32 year old from 1972 is launched.

2007 A Ledaig 10 year old is released.

2008 A limited edition Tobermory 15 year old is released.

2013 Burn Stewart Distillers is sold to Distell Group Ltd. A 40 year old Ledaig is released.

2015 Ledaig 18 years and 42 years are released together with Tobermory 42 years.

2018 Two 19 year old Ledaig are released.

2019 A 12 year old Tobermory is released, replacing the 10 year old.

Tasting notes Tobermory 12 years old:

IR – Butterscotch and heather honey on the nose with peaches and a hint of orange peel. Mouthcoating, rich and malty with notes of fudge, Danish pastry, citrus, pineapple and a hint of pepper. The finish is slightly salty.

Tasting notes Ledaig 10 years old:

GS – The nose is profoundly peaty, sweet and full, with notes of butter and smoked fish. Bold, yet sweet on the palate, with iodine, soft peat and heather. Developing spices. The finish is medium to long, with pepper, ginger, liquorice and peat.

12 years old

Tomatin

[to•mat•in]

Owner: **Region/district:**
Tomatin Distillery Co   Highland
(Takara Shuzo Co., Kokubu & Co., Marubeni Corp.)

Founded: **Status:**   **Capacity:**
1897   Active (vc)   5 000 000 litres

Address: Tomatin, Inverness-shire IV13 7YT

Website: **Tel:**
tomatin.com   01463 248144 (vc)

In 2005, the owners of Tomatin decided to do something unusual. The very last week of production that year, they distilled a batch of spirit from Optic barley peated to 15ppm. The 60,000 litres were filled into a variety of different cask types.

Eight years later, a completely different range of Tomatin single malt saw the light of day – Cù Bòcan was born. Presented as a stand alone range, a number of differemnt bottlings were released in the following six years. Eventhough the brand was popular, it was decided that a relaunch was due. In summer 2019 three new expressions were launched with Signature being the core bottling complemented by two versions with an unusual maturation history. Creation #1 had been filled into Black Isle Brewery Imperial Stout casks and Bacalhôa Moscatel de Setúbal wine casks while Creation #2 had been maturing in a combination of Japanese Shochu and European Virgin oak casks.

The distillery is equipped with a nine ton stainless steel, full lauter mash tun, 12 stainless steel washbacks with a fermentation time from 54 to 110 hours and six pairs of stills. The goal is to produce 1.8 million litres in 2019, including a couple of weeks of peated production at 40ppm.

The core range consists of **Legacy, 12, 18, 30** and **36 year old**. Included are also **Cask Strength, 14 year old port finish** and a **First Fill Bourbon** exclusive to the UK. In 2018, a **30 year old** was added to the range. Recent limited releases include **Five Virtues (Wood, Fire, Earth, Water** and **Metal)** focusing on the effect of different casks. In July 2018, a **15 year old Tomatin** with a second maturation in Moscatel casks was released. Warehouse 6 Collection is a range of 40-year-old-plus whiskies with **1975 vintage** matured in ex-oloroso casks being the latest. The piece de resistance appeared in late 2018 when a **50 year old** was released. The distillery's duty free range consists of **8, 12, 15** and **40 year olds** while the smoky side of Tomatin is represented by the relaunched **Cù Bòcan Signature, Cù Bòcan Creation #1** and **Cù Bòcan Creation #2**.

History:

1897 The distillery is founded by Tomatin Spey Distillery Company.

1906 Production ceases.

1909 Production resumes through Tomatin Distillers.

1956 Stills are increased from two to four.

1958 Another two stills are added.

1961 The six stills are increased to ten.

1974 The stills now total 23 and the maltings closes.

1985 The distillery company goes into liquidation.

1986 Takara Shuzo Co. and Okara & Co., buy Tomatin through Tomatin Distillery Co.

1998 Okara & Co is liquidated and Marubeni buys out part of their shareholding.

2004 Tomatin 12 years is launched.

2005 A 25 year old and a 1973 Vintage are released.

2006 An 18 year old and a 1962 Vintage are launched.

2008 A 30 and a 40 year old as well as several vintages from 1975 and 1995 are released.

2009 A 15 year old, a 21 year old and four single casks (1973, 1982, 1997 and 1999) are released.

2010 The first peated release - a 4 year old exclusive for Japan.

2011 A 30 year old and Tomatin Decades are released.

2013 Cù Bòcan, the first peated Tomatin, is released.

2014 14 year old port finish, 36 year old, Vintage 1988, Tomatin Cuatro, Cù Bòcan Sherry Cask and Cù Bòcan 1989 are released.

2015 Cask Strength and Cù Bòcan Virgin Oak are released.

2016 A 44 year old Tomatin and two Cù Bòcan vintages (1988 and 2005) are released.

2017 New releases include Wood, Fire and Earth as well as a 2006 Cù Bòcan.

2018 A 30 year old and a 50 year old are released.

2019 The entire range of Cù Bòcan is relaunched with three new expressions.

Tasting notes Tomatin 12 years old:

GS – Barley, spice, buttery oak and a floral note on the nose. Sweet and medium-bodied, with toffee apples, spice and herbs in the mouth. Medium-length in the finish, with sweet fruitiness.

Cù Bòcan Creation #2

Tomintoul

[tom•in•towel]

Owner:	**Region/district:**
Angus Dundee Distillers	Speyside
Founded: **Status:**	**Capacity:**
1965 Active	3 300 000 litres

Address: Ballindalloch, Banffshire AB37 9AQ

Website:	**Tel:**
tomintouldistillery.co.uk	01807 590274

In the last two decades, Angus Dundee Distillers have done an excellent job in promoting two distilleries that were hardly known to the public before. Affordable and well-made single malts have been made available to consumers and sales figures are rising.

But there are also other sides to the company's business. The major part of the production is used either for its blends, or as part of whiskies specially blended for other companies, so called own-label whiskies. These companies include super markets around the world. A third side of the business is exporting in bulk. This means selling the whisky as Scotch blend, blended malt or blended grain. According to the rules of the SWA, single malt Scotch must be bottled in Scotland and cannot be sold in bulk. Angus Dundee Distillers was founded by Terry Hillman in 1988 and now, aged 86, he is still a member of the board. The day to day business however is run by his two children, Aaron and Tania.

Tomintoul is equipped with a 12 ton semi lauter mash tun, six stainless steel washbacks with a fermentation time of 54-60 hours and two pairs of stills. There are currently 15 mashes per week, which means that capacity is used to its maximum, a total of 13 warehouses (racked and palletised) have a storage capacity of 120,000 casks. The malt used for mashing is unpeated, but every year since 2001, heavily peated (55ppm) spirit is produced. On the site there is also a blending centre with 14 large blending vats.

The core range consists of **Tlàth** without age statement, **10, 14, 16, 21** and **25 year old**. There are also two finishes; a **12 year old oloroso sherry cask** and a **15 year old port finish**. The peaty side of Tomintoul is represented by **Peaty Tang**, both **without age statement** and as a newly released **15 year old**. As a stand alone range, there is also the heavily peated **Old Ballantruan without age statement** as well as a **10** and **15 year old**. Recent limited releases include a **1977 single sherry cask**, a **Vintage 1973 double wood matured** and **Tomintoul 1965 The Ultimate cask.**

History:

1965 The distillery is founded by Tomintoul Distillery Ltd, which is owned by Hay & MacLeod & Co. and W. & S. Strong & Co.

1973 Scottish & Universal Investment Trust, owned by the Fraser family, buys both the distillery and Whyte & Mackay.

1974 The two stills are increased to four and Tomintoul 12 years is launched.

1978 Lonhro buys Scottish & Universal Investment Trust.

1989 Lonhro sells Whyte & Mackay to Brent Walker.

1990 American Brands buys Whyte & Mackay.

1996 Whyte & Mackay changes name to JBB (Greater Europe).

2000 Angus Dundee plc buys Tomintoul.

2002 Tomintoul 10 year is launched.

2003 Tomintoul 16 years is launched.

2004 Tomintoul 27 years is launched.

2005 The peated Old Ballantruan is launched.

2008 1976 Vintage and Peaty Tang are released.

2009 A 14 year old and a 33 year old are released.

2010 A 12 year old Port wood finish is released.

2011 A 21 year old, a 10 year old Ballantruan and Vintage 1966 are released.

2012 Old Ballantruan 10 years old is released.

2013 A 31 year old single cask is released.

2015 Five Decades and a 40 year old are released.

2016 A 40 year old and Tlàth without age statement are launched.

2017 15 year old Peaty Tang and 15 year old Old Ballantruan are launched.

2018 Tomintoul 1965 The Ultimate Cask is released.

Tasting notes Tomintoul 10 years old:

GS – A light, fresh and fruity nose, with ripe peaches and pineapple cheesecake, delicate spice and background malt. Medium-bodied, fruity and fudgy on the palate. The finish offers wine gums, mild, gently spiced oak, malt and a suggestion of smoke.

10 years old

Tormore

THE TORMORE DISTILLERY

[tor•more]

Owner:
Chivas Bros (Pernod Ricard)

Region/district:
Speyside

Founded: 1958

Status: Active

Capacity: 4 800 000 litres

Address: Tormore, Advie, Grantown-on-Spey, Morayshire PH26 3LR

Website: tormoredistillery.com

Tel: 01807 510244

In the mid 1970s it became fashionable amongst brewing companies to also get involved in whisky production. At the same time several of them made it into the retail side of the business by opening restaurants and pubs.

It all started in 1960 when Scottish & Newcastle took over the newly reopened Jura distillery. In 1972, Watney Mann bought IDV (International Distillers & Vintners) which included Auchroisk, Glen Spey, Knockando and Strathmill and two years later, Whitbread acquired Long John International and with that took control of Tormore, Laphroaig and three other distilleries. Allied Breweries followed the competitors' example in 1976 when they bought William Teacher & Sons and their two malt whisky distilleries, Ardmore and Glendronach. And then, in 1986, perhaps the biggest takeover of them all – the famous Guinness brewery acquired the giant DCL which later became what we know today as Diageo.

Before a 2014 re-launch, there had been two attempts to establish Tormore single malt as a brand. In 1991 it became part of Caledonian Malts, a range introduced by Allied Distillers. It was later replaced in the range by Scapa. The second time was in 2004 when a 12 year old was launched under the name "The Pearl of Speyside". The name had reference to the generous stock of freshwater pearl mussels which you can find in the river Spey that flows close to the distillery.

From the outside, Tormore is without competition the most unusual looking distillery in Scotland, at least until the new Macallan was opened. Following an upgrade in 2012, Tormore is now equipped with a stainless steel full lauter mash tun, 11 stainless steel washbacks and four pairs of stills. Tormore´s fruity and light character is achieved by a clear wort, slow distillation and by using purifiers on all the stills.

The only official bottlings are a **14 year old** bottled at 43% and a **16 year old**, non chill-filtered, bottled at 48%. There is also a **16 year old cask strength** bottling in the Distillery Reserve Collection, available at all Chivas´ visitor centres.

History:

1958 Schenley International, owners of Long John, founds the distillery.

1960 The distillery is ready for production.

1972 The number of stills is increased from four to eight.

1975 Schenley sells Long John and its distilleries (including Tormore) to Whitbread.

1989 Allied Lyons (to become Allied Domecq) buys the spirits division of Whitbread.

1991 Allied Distillers introduce Caledonian Malts where Miltonduff, Glendronach and Laphroaig are represented besides Tormore. Tormore is later replaced by Scapa.

2004 Tormore 12 year old is launched as an official bottling.

2005 Chivas Brothers (Pernod Ricard) becomes new owners through the acquisition of Allied Domecq.

2012 Production capacity is increased by 20%.

2014 The 12 year old is replaced by two new expressions - 14 and 16 year old.

Tasting notes Tormore 14 years old:

GS – Vanilla, butterscotch, summer berries and light spice on the nose. Milk chocolate and tropical fruit on the smooth palate, with soft toffee. Lengthy in the finish, with a sprinkling of black pepper.

14 years old

Tullibardine

[tully•bar•din]

Owner:
Terroir Distillers
(Picard Vins & Spiritueux)

Region/district:
Highlands

Founded: 1949

Status: Active (vc)

Capacity: 3 000 000 litres

Address: Blackford, Perthshire PH4 1QG

Website:
tullibardine.com

Tel:
01764 682252

After a ten year closure, Tullibardine was bought in 2003 by former directors of DCL and United Spirits, Michael Beamish and Doug Ross. They made an excellent job resurrecting the distillery.

Their business plan was to combine the distillery with a large commercial outlet next door. In the following years however, most of the shops were closed not least due to economical difficulties and the property crash which followed the financial crisis in 2008. Even though the owners managed to show positive figures in the balance sheet already in 2007, they confirmed in 2008 that they were considering offers from other companies to buy the distillery.

It would take another six years before the perfect buyer turned up, Picard Vins & Spiritueux. Tullibardine had always struggled with a blurry range of whiskies but the new owners quickly created a core range which seems to have attracted the consumers attention. The remains of the closed outlet have been turned into a visitor centre, a bottling line, a vatting hall, more warehouses and a small cooperage and the distillery also serves as the spiritual home of the owners blended Scotch – Highland Queen.

Tightly fitted into a cramped production area, the equipment consists of a 6.2 ton stainless steel semi-lauter mash tun, nine stainless steel washbacks with a fermentation of 55-60 hours, two 21,000 litre wash stills and two 16,000 litre spirit stills. During 2019, they will be working 26-27 mashes per week which will result in 2.8 million litres of pure alcohol. Initially, the whole production is filled into first fill bourbon.

The core range consists of **Sovereign,** bourbon matured and without age statement, **225 Sauternes finish, 228 Burgundy finish** and **500 Sherry finish** and two older bottlings – a **20 year old and a 25 year old.** A new range called Custodian's Collection was introduced in 2015. The latest in that range, released in 2019, was **Vintage 1964.** The first cask strength Tullibardine made this century was The Murray 2004 vintage. The latest expression, **The Murray Marsala Finish,** was released in October 2018.

History:

1949 The architect William Delmé-Evans founds the distillery.

1953 The distillery is sold to Brodie Hepburn.

1971 Invergordon Distillers buys Brodie Hepburn Ltd.

1973 The number of stills increases to four.

1993 Whyte & Mackay buys Invergordon Distillers.

1994 Tullibardine is mothballed.

1996 Whyte & Mackay changes name to JBB (Greater Europe).

2001 JBB (Greater Europe) is bought out from Fortune Brands by management and changes name to Kyndal (Whyte & Mackay from 2003).

2003 A consortium buys Tullibardine for £1.1 million. The distillery is in production again.

2005 Three wood finishes from 1993, Port, Moscatel and Marsala, are launched together with a 1986 John Black selection.

2006 Vintage 1966, Sherry Wood 1993 and a new John Black selection are launched.

2007 Five different wood finishes and a couple of single cask vintages are released.

2008 A Vintage 1968 40 year old is released.

2009 Aged Oak is released.

2011 Three vintages and a wood finish are released. Picard buys the distillery.

2013 A completely new range is launched – Sovereign, 225 Sauternes, 228 Burgundy, 500 Sherry, 20 year old and 25 year old.

2015 A 60 year old Custodian Collection is released.

2016 A Vintage 1970 and The Murray from 2004 are released.

2017 Vintage 1962 and The Murray Chateauneuf-du-Pape are released.

2018 The Murray Marsala Finish is released.

2019 A Vintage 1964 is released.

Tasting notes Tullibardine Sovereign:

GS – Floral on the nose, with new-mown hay, vanilla and fudge. Fruity on the palate, with milk chocolate, brazil nuts, marzipan, malt, and a hint of cinnamon. Cocoa, vanilla, a squeeze of lemon and more spice in the finish.

Sovereign

Lagg Distillery on Arran

New
distilleries

New distilleries are being opened in Scotland
at a rate we haven´t experienced since the great whisky boom
of the late 19th century. In the first twelve years of the new millennium,
six new malt whisky distilleries opened up in Scotland. In the next six
years, another twentyfour distilleries came on stream and in 2019,
three more distilleries were added! There are at least another
15-20 distilleries in different stages of construction or planning.
In theory, this means that in a couple of years there could be a
total of 140 malt whisky distilleries operating in Scotland.

Wolfburn

[wolf•burn]

Owner:
Aurora Brewing Ltd.

Region/district:
Northern Highlands

Founded: **Status:** **Capacity:**
2013 Active (vc) 135 000 litres

Address: Henderson Park, Thurso,
Caithness KW14 7XW

Website: **Tel:**
wolfburn.com 01847 891051

The most northerly distillery on the Scottish mainland, Wolfburn, is situated in an industrial area on the outskirts of Thurso.

The owners have chosen a site that is situated 350 metres from the ruins of the old Wolfburn Distillery. Construction work commenced in August 2012 and the first newmake came off the stills at the end of January 2013. The distillery manager from the very start, Shane Fraser who used to manage Glenfarclas, has recently left the company to work for another distillery in the USA.

The distillery is equipped with a 1.1 ton semi-lauter stainless steel mash tun with a copper canopy, four stainless steel washbacks with a fermentation time of 70-92 hours, holding 5,500 litres each, one wash still (5,500 litres) and one spirit still (3,600 litres). Wolfburn uses a mix of casks: ex-bourbon quarter casks, ex-bourbon hogsheads as well as barrels and ex-sherry butts.

The main part of the malt is unpeated but since 2014, a lightly peated (10 ppm) spirit has also been produced. The inaugural bottling from the distillery appeared in early 2016 and had a smoky profile due to the fact that it had partly been matured in quarter casks from Islay. This limited release was followed by a more widely available bourbon matured whisky which in September 2016 was re-named Northland. At the same time a second bottling appeared, Aurora, which had been partly matured in oloroso sherry casks. The core range was expanded in 2017 with Morven, the distillery´s first peated whisky and currently their biggest seller, and September 2018 saw the fourth expression being released - Langskip, matured in ex-bourbon barrels and bottled at 58%. A range of limited bottlings started in 2017 with the lightly peated Batch 128 and was followed in 2018 by the unpeated Batch 270 and finally in spring 2019, Batch 375 was launched. The unpeated 4 year old whisky had been matured in a combination of first fill bourbon and second fill oloroso sherry casks.

Kingsbarns

[kings•barns]

Owner:
Wemyss family

Region/district:
Lowlands

Founded: **Status:** **Capacity:**
2014 Active (vc) 600 000 litres

Address: East Newhall Farm, Kingsbarns,
St Andrews KY16 8QE

Website: **Tel:**
kingsbarnsdistillery.com 01333 451300

The plans for this distillery near St Andrews in Fife, were drafted in 2008 and came to fruition in 2014 when the distillery was opened.

The idea was to restore a derelict farm-steading from the late 18th century and turn it into a modern distillery. Planning permission was received in March 2011 and in September 2012, the Scottish government awarded a grant of £670,000. This, in turn, led to the Wemyss family agreeing to inject £3m into the project and becoming the new owners. The family-owned company owns and operates the independent bottling company, Wemyss Malts, and also owns other companies in the field of wine and gin.

Construction began in June 2013 and the distillery was officially opened on 30th November 2014 on St Andrew´s Day. Commissioning of the distillery began in January 2015 with the first casks being filled early in March.

The distillery is equipped with a 1.5 ton stainless steel mash tun, four 7,500 litre stainless steel washbacks with a fermentation time of 72-120 hours, one 7,500 litre wash still and one 4,500 litre spirit still. A slow distillation and an early cut are important to achieve the fruity character. Mainly first fill bourbon barrels are used for maturation together with STR casks (wine barriques that have been shaved, toasted and re-charred). The current yearly production is 200,000 litres of alcohol.

The first release of Kingsbarns single malt was in 2018 when a limited number of bottles were made available to the members of the Founder´s Club. Their first generally available flagship malt, Dream to Dram, was released in early 2019. Since then, various first fill bourbon single casks have been released as well. In summer 2017, a designated gin distillery was opened on site to produce the already succesful Darnley´s Gin.

Ballindalloch

[bal•lin•da•lock]

Owner:
The Macpherson-Grant family

Region/district:
Speyside

Founded:
2014

Status:
Active (vc)

Capacity:
100 000 litres

Address: Ballindalloch, Banffshire AB37 9AA

Website:
ballindallochdistillery.com

Tel:
01807 500 331

In the heart of Speyside, the owners of Ballindalloch Castle, the Macpherson-Grant family, decided in 2012 to turn a steading from 1820 into a whisky distillery.

Previous generations of the family had been involved in distilling from the 1860s and from 1923 to 1965, they owned part of Cragganmore distillery, not far away from the castle. The old farm building was meticulously renovated with attention given to every little detail and the result is an amazingly beautiful distillery which can be seen from the A95 between Aberlour and Grantown-on-Spey.

Ballindalloch distillery takes its water from the nearby Garline Springs and all the barley is grown on the Estate. All of the distillery equipment are gathered on the second floor which makes it easy for visitors to get a good view of the production. The equipment consists of an extraordinary 1 ton semi lauter, copper clad mash tun with a copper dome. There are four washbacks made of Oregon pine where the fermentation time was increased a while ago to increase the fruity character of the spirit. There are now four long fermentations (140 hours) and on short (92 hours). Finally there is a 5,000 litre lantern-shaped wash still and a 3,600 litre spirit still with a reflux ball. Both stills are connected to two wooden worm tubs for cooling the spirit vapours. The distillery came on stream in September 2014 and was officially opened 16th April 2015 by Prince Charles. The distillery is working 5 days a week, making 100,000 litres of alcohol. The idea is to produce a robust and bold whisky, enhanced not least by the use of worm tubs. The first single malt release is expected in 2022. On the 7th of March 2019, the owners filled their 3000th cask.

The distillery is open for visitors by appointment and there is also the opportunity to take part in The Art of Whisky Making, which means spending a day with the crew and learning about whisky from mashing to warehousing.

Ardnamurchan

[ard•ne•mur•ken]

Owner:
Adelphi Distillery Ltd

Region/district:
Western Highlands

Founded:
2014

Status:
Active (vc)

Capacity:
500 000 litres

Address: Glenbeg, Ardnamurchan, Argyll PH36 4JG

Website:
adelphidistillery.com

Tel:
01972 500 285

It takes a good 90 minutes to go from Fort William to the distillery on the Ardnamurchan peninsula north of Mull by car but it's well worth the journey.

Winding single track roads beg for careful driving but the stunning scenery it also a good reason for driving slowly. The distillery is owned by independent bottler Adelphi Distillery. In 2007, they realised that the supply of good whisky could become scarce in years to come for those companies not having a distillery of their own. They decided to build one and part of the reason for choosing this remote site was that the land was owned by one of the directors of Adelphi Distillery, Donald Houston.

Ardnamurchan distillery came on stream on 11th July 2014 and is equipped with a two tonne semi lauter mash tun made of stainless steel with a copper canopy, four wooden washbacks and three made of stainless steel. The initial wooden washbacks were made from oak, having been used as cognac vats in France but two of them were exchanged in 2018 to Oregon pine. The fermentation time is 72-96 hours. There is one wash still (10,000 litres) and one spirit still (6,000 litres) with condensers placed outside and quite recently, sub-coolers made from stainless steel were also fitted. Two different styles of whisky are produced; peated and unpeated and for the peated spirit, the barley has a phenol specification of 30-35ppm. The production goal for 2019 is 350,000 litres of alcohol. There are also plans to malt some of their barley themselves and a malting floor is already in place but hasn't been used so far.

The owners expect to launch their first single malt in 2021 but since 2016 they have released spirit from the distillery under the name AD that is still not legally whisky. The 2018 version was the first to actually contain some 3 year old whisky.

Annandale

[ann•an•dail]

Owner: **Region/district:**
Annandale Distillery Co.   Lowlands

Founded: **Status:**   **Capacity:**
2014   Active (vc)   500 000 litres

Address: Northfield, Annan, Dumfriesshire DG12 5LL

Website: **Tel:**
annandaledistillery.com   01461 207817

In 2010 Professor David Thomson and his wife, Teresa Church, obtained consent from the local council for the building of the new Annandale Distillery in Dumfries and Galloway in the south-west of Scotland.

The old one had been producing since 1836 and was owned by Johnnie Walker from 1895 until it closed down in 1918. From 1924 to 2007, the site was owned by the Robinson family, who were famous for their Provost brand of porridge oats. David Thomson began the restoration of the site in June 2011 with the two, old sandstone warehouses being restored to function as two-level dunnage warehouses. The distillery was in a poor condition and the mash house and the tun room was largely reconstructed while the other buildings were refurbished substantially. The old maltings, with the kiln and original pagoda roof, have been turned into an excellent visitor centre.

Entering the production area of the new distillery is like walking into a beautiful village church. First you run into the 2.5 ton semi-lauter mash tun with an elegant copper dome. Then, with three wooden washbacks (a fermentation time of 72-96 hours) on each side, you are guided up to the two spirit stills (4,000 litres). Once you have reached them, you find the wash still (12,000 litres) slightly hidden behind a wall. The capacity is 500,000 litres per annum.

The first cask was filled on 15 November 2014 and both unpeated and peated (45ppm) whisky is distilled. Finally, in June 2018, two single malts were released, both matured in ex-Buffalo Trace barrels. In both cases the whiskies, the un-peated Man O´Words and the peated (18ppm in the bottle) Man O´Swords were bottled at cask strength. More bottlings, including single casks, have followed. The distillery has also launched two blended whiskies - Nation of Scots and, most recently, Outlaw King celebrating king Robert the Bruce and named after the Netflix film.

Inchdairnie

[inch•dairnie]

Owner: **Region/district:**
John Fergus & Co. Ltd   Lowlands

Founded: **Status:**   **Capacity:**
2015   Active   2 000 000 litres

Address: Whitecraigs Rd, Glenrothes, Fife KY6 2RX

Website: **Tel:**
inchdairniedistillery.com   01595 510010

For Inchdairnie´s distillery manager Ian Palmer there are three key words that govern operation – flavour, innovation and experimentation.

Opened in May 2016, a few miles west of Glenrothes in Fife, the distillery has a capacity of two million litres per year with a possibility of expanding to four million. It is owned by John Fergus & Co. which was founded by Ian Palmer in 2011. Palmer has 40 years of experience in the Scotch whisky industry and his latest position was general manager for Glen Turner. He is a minority share holder, with CES Whisky holding the rest of the shares.

Ian Palmer's unorthodox ideas start already with the equipment. Unusually, the distillery has a hammer mill and a Meura mash filter, instead of a traditional mash tun. There are four washbacks with a fermentation time of 72 hours and one pair of traditional pot stills with double condensers

and after-coolers to increase the copper to spirit ratio. The two stills are complemented by a Lomond still with six plates to provide the opportunity for triple distillation and experimental distillation. A unique yeast recipe is used and high gravity fermentation will create a fruitier character of the newmake. Furthermore, Palmer is working with both the standard spring barley as well as winter barley to give the possibility for a broader palette of flavours.

And it does not stop at that. In November 2017, it was revealed that the distillery was working on a rye whisky which will be called Ryelaw once bottled and in June 2019, the first distillation of a whisky made from oat was made. This was the first time in over a century that whisky made from oat was produced in Scotland. Two main styles of whisky will be produced. Strathenry (80% of the production both unpeated and peated) will be used for blended whisky while Inchdairnie will be sold as a single malt.

Daftmill

[daf•mil]

Owner:
Francis Cuthbert

Region/district:
Lowlands

Founded: 2005   **Status:** Active   **Capacity:** c 65 000 litres

Address: By Cupar, Fife KY15 5RF

Website:
daftmill.com

Tel:
01337 830303

The distillery may be one of the smallest in Scotland but few single malt releases have been more eagerly awaited by the whisky enthusiasts than the inaugural release from Daftmill.

Ever since December 2008, when the spirit legally became whisky, questions to the owners Francis and Ian Cuthbert about when the first whisky would be launched have always been answered by "when it's ready". In 2017, they signed a distribution agreement with Berry Brothers and in May 2018, a ballot was opened for buying one of the first 629 bottles of a 12 year old matured in ex-bourbon casks. The first release was followed by a Summer Relase in June where seven casks rendered 1665 bottles. More bottlings followed and in May 2019, six different single casks were released - all distilled in 2006.

Daftmill's first distillation was on 16th December 2005 and it is run as a typical farmhouse distillery. The barley is grown on the farm and they also supply other distilleries. The malting is done without peat at Crisp's in Alloa. The equipment consists of a one tonne semi-lauter mash tun with a copper dome, two stainless steel washbacks with a fermentation between 72 and 100 hours and one pair of stills with slightly ascending lyne arms. The equipment is designed to give a lot of copper contact, a lot of reflux. The wash still has a capacity of 3,000 litres and the spirit still 2,000 litres and around 100 casks are filled very year.

The Cuthbert's aim is to do a light, Lowland style whisky. In order to achieve this they have very short foreshots (five minutes) and the spirit run starts at 78% to capture all of the fruity esters and already comes off at 73%. Taking care of the farm obviously prohibits Francis from producing whisky full time. His silent season is during spring and autumn when work in the fields take all of his time. Whisky production is therefore reserved for two months in the summertime and two in the winter.

Abhainn Dearg

[aveen jar•rek]

Owner:
Mark Tayburn

Region/district:
Highlands (Isle of Lewis)

Founded: 2008   **Status:** Active   **Capacity:** c 20 000 litres

Address: Carnish, Isle of Lewis,
Na h-Eileanan an Iar HS2 9EX

Website:
abhainndearg.co.uk

Tel:
01851 672429

In September 2008, spirit flowed from a newly constructed distillery in Uig on the island of Lewis in the Outer Hebrides.

This was the first distillery on the island since 1840 when Stornoway distillery was closed. The conditions for new distilleries being built at that time were not improved when James Matheson, a Scottish tradesman, bought the entire island in 1844. Even though he had made his fortune in the opium trade, he was an abstainer and a prohibitionist and did not look kindly on the production or use of alcohol.

The Gaelic name of the new distillery is Abhainn Dearg which means Red River, and the founder and owner is Mark "Marko" Tayburn who was born and raised on the island. Part of the distillery was converted from an old fish farm while some of the buildings are new. There are two 500 kg mash tuns made of stainless steel and two 7,500 litre washbacks made of Douglas fir with a fermentation time of 4 days. The wash still has a capacity of 2,112 litres and the spirit still 2,057 litres. Both have very long necks and steeply descending lye pipes leading out into two wooden worm tubs. Both bourbon and sherry casks are used for maturation. The plan is to use 100% barley grown on Lewis and in 2013 the first 6 tonnes of Golden Promise (15% of the total requirement) were harvested. Over the years, production has been limited to around 10,000 litres of pure alcohol yearly even though the distillery has the capacity to do more.

The first release from the distillery was The Spirit of Lewis (matured for a short time in sherry casks) in 2010 and the first single malt was a limited release of a 3 year old in October 2011, followed up by a cask strength version (58%) in 2012. The distillery's first 10 year old appeared in late 2018 when 10,000 bottles were released, bottled at 46%. At the same time 100 bottles of a limited 10 year old single cask, also bottled at 46% were launched.

Ailsa Bay

[ail•sah bey]

Owner:
William Grant & Sons

Region/district:
Lowlands

Founded: **Status:**
2007 Active

Capacity:
12 000 000 litres

Address: Girvan, Ayrshire KA26 9PT

Website:
-

Tel:
01465 713091

Commissioned in September 2007, it only took nine months to build this distillery on the same site as Girvan Distillery near Ayr on Scotland's west coast.

Initially, it was equipped with a 12,1 tonne full lauter mash tun, 12 washbacks made of stainless steel and eight stills. In 2013 however, it was time for a major expansion when yet another mash tun, 12 more washbacks and eight more stills were commissioned, doubling the capacity to 12 million litres of alcohol.

Each washback will hold 50,000 litres and fermentation time is 60 hours for the heavier styles and 72 hours for the lighter "Balveniestyle". The stills are made according to the same standards as Balvenie's and one of the wash stills and one of the spirit stills have stainless steel condensers instead of copper. That way, they have the possibility of

making batches of a more sulphury spirit if desired. To increase efficiency and to get more alcohol, high gravity distillation is used. The wash stills are heated using external heat exchangers but they also have interior steam coils. The spirit stills are heated by steam coils. In 2019, the distillery will be doing around 50 mashes per week, producing 10 million litres of alcohol.

Five different types of spirit are produced. The most common is a light and rather sweet spirit. Then there is a heavy, sulphury style and three peated with the peatiest having a malt specification of 50ppm. The production is destined to become a part of Grant's blended Scotch but in 2016, a peated single malt Ailsa Bay was released. In September 2018 Ailsa Bay Sweet Smoke was launched. Definitely sweeter and slighly smokier, it replaced the inaugural bottling. The ppm on the label (22) is the actual phenol content of the liquid itself and not the barley.

Roseisle

[rose•eyel]

Owner:
Diageo

Region/district:
Highlands

Founded: **Status:**
2009 Active

Capacity:
12 500 000 litres

Address: Roseisle, Morayshire IV30 5YP

Website:
-

Tel:
01343 832100

Roseisle distillery is located on the same site as the already existing Roseisle maltings just west of Elgin. The distillery has won several awards for its ambition towards sustainable production.

The distillery is equipped with two stainless steel, full lauter mash tuns with a 12.5 tonne charge each. There are 14 huge (115,500 litres) stainless steel washbacks and 14 stills with the wash stills being heated by external heat exchangers while the spirit stills are heated using steam coils. The spirit vapours are cooled through copper condensers but on three spirit stills and three wash stills there are also stainless steel condensers attached, that you can switch to for a more sulphury spirit. The fermentation time for a Speyside style of whisky is 90-100 hours and for a heavier style it is 50-60 hours. The plan for 2019 is to do 23 mashes per week and a total of 12 million litres of alcohol.

The total cost for the distillery was £40m and how to use the hot water in an efficient way was very much a focal point from the beginning. For example, Roseisle is connected by means of two long pipes with Burghead maltings, 3 km north of the distillery. Hot water is pumped from Roseisle and then used in the seven kilns at Burghead and cold water is then pumped back to Roseisle. The pot ale from the distillation will be piped into anaerobic fermenters to be transformed into biogas and the dried solids will act as a biomass fuel source. The biomass burner on the site, producing steam for the distillery, covers 72% of the total requirement. Furthermore, green technology has reduced the emission of carbon dioxide to only 15% of an ordinary, same-sized distillery.

Destined to be used for blends, Roseisle single malt was in autumn 2017, for the first time used in a different role. It was part of the blended malt Collectivum XXVIII where Diageo had used whiskies from all 28 malt distilleries.

Strathearn

[strath•earn]

Owner:
Tony Reeman-Clark

Region/district:
Southern Highlands

Founded: 2013
Status: Active
Capacity: c 30 000 litres

Address: Bachilton Farm Steading, Methven PH1 3QX

Website:
strathearndistillery.com

Tel:
01738 840 100

This is something as unique as Scotland's first micro-distillery. Abhainn Dearg on the Isle of Lewis has the same capacity, but the stills at Strathearn are considerably smaller.

The brainchild of Tony Reeman-Clark, it is situated a couple of miles west of Methven near Perth. Gin production was started in August 2013 and the first whisky was filled into casks in October. The distillery uses the Maris Otter barley which was abandoned by other distillers years ago due to the low yield. Reeman-Clark prefers it though, because of the flavours that it contributes. All the equipment is fitted into one room and consists of a stainless steel mash tun, two stainless steel washbacks with a fermentation time of 4-5 days, one 1,000 litre wash still and a 500 litre spirit still. Both stills are of the Alambic type with vertical tube copper condensers. When they are producing gin, they simply detach the lyne arm and mount a copper basket

to the still to hold the botanicals. On the whisky side, both peated (35ppm) and un-peated whisky is produced and for maturation a variety of 50-100 litre casks are used; virgin French oak, virgin American oak and ex-sherry casks.

Reeman-Clark has also been experimenting with other types of wood like chestnut, mullberry and cherry. According to the rules, spirit matured in anything other than oak, cannot be called Scotch whisky. This problem was solved by labelling the content Uisge Beatha – the ancient name for Scotch. The first single malt Scotch from the distillery was released in December 2016. One hundred 50cl bottles were put up for auction and they were sold for a median price of £333. Another batch was released in September 2017. Several gins have been released over the years with Scottish Gin and Heather Rose being the big sellers and in autumn 2017, Dunedin rum was released. The distillery also runs a private cask club where you can buy your own cask of maturing whisky.

Eden Mill

[eden mill]

Owner:
Paul Miller

Region/district:
Lowlands

Founded: 2014
Status: Active (vc)
Capacity: 100 000 litres

Address: St Andrews, Fife, KY16 0UU

Website:
edenmill.com

Tel:
01334 834038

In 2012, Paul Miller, the former Molson Coors Sales Director, with a background in the whisky industry, opened up the successful Eden Brewery in Guardbridge, west of St Andrews.

The site was an old paper mill and only 50 metres away, there was a distillery called Seggie which was operative between 1810 and 1860 and owned by the Haig family. As an extension of the brewery, Paul decided to build a distillery called Eden Mill Distillery. The distillery, with a capacity of 80,000 litres per year, mainly produces malt whisky, but gin is also on the map. The distillery is equipped with two wash stills and one spirit still of the alembic type. Made by Hoga in Portugal, all three stills are of the same size – 1,000 litres. Eden Mill is the first combined brewery and distillery in Scotland. The brewery/distillery also has a visitor centre which already attracts 20,000 visitors a year.

Whisky production started in 2014 and the first release of a single malt appeared in 2018. Matured in a combination of French virgin oak, American virgin oak, and Pedro Ximenez casks it sold out instantly with bottle No. 1 going for £7,100 at an auction. The second bottling, aged in a variety of casks, appeared in November 2018. Meanwhile a series of seven different 20cl bottlings called the Hip Flask Series was launched. All of them had been made from different mashbills and had matured in different types of casks.

In 2018, the owners announced that they were about to move the entire operation to another building on the same site. The shop and tasting rooms moved to the Gatehouse in summer 2018 as a temporary pop-up and in spring 2019 a new, permanent visitor centre opened. In June 2019, the owners said that £3.1m had been set aside for an expansion of the production side of the distillery which would double the capacity for whisky distilling to 200,000 litres.

Dalmunach

[dal•moo•nack]

Owner:	**Region/district:**
Chivas Brothers	Speyside
Founded:	**Status:**	**Capacity:**
2015	Active	10 000 000 litres

Address: Carron, Banffshire AB38 7QP

Website:	**Tel:**
-	-

One of the newest distilleries in Scotland, and one of the most beautiful, has been built on the site of the former Imperial distillery.

Imperial distillery was founded in 1897, the year of Queen Victoria´s Diamond Jubilee and on the top of the roof there was even a large cast iron crown to mark the occasion. The founder was Thomas Mackenzie who at the time already owned Dailuaine and Talisker. The timing was not the best though. One year after the opening, the Pattison crash brought the whisky industry to its knees and the distillery was forced to close. Eventually it came into the hands of DCL who owned it from 1916 until 2005, when Chivas Brothers took over. It was out of production for 60% of the time until 1998 when it was mothballed. The owners probably never planned to use it for distillation again as it was put up for sale in 2005 to become available as residential flats. Soon after, it was withdrawn from the market and, in 2012, a decision was taken to tear down the old distillery and build a new. Demolition of the old distillery began in 2013 and by the end of that year, nothing was left, except for the old warehouses.

Construction on the new Dalmunach distillery started in 2013 and it was commissioned in October 2014. The exceptional and stunning distillery is equipped with an efficient (4 hour mash) 13 ton Briggs full lauter mash tun and 16 stainless steel washbacks charged with 56,000 litres and with a fermentation time of 56-62 hours. There are four pairs of stills of a considerable size - wash stills 28,000 litres and spirits stills 18,000. They are all positioned in a circle with a hexagonal spirit safe in the middle. The distillery, which cost £25m to build, is the company´s most ennergy efficient distillery and uses 38% less energy than the indutsry average. has a capacity of 10 million litres. In autumn 2019, the first official release of Dalmunach single malt appeared – a 4 year old bottled at cask strength.

Glasgow

[glas•go]

Owner:	**Region/district:**
Liam Hughes, Ian McDougall	Lowlands
Founded:	**Status:**	**Capacity:**
2015	Active	270 000 litres

Address: Deanside Rd, Hillington, Glasgow G52 4XB

Website:	**Tel:**
glasgowdistillery.com	0141 4047191

When Glasgow Distillery was opened in Hillington Business Park, it became the first new whisky distillery in Glasgow in modern times.

There were stills within the Strathclyde grain distillery producing the malt whisky Kinclaith from 1958-1975 but Liam Hughes, Mike Hayward and Ian McDougall had the intention of building the first proper malt distillery in Glasgow in more than hundred years. Backed up by Asian investors, the distillery was ready to start production in February 2015.

The first product to be bottled was the Makar gin which now exists in several versions. The owners have also bottled old (26-28 years), sourced single malts under the name Prometheus. The first single malt from their own production appeared in June 2018. Aged in ex-bourbon barrels and finished in virgin oak, the whisky was called 1770 Glasgow Single Malt, named after Glasgow´s first distillery which was founded at Dundashill in 1770. This was followed by a second edition in spring 2019. The 1770 peated version, matured in ex-sherry butts with a finish in virgin oak was launched in September 2019 with a triple distilled expression planned for early 2020.

The first distillation of whisky was unpeated but since then peated spirit (50ppm) is also part of the production and since January 2017, triple distillation is also practised one month per year. The distillery is equipped with a one ton mash tun, seven wash backs (5,400 litres each) with a minimum fermentation of 72 hours, one 2,500 litre wash still, one 1,400 litre spirit still and one 450 litre gin still - all from Firma Carl in Germany. Starting with 75,000 litres, since 2017 they are more or less on full production and an already planned expansion is due to be finished by 2019. This means yet another pair of stills, seven more washbacks and a total capacity of 500,000 litres.

Harris

[har•ris]

Owner:		Region/district:
Isle of Harris Distillers Ltd.		Highlands (Isle of Harris)
Founded:	**Status:**	**Capacity:**
2015	Active (vc)	399 000 litres

Address: Tarbert, Isle of Harris,
Na h-Eileanan an Iar HS3 3DJ

Website:	Tel:
harrisdistillery.com	01859 502212

More than ten years ago, Anderson Bakewell had conjured up an idea to build a distillery on the Isle of Harris.

Joining Bakewell, who had been connected to the island for more than 40 years, was Simon Erlanger, a former marketing director for Glenmorangie and now the MD of the new distillery. Construction started in 2014 and the distillery came into production in September 2015. The total cost for the whole project was £11.4m. The distillery, located in Tarbert, was the second distillery after Abhainn Dearg on Lewis to be located in the Outer Hebrides.

Together with three other distilleries (Talisker and Torabhaig on Skye and Isle of Raasay), Harris distillery launched a new whisky route called Hebridean Whisky Trail on 15th August 2018 – www.hebrideanwhisky.com. In 2018, the distillery had no less than 91,000 visitors!

The equipment consists of a 1.2 tonne semi lauter mash tun made of stainless steel but clad with American oak and 8 washbacks made of Oregon pine with a fermentation time of 72-96 hours. Three of the washbacks were installed recently, increasing the capacity by 75%. There are also one 7,000 litre wash still and a 5,000 litre spirit still - both with descending lyne arms and made in Italy. In 2018, the distillery increased production from 5 to 7 mashes per week and the plan is to go to 9 mashes during 2019.

The style of the whisky, which will be called Hearach (the Gaelic word for a person living on Harris), will be medium peated with a phenol specification in the barley of 12-14ppm although in 2018 they distilled a batch of heavily peated malt (30ppm) made with Isle of Harris peat. The first spirit to be distilled in September 2015 was gin and this was followed by whisky in December. The gin has already been released and apart from traditional botanicals, local ingredients are also used such as sugar kelp.

Brew Dog

[bru•dog]

Owner:		Region/district:
Brewdog plc.		Highlands
Founded:	**Status:**	**Capacity:**
2016	Active	450 000 litres

Address: Balmacassie Commercial Park, Ellon,
Aberdeenshire AB41 8BX

Website:	Tel:
lonewolfspirits.com	01358 724924

In spring 2019, the distillery changed the name from Loan Wolf to Brew Dog in order to tap into the name and fame of the well-known brewery

Founded in 2007 by James Watt and Martin Dickie, Brew Dog grew to become the biggest independent brewery in the UK and in 2014 a decision was taken to open up also a distillery. It is situated next to the brewery in Ellon outside of Aberdeen and as distillery manager, Steven Kearsley who had a background at several Diageo distilleries, was called in. In autumn 2018, David Gates who previously ran Diageo Futures and worked as brand director for Johnnie Walker, joined the company as managing director.

The adjacent brew house provides the wash for the distillery which has the following equipment; one 3,000 litre pot still with an 8 plate rectification column which will be used for stripping the wash for vodka, whisky and rum, another 3,000 litre still with a 60-plate column is used for the final distillation of vodka and whisky, a 600 litre pot still is dedicated to gin and brandy production, while a 50 litre pot still is used for research and experimentation. Apart from malt whisky, grain and rye, bourbon style whiskey, vodka, gin and rum are produced.

The initial production was gin and vodka and the first bottles were launched in spring 2017. Whisky and rum production has also commenced. In 2017, Lone Wolf became one of the first Scottish distilleries in modern times to distill a rye whisky. In spring 2019, the company entered into a collaboration with three other whisky makers who all designed one whisky each to be paired with Brew Dog beers. The Boilermaker series is made up by Transistor blended Scotch from Compass Box, Torpedo Tulip, a 100% rye whisky from Millstone and a blended Scotch named Skeleton Key from Duncan Taylor.

Arbikie

[ar•bi•ki]

Owner:
The Stirling family

Region/district:
Eastern Highlands

Founded: 2015
Status: Active
Capacity: 200 000 litres

Address: Inverkeilor, Arbroath, Angus DD11 4UZ

Website:
arbikie.com

Tel:
01241 830770

The Stirling family has been farming since the 17th century and the 2000-acre Arbikie Highland Estate in Angus has now been in their possession for four generations.

The three brothers (John, Iain and David) started their careers within other fields but have now returned to the family lands to open up a single-estate distillery. The definition of a single-estate distillery is that, not only does the whole chain of production take place on site, but all the ingredients are also grown on the farm. Ballindalloch is one example but Arbikie is the first to produce both brown and white spirits.

The first vodka from potatoes was distilled in October 2014 which was followed by gin in May 2015. Trials with malt whisky, started in March 2015, went over to full production in October 2015. Responsible for the production side at the distillery is master distiller Kirsty Black.

The barley is grown in fields of their own and then sent to Boorts malt in Montrose. The distillery is equipped with a stainless steel, semi-lauter mash tun with a 0.75 ton charge and four washbacks (two 4,400 litre and two 9,000 litre) with a fermentation time of 96-120 hours. There is also one 4,000 litre wash still and one 2,400 litre spirit still. For the final stage of vodka and gin production, there is a 40 plate rectification column. The whisky is matured in ex bourbon barrels and ex sherry hogsheads. The Stirlings don't intend to launch their first single malt whisky until 2029/2030.

In common with a few other distilleries in Scotland, Arbikie started trials with rye whisky production in December 2015. It was made from 52% unmalted rye, 33% unmalted wheat and 15% malted barley grown on their own farm. Matured for three years in American oak and finished in ex-PX casks, it was released in December 2018 as the first rye whisky made in Scotland for more than 100 years.

Dornoch

[dor•nock]

Owner:
Phil and Simon Thompson

Region/district:
Northern Highlands

Founded: 2016
Status: Active
Capacity: 30 000 litres

Address: Castle Street, Dornoch, Sutherland, IV25 3 SD

Website:
dornochdistillery.com

Tel:
01862 810 216

Along with their parents, Phil and Simon Thompson have been running the Dornoch Castle Hotel in Sutherland for many years.

The hotel is famous for its outstanding whisky bar and the two brothers are passionate about whisky and other spirits. So passionate in fact that they decided to convert a 135-year old fire station into a distillery. The building is only 47 square metres and the brothers have struggled to fit all the equipment into the limited space. That is why they have plans to expand into a larger site in the near future. Currently, the distillery is equipped with a 300 kg stainless steel, semi-lauter mash tun from China, seven washbacks made of oak with a minimum fermentation time of seven days, a 1,000 litre wash still and a 600 litre spirit still. Both stills, made by Hoga in Portugal, have shell and tube condensers. The stills are directly fired using gas but they are also equipped with steam coils as an alternative heating method. There is also a 2,000 litre still with a column from Holland for the production of gin and other spirits. The distillery has a yearly capacity of 30,000 litres of pure alcohol of which approximately 15,000 litres are dedicated to whisky. The first distillation was gin in October 2016 and whisky production commenced in January 2017. A range of experimental batches of the gin were released during spring 2017 and in November the same year, the brothers finally launched their key expression - Thompson Bros Organic Highland Gin. Since then, and inspired by a trip to Japan, they have also released a very limited volume of new make spirit.

Their interest in "old-style" whiskies produced in the 1960s and earlier also has an influence on the production. All the barley is floor malted, often using old heritage varieties, predominantly Plumage Archer and Maris Otter, and different strains of brewer's yeast is used instead of distiller's yeast.

Torabhaig

[tor•a•vaig]

Owner: Mossburn Distillers

Region/district: Highlands (Skye)

Founded: 2016   **Status:** Active (vc)

Capacity: 500 000 litres

Address: Teangue, Sleat, Isle of Skye IV44 8RE

Website: www.torabhaig.com

Tel: 01471 833447

The idea to build a second distillery on Skye (with Talisker being the first) was presented several years ago by the late Sir Iain Noble.

A Skye landowner, he became known for his interest in and dedication to the Gaelic language. He was also the founder of Sabhal Mòr Ostaig, a Gaelic college on the Sleat peninsula, just a mile south of the distillery. When he died in 2010, Mossburn Distillers took over and finalised the plans for a distillery.

Located in a farmstead from the 1820s, the owners meticulously restored some of the buildings and added new ones in the same style resulting in a distillery with a grand view across the sea to the mainland. Production started in January 2017 and, in March 2018, an excellent visitor centre opened, attracting 12,000 people in its first year.

The entire set of production equipment is conveniently

(not least for visitors) situated on one level with a 1.5 ton stainless steel semi lauter mash tun with a copper top. There are eight washbacks made of Douglas fir (10,000 litres) with a fermentation time between 80 and 120 hours, one 8,000 litre wash still (named Sir Iain) and one 5,000 litre spirit still (Lady Noble). In 2019, 350,000 litres will be produced. The owners aim to produce a heavily peated whisky with an unusually high phenol specification of 75ppm in the malted barley although coming off spirit at 63%, they are not looking to catch the heaviest phenols.

The owner of Torabhaig, Mossburn Distillers, was founded in 2013 and is owned by Marussia Beverages, a Dutch company specialising in spirits and fine wine. That company in turn is actually a part of the privately owned Swedish group Haydn Holding. So, for the first time in history, we now have a Swedish-owned malt distillery in Scotland!

Isle of Raasay

[ajl ov rassay]

Owner: R&B Distillers

Region/district: Highlands (Raasay)

Founded: 2017   **Status:** Active (vc)

Capacity: 200 000 litres

Address: Borodale House, Raasay, By Kyle IV40 8PB

Website: rbdistillers.com

Tel: 01478 470177

The owners, R&B Distillers, were working on establishing a distillery in The Borders when a new plan surfaced – to build a distillery on the small island of Raasay, east of Skye.

Alasdair Day, with an ancestral interest in Scotch whisky, teamed up with entrepreneur Bill Dobbie and bought Borodale House. With more buildings added for the whisky production, the old Victorian house is now the hotel part of the distillery. With a stunning view towards the Cuillin Mountains on the Isle of Skye, this is now an excellent way of spending a night on Raasay.

The distillery started production in September 2017 and is equipped with a one ton mash tun and six stainless steel (5,000 litre) washbacks currently with four short fermentations (67 hours) and six long (118 hours) adding up to 200,000 litres of alcohol. The 5,000 litre wash still has a

cooling jacket around the lyne arm which, unusually, is used to produce a heavy spirit and there's also a 3,600 litre spirit still with a copper column attached should they want to use it for special runs. Added to that is a vapour basket for future gin production. The production is a combination of peated (45 ppm) and unpeated spirit. Starting with the one warehouse, another three were added in 2019. Eventually, the first warehouse will be used as a bottling hall.

Meanwhile, the distillery is also working on growing their own barley on the island. So far they've tried four different grains – the classic Golden Promise and three different Scandinavian types including a 6-row barley named Brage.

A wide variety of casks are used for maturation. The first release will be made up by first fill bourbon and ex-Tuscany casks while the core range going forward will be a combination of heavily charred *muehlenbergi* American oak, Bordeaux wine casks and ex-rye whiskey casks.

Lindores Abbey

[linn•doors aebi]

Owner:
The Lindores Distilling Co.

Region/district:
Lowlands

Founded: **Status:** **Capacity:**
2017 Active (vc) 260 000 litres

Address: Lindores Abbey House, Abbey Road,
Newburgh, Fife KY14 6HH

Website: **Tel:**
lindoresabbeydistillery.com 01337 842547

The famous, first written record of whisky was a letter to Friar John Cor, a monk at the Abbey of Lindores, dated 1494 where, by order of King James IV, he was instructed to make "aqua vitae, VIII bolls of malt".

The archeologial and historical evidence for Lindores being the birthplace of Scotch whisky are by no way inconclusive but further excavations of the site may reveal some evidence. Be that as it may, the current owners of the abbey in ruins are Drew and Helen McKenzie Smith and in December 2017 they commissioned a distillery next to the old monastery. The location is stunning and with all the production equipment on one level you have a spectacular view of the surroundings. Behind the washbacks you catch a glimpse of Dundee and from the stills you look down on the abbey ruins with the river Tayne in the background.

The owners idea is in 2020 to make Lindores Abbey one of few producers in Scotland (in the same way as Daftmill, Ballindalloch and Arbikie) of a single estate whisky in the sense of using locally produced barley. The equipment consists of a 2 ton semi lauter mash tun with a copper lid, four Oregon pine washbacks (with space for another four in the future) with a fermentation time between 90 and 115 hours, one 10,000 litre wash still and two 3,500 litre spirit stills. The foreshots are 15-20 minutes and the spirit cut starts at 75% and goes down to 67%. The idea behind having two spirit stills is to allow for more copper contact during distillation. The production goal for 2019 is four mashes per week and 150,000 litres of pure alcohol. The forthcoming style of the whisky will be light and fruity

An excellent visitor centre with a wide range of activities, including whisky and champagne afternoon teas and an apothecary experience where you you can create your own version of aqua vitae, is also a part of the distillery.

The Clydeside

[klajdsajd]

Owner:
Morrison Glasgow Distillers

Region/district:
Lowlands

Founded: **Status:** **Capacity:**
2017 Active (vc) 500 000 litres

Address: 100 Stobcross Road, Glasgow G3 8QQ

Website: **Tel:**
theclydeside.com 0141 2121401

If Tim Morrison, the owner of indepednent bottler AD Rattray. ever wanted to found a distillery he couldn´t have picked a better spot.

The queens Docks in Glasgow oozes of whisky history with ships coming in with barley and coal and going out with barrels of whisky, To add to the picture, Tim´s great grandfather designed the pumphouse which was used to power the hydraulic gates allowing ships in and out of the Queens Dock and which is now the site of Clydeside distillery. Tim Morrison represents the fourth generation of one of Scotland´s best known whisky families. Eventually he took over the independent bottler AD Rattray and expanded the business by opening up a first class shop and whisky centre in Kirkoswald in Ayrshire.

The distillery is beautifully situated on the river Clyde with well-known attractions such as the Riverside

Museum, Glasgow Science Centre and the SEC Centre as its closest neighbours. The equipment consists of a 1.5 ton semi lauter mash tun made of stainless steel, 8 stainless steel washbacks with a fermentation time of 72 hours, a 7,500 litre wash still and a 5,000 litre spirit still. The foreshots are 15 minutes with a slow distillation and the cutpoints for the spirit run are 76-71%. Production started in autumn 2017 and the aim for 2019 is to produce 365,000 litres of alcohol on 13 mashes per week.

An excellent visitor centre has been constructed within the old Pump House building from 1877 while an adjacent, modern building houses the distillery. Apart from a variety of tours, the distillery shop also offers a wide range of whiskies including new make spirit from the distillery itself. In their first full year, 2018, the distillery received an astonishing 25,000 visitors.

Ncn´ean

[nook•knee•anne]

Owner: **Region/district:**
Ncn´ean Distillery Ltd.   Western Highlands

Founded: **Status:**   **Capacity:**
2017   Active (vc)   100 000 litres

Address: Drimnin, By Lochaline PA80 5XZ

Website: **Tel:**
ncnean.com   01967 421698

Standing in Tobermory on Mull, you can actually see the distillery on the Morvern peninsula. It's a mere 20 minute trip across the sound but unfortunately there are no boat connections operating at the moment.

Instead, if you travel from Mull, you have to take the ferry from Fishnish to Lochaline and then drive for 45 minutes to reach the distillery. The reward when you arrive, however, is gratifying. Beautifully situated on the Drimnin estate, the distillery has an astounding view towards Mull. The estate was bought by Derek and Louise Lewis in 2001 and their, daughter, Annabel Thomas, is the founder of the distillery.

One of the fundamentals when it was built was to make it as sustainable as possible. A boiler, fired by wood chips, was brought in from Germany. The ashes from the boiler is used to fertilise the distillery garden. The barley is Scottish and certified organic and the waste heat is recycled through the temperature-controlled warehouse.

The distillery came on stream in March 2017 and is equipped with a one ton semi lauter mash tun with a one hour rest to get as clear wort as possible. Trials with different types of yeast (champagne, red wine, etc.) to create different flavour profiles is part of the work. There are four stainless steel washbacks (with space for one more) and the fermentation time is between 65 and 115 hours. Furthermore there is a 5,000 litre wash still and a 3,500 litre spirit still, both with slightly descending lyne arms. The owner is working on two basic recipes of fruity whisky – "new style" to be enjoyed young and "old style", with lower cut points and destined for a longer maturation.

The first whisky release from the distillery is planned for spring 2020 but a botanical spirit including wild herbs and flowers foraged from the surroundings of the distillery, was launched already in autumn 2018.

The Borders

[boar•ders]

Owner: **Region/district:**
The Three Stills Co. Ltd.   Lowlands

Founded: **Status:**   **Capacity:**
2017   Active (vc)   2 000 000 litres

Address: Commercial Road, Hawick TD9 7AQ

Website: **Tel:**
thebordersdistillery.com   01450 374330

On the 6th of March 2018, the first whisky distillery in the Borders in 180 years started production and the distillery opened to the public a few weeks later.

Behind the Borders Distillery in Hawick is a company called The Three Stills Company which was founded in 2013. The owners include four men who had all previously worked for William Grant & Sons – George Tait, Tony Roberts, John Fordyce and Tim Carton. The owners also include private investors as well as companies in the UK and abroad. In 2016, the company started to renovate the beautiful buildings dating from the late 1880s and which used to be an electric company and turned it into a distillery. The river Teviot is running just behind the distillery and like the textile companies that Hawick is renowned for were using the water for dyeing and power, the distillery now uses it for cooling the spirit vapours.

The distillery is equipped with a 5 ton mash tun, eight stainless steel washbacks, two wash stills and two spirit stills with all equipment provided by Forsyths. The capacity is quite large, 2 million litres, and the aim is to produce an un-peated, floral whisky. Other spirits will also be produced, including gin using local botanicals and there is also a dedicated Carterhead gin and vodka still on site.

The owners also have plans to install a bio plant on the site. Using anaerobic digestion technique, by-products from the distillation will be converted into biogas which will help power the distillery. The company has already released a blended Scotch from sourced whisky called Clan Fraser and a blended malt named Lower East Side with almost 100% sold abroad. The first bottling of spirit actually made at the distillery appeared in July 2018 when William Kerr´s Borders Gin was launched.

Aberargie

[aber•ar•jee]

Owner:	**Region/district:**
The Perth Distilling Co.	Lowlands
Founded: **Status:**	**Capacity:**
2017 Active	750 000 litres
Address: Aberargie, Perthshire PH2 9LX	
Website:	**Tel:**
-	01738 787044

The distillery was built on the same grounds in Fife as Morrison & Mackay, independent bottler and producer of Scottish liqueurs, and a company which can trace it´s roots back to 1982.

Founded as John Murray & Co., the company was taken over in 2005 by Kenny Mackay and Brian Morrison, once the chairman of Morrison Bowmore, and his son Jamie. The production of liqueurs, especially Columba Cream, continued while bottling of Scotch single malts (The Carn Mor) was added to the business. Later on, they also took over the Old Perth brand from Whyte & Mackay and relaunched it as a blended malt. The company name was changed to Morrison & Mackay in 2014.

At the same time, the Morrison´s of the company decided to build a distillery on the premises and founded a company called The Perth Distilling Company. Construction work started in summer 2016 and the first spirit was distilled in November 2017.

The distillery is equipped with a 2 ton semilauter mash tun, six stainless steel washbacks with a fermentation time of 72 hours, one 15,000 litre wash still and one 10,000 litre spirit still. The stills, both with steeply descending lyne arms, were made by Forsyths and are heated with panels instead of coils or pans. With a maturation in a mixture of first fill sherry butts, first fill bourbon barrels and second fill sherry/bourbon casks, the owners are aiming for a fruity character which will be enhanced by occasional peated spirit runs. Different barley varieties are being used, including Golden Promise, and they are all grown in 300 acres of field owned by the Morrison family and that surround the distillery. With the Morrison & Mackay blending and bottling facility next to the distillery, every step of the production (except malting) will take place on site.

GlenWyvis

[glen•wivis]

Owner:	**Region/district:**
GlenWyvis Distillery Ltd.	Highlands
Founded: **Status:**	**Capacity:**
2017 Active (vc)	140 000 litres
Address: Upper Docharty, Dingwall IV15 9UF	
Website:	**Tel:**
glenwyvis.com	01349 862005

In 2015, the local farmer John McKenzie, came up with the idea to establish a distillery that was owned by the local people – the first ever 100% community-owned distillery.

A planning application was submitted to the local council in March 2016 and by summer more than £2.5 million had been raised via a community share offer with more than 3,000 people investing. Construction started in January 2017 and later that year, the owners managed to hire one of the most experienced distillers in Scotland as the manager – Duncan Tait – who over the years had been managing several of the Diageo distilleries. The first distillation was on the 30th of January 2018 and the production goal for 2019 is to do 3 mashes per week and 24,000 litres of pure alcohol which will allow the owners time to build a second warehouse.

The distillery is equipped with a 0.5 ton semi lauter mash tun, six washbacks (4,400 litres each) made of stainless steel with a fermentation time of 72-120 hours, one 2,500 litre wash still and one 1,700 litre spirit still. The unpeated spirit, which will mainly be filled into American oak, will be matured in dunnage warehouses on site. The style of the newmake is a combination o fruity and green/grassy and a small batch of the newmake was released in summer 2019. A dedicated 400 litre gin still was installed in spring 2018 and the distillery has been selling GoodWill Gin for a year now.

The distillery is located in Dingwall, north of Inverness but this is not the first distillery in the town. In 1879, Ben Wyvis was founded and it went on producing until 1926 when it was closed. The area, however, is famous for yet another distillery – namely Ferintosh. It was built across the Cromarty Firth from Dingwall in 1689 by members of the Forbes family.

Ardnahoe

[ard•na•hoe]

Owner:
Hunter Laing & Company

Region/district:
Islay

Founded: 2017
Status: Active (vc)
Capacity: 1,000,000 litres

Address: Isle of Islay, Port Askaig PA46 7RU

Website: ardnahoedistillery.com

Tel: 01496 840711

Ardnahoe, the newest distillery on Islay, came on stream in November 2018. The location, between Caol Ila and Bunnahabhain and over-looking Jura, is absolutely stunning!

The distillery, owned by independent bottler Hunter Laing, is equipped with a 2.5 ton semi lauter mash tun with a copper lid. The lauter gear is used as little as possible to get a clear wort. There are four washbacks made from Oregon pine with a fermentation time between 60 and 70 hours, one wash still (12,500 litres) and one spirit still (7,500 litres) with a slow distillation, both with extremely long lyne arms (7,5 metres). Actually, these are the longest in Scotland. The distillery is equipped with wooden worm tubs (the only ones on Islay) with a 77 metre copper tube in each. Another unusual piece of equipment is the 4-roller Boby mill from the 1920s which was brought in from

Fettercairn. The aim is to produce a variety of single malts from unpeated to peated whiskies on several levels (from 5ppm up to 40ppm) with the malt being bought from Port Ellen maltings. The distillery has an annual capacity of one million litres of alcohol and in 2019 the production will be 14 mashes per week which amounts to 600,000 litres. Eighty percent of the newmake goes into first fill bourbon and the rest is matured in sherry casks. There is currently one warehouse on site but more will be built – a mix of dunnage and racked. In 2017, the legendary Jim McEwan joined the team as Production Director.

Ardnahoe is the first new distillery on the island since Kilchoman was opened in 2005 and the 9th on Islay. The distillery visitor centre is surprisingly huge (given the size of the distillery) with a large shop and a café/whisky bar with an excellent view towards the paps of Jura and with Mull in the distance.

Ardross

[ard•ross]

Owner:
Greenwood Distillers

Region/district:
N Highland

Founded: 2019
Status: Active (vc)
Capacity: 1,000,000 litres

Address: Ardross Mains, Ardross, Alness

Website: greenwooddistillers.com

Tel: -

An old, derelict 19th century farm site north of Inverness, has recently been transformed into one of Scotland's newest distilleries.

The nearest distillery neighbors are Teaninich and Dalmore, some 5 km to the southeast. A planning application was filed and approved in 2017 and behind the project lies Greenwood Distillers Ltd., incorporated in 2018 and that company, in turn, is an affiliate of Vevil International, owner of Ned Hotel and the Wolseley restaurant in London. The CEO of Greenwood Distillers is Barthelemy Brosseau and one of the directors is Andrew Rankin who was Operations Director and Chief Blender at Morrison Bowmore for almost 25 years.

The distillery will probably be commissioned in autumn 2019 and according to the plans submitted with the planning application, it will be equipped with one pair of stills

and ten washbacks. The capacity is an impressive one million litres of pure alcohol. The cost for the entire project amounts to £15-18m

The first release from the distillery was Theodore gin which appeared in August 2019 (although having been pre-released already in February). It features 16 botanicals inspired by those that the Picts may have encountered on their travels to Scotland, including pine, damask rose and honey. The gin will not just be a way of creating cash flow until the future whisky has been released, as a designated gin distillery has been built on the site. Ardross is also said to be on of few distilleries in Scotland to own and manage its own loch (Loch Dubh).

As Distillery Manager, the company hired Sandy Jamieson with a long career in the Scotch whisky business, most recently as manager of Speyside Distillery near Kingussie.

Lagg

[laag]

Owner:		**Region/district:**
Isle of Arran Distillers Ltd.		Islands
Founded:	**Status:**	**Capacity:**
2019	Active (vc)	750 000 litres

Address: Kilmory, Isle of Arran KA27 8PG

| **Website:** | **Tel:** |
| laggwhisky.com | 01770 870565 |

The success for Arran distillery, which was opened in 1993, has now encouraged the owners to open yet another distillery on the island. Work on Lagg distillery began in February 2017 and the first distillation took place in March 2019.

Three months later the distillery opened to the public. With Arran distillery located in Lochranza on the northern tip of the island, Lagg is situated in the south. Already now, Arran is by far the most visited distillery in Scotland with more than 100,000 visitors last year and the owners anticipate that the combined distilleries will see more than 200,000 visitors in 2020. For those wanting to do more than just visiting the distillery, there is the possibility of buying entire casks for future bottling.

The distillery is equipped with a four ton semilauter mash tun, four Oregon pine washbacks holding 20,000 litres, one wash still (10,000 litres) and one spirit still (7,000 litres). However, there is space for an additional four washbacks and one more pair of stills in the future. The idea is to move all the peated production from Lochranza to Lagg and the first 6 months will probably see 250,000 litres produced from 50ppm barley. The peat used to dry the barley will be sourced from different places in Scotland and also from abroad. This is part of the owners ambition to explore the impact of terroir on whisky flavour. Also, Lagg will act as an experimental plant with trials of different yeast strains and types of barley and they have plans to produce their own cider and apple brandy in the future.

The distillery manager for Lagg is Graham Omand, raised on Islay and working as a stillman and mashman at Arran for nearly a decade. He will work together with his uncle, James Mactaggart, the Master Distiller for both of the island's distilleries.

Holyrood

[holly•rude]

Owner:		**Region/district:**
The Holyrood Distillery Ltd.		Lowlands
Founded:	**Status:**	**Capacity:**
2019	Active (vc)	250 000 litres

Address: 19 St Leonard´s Lane, Edinburgh EH8 9SH

| **Website:** | **Tel:** |
| holyrooddistillery.co.uk | 0131 2858977 |

A number of distillery projects are going on in Edinburgh at the moment and the first single malt distillery to open in the city for almost 100 years was Holyrood.

Planning permission was granted in August 2016, construction work commenced two years later and in summer 2019 the production started. Behind the project lie whisky veteran David Robertson (ex Macallan master distiller) and the Canadian couple Kelly and Rob Carpenter who together with 60 other investors managed to raise the £7.3m needed for the project. The distillery is located in a listed building from 1835 and as Distillery Manager the company has hired Jack Mayo who used to work for Glasgow Whisky Company.

The owners have decided on five core whisky styles; Floral where wine yeast has been used, with a small spirit cut and maturing in American and virgin oak, Fruity using three types of yeast and maturing in American oak, ex-wine and ex-sherry casks, Sweet including speciality malt and a mix of yeast and maturation in American oak, Spicy where, again, a mix of yeast (brewers and distillers) are used and maturation takes place in European ex-sherry and finally Smoky where peated newmake is filled into American and European oak. Apart from whisky, the distillery will also produce other sprits including gin which was the first to be released.

The distillery is equipped with a one ton lauter mash tun and six 5,000 litre washbacks made of stainless steel with a fermentation time of 48-168 hours depending on the style they're making. There are one 5,000 litre wash still and one 3,750 litre spirit still – both of them very tall (7 metres)! Foreshots vary from 6 to 30 minutes, also depending on the spirit style. The distillery was commissioned in summer 2019 and opened to the public in July that year.

Distilleries per owner

c = closed, d = demolished, mb = mothballed, dm = dismantled

Diageo
Auchroisk
Banff (d)
Benrinnes
Blair Athol
Brora (c)
Caol Ila
Cardhu
Clynelish
Coleburn (dm)
Convalmore (dm)
Cragganmore
Dailuaine
Dallas Dhu (c)
Dalwhinnie
Dufftown
Glen Albyn (d)
Glendullan
Glen Elgin
Glenesk (dm)
Glenkinchie
Glenlochy (d)
Glenlossie
Glen Mhor (d)
Glen Ord
Glen Spey
Glenury Royal (d)
Inchgower
Knockando
Lagavulin
Linkwood
Mannochmore
Millburn (dm)
Mortlach
North Port (d)
Oban
Pittyvaich (d)
Port Ellen (dm)
Roseisle
Royal Lochnagar
St Magdalene (dm)
Strathmill
Talisker
Teaninich

Pernod Ricard
Aberlour
Allt-a-Bhainne
Braeval
Caperdonich (d)
Dalmunach
Glenburgie
Glen Keith
Glenlivet
Glentauchers
Glenugie (dm)
Imperial (d)
Inverleven (d)

Kinclaith (d)
Lochside (d)
Longmorn
Miltonduff
Scapa
Strathisla
Tormore

Edrington Group
Glenrothes
Highland Park
Macallan

Inver House (Thai Beverage)
Balblair
Balmenach
Glen Flagler (d)
Knockdhu
Pulteney
Speyburn

John Dewar & Sons (Bacardi)
Aberfeldy
Aultmore
Craigellachie
Macduff
Royal Brackla

William Grant & Sons
Ailsa Bay
Balvenie
Glenfiddich
Kininvie
Ladyburn (dm)

Whyte & Mackay (Emperador)
Dalmore
Fettercairn
Jura
Tamnavulin

Beam Suntory
Ardmore
Auchentoshan
Bowmore
Glen Garioch
Laphroaig

Distell International
Bunnahabhain
Deanston
Tobermory

Benriach Dist. Co. (Brown Forman)
Benriach
Glendronach
Glenglassaugh

Loch Lomond Group
Glen Scotia
Littlemill (d)
Loch Lomond

J & A Mitchell
Glengyle

Distilleries per owner

c = closed, d = demolished, mb = mothballed, dm = dismantled

Springbank

Glenmorangie Co. (LVMH)
Ardbeg
Glenmorangie

Angus Dundee Distillers
Glencadam
Tomintoul

Ian Macleod Distillers
Glengoyne
Rosebank (c)
Tamdhu

Campari Group
Glen Grant

Isle of Arran Distillers
Arran
Lagg

Signatory
Edradour

Tomatin Distillery Co.
Tomatin

J & G Grant
Glenfarclas

Rémy Cointreau
Bruichladdich

David Prior
Bladnoch (c)

Gordon & MacPhail
Benromach

La Martiniquaise
Glen Moray

Ben Nevis Distillery Ltd (Nikka)
Ben Nevis

Picard Vins & Spiritueux
Tullibardine

Harvey's of Edinburgh
Speyside

Kilchoman Distillery Co.
Kilchoman

Cuthbert family
Daftmill

Mark Tayburn
Abhainn Dearg

Aurora Brewing Ltd
Wolfburn

Strathearn Distillery Ltd
Strathearn

Annandale Distillery Co.
Annandale

Adelphi Distillery Co.
Ardnamurchan

Wemyss
Kingsbarns

Mcpherson-Grant family
Ballindalloch

Paul Miller
Eden Mill

Isle of Harris Distillers
Harris

The Glasgow Distillery Company
Glasgow Distillery

John Fegus & Co. Ltd
Inchdairnie

Stirling family
Arbikie

Brewdog plc
Lone Wolf

Thompson family
Dornoch

Mossburn Distillers
Torabhaig

R & B Distillers
Isle of Raasay

The Lindores Distilling Company
Lindores Abbey

Morrison Glasgow Distillers
Clydeside

Ncn´ean Distillery Ltd.
Ncn´ean

The Three Stills Co.
The Borders

The Glenallachie Distillers Co.
Glenallachie

The Perth Distilling Company
Aberargie

GlenWyvis Distillery Ltd.
GlenWyvis

Hunter Laing
Ardnahoe

The Holyrood Distillery Ltd.
Holyrood

Greenwood Distillers
Ardross

Lalique Group/Hansjörg Wyss
Glenturret

Coleburn Distillery, closed in 1985 but still used for warehousing

Closed
distilleries

The distilleries on the following pages
have all been closed and some of them even demolished.
New releases from a few of them appear on a regular basis
but for most of them chances are very slim of ever finding another bottling.
One is also tempted to say that none of the distilleries will ever be opened
again but recent developments clearly show that you can never
be certain. In October 2017, Diageo announced that they had
plans to re-start Brora and Port Ellen and the following day,
Ian Macleod Distillers declared that Rosebank would
be reinstated as a working distillery as well.

Brora

[bro•rah]

Owner: **Region/district:**
Diageo   Northern Highlands

Founded: **Status:**   **Capacity:**
1819   Closed   800,000 litres

Address: Brora, Sutherland KW9 6LR

Website: **Tel:**
malts.com   01408 623003 (vc)

Plans to rebuild and open up Brora distillery were announced in autumn 2017 by Diageo. In October 2018, planning approval from the Highland Council was received and soon after the work began.

Many of the original buildings can be used as they are but the stillhouse will be demolished and rebuilt brick for brick. The original two stills were in fairly good condition but were sent to Diageo's Abercrombie coppersmiths in Alloa for inspection and refurbishing. Other equipment that was still in place after 35 years was the feints receiver, the spirit receiver and the brass safe. With a capacity of 800,000 litres of alcohol, the plan is to start production at Brora sometime in 2020.

The distillery was built in the time referred to as the Highland Clearances. Many land-owners wished to increase the yield of their lands and consequently went into large-scale sheep farming. Thousands of families were ruthlessly forced away and the most infamous of the large land-owners was the Marquis of Stafford who founded Clynelish (Brora) in 1819. The distillery had a chequered history until 1896 when the brewer and whisky broker James Ainslie assumed ownership. He rebuilt the distillery and soon Clynelish single malt enjoyed a good reputation amongst blenders. In 1967 the owners, DCL, decided to build a new, modern distillery on the same site. This was given the name Clynelish and it was decided the old distillery, with a capacity of 1 million litres of alcohol, should be closed. Shortly after, the demand for peated whisky, especially for the blend Johnnie Walker, increased and the old site re-opened but now under the name Brora and the "recipe" for the whisky was changed to a heavily peated malt.

Brora was closed in 1983 and from 1995 United Distillers regularly released different expressions of Brora in the Rare Malts series. In 2002 a new range was created, Special Releases, and bottlings of Brora have appeared ever since. Very little stock remains and from 2018, this will not be released in connection with the Special Releases. The latest bottling, in August 2019, was a 40 year old to celebrate the distillery's 200th anniversary.

History:

1819 The Marquis of Stafford, 1st Duke of Sutherland, founds the distillery as Clynelish Distillery.

1827 The first licensed distiller, James Harper, files for bankruptcy and John Matheson takes over.

1828 James Harper is back as licensee.

1833 Andrew Ross takes over the license.

1846 George Lawson & Sons takes over.

1896 James Ainslie & Heilbron takes over and rebuilds the facilities.

1912 Distillers Company Limited (DCL) takes over together with James Risk.

1925 DCL buys out Risk.

1930 Scottish Malt Distillers takes over.

1931 The distillery is mothballed.

1938 Production restarts.

1960 The distillery becomes electrified (until now it has been using locally mined coal from Brora).

1967 A new distillery is built adjacent to the first one, it is also named Clynelish and both operate in parallel from August with the new distillery named Clynelish A and the old Clynelish B.

1969 Clynelish B is closed in April but reopened shortly after as Brora and starts using a heavily peated malt until 1973.

1975 A new mashtun is installed.

1983 Brora is closed in March.

1995 Brora 1972 (20 years) and Brora 1972 (22 years) are launched as Rare Malts.

2002 A 30 year old is the first bottling in the Special Releases.

2014 The 13th release of Brora – a 35 year old.

2015 The 14th release of Brora – a 37 year old.

2016 The 15th release of Brora – a 38 year old.

2017 The 16th release of Brora - a 34 year old. Diageo announces that the distillery will re-open in 2020.

2019 A 40 year old is released and the reconstruction of the distillery begins.

40 years old

Port Ellen

[port ell•en]

Owner:
Diageo

Region/district:
Islay

Founded: 1825

Status: Dismantled

Capacity: -

Address: Port Ellen, Isle of Islay, Argyll PA42 7AJ

At the same time, autumn 2017, that Diageo announced that the closed Brora distillery would be re-opened they also revealed that they had the same plans for Port Ellen distillery on Islay.

Both distilleries stopped producing in 1983 but the main difference between the two is that no equipment remains at Port Ellen and very few of the buildings can be used. Actually it's only the kiln and the warehouses that are functional. A new distillery will have to be built in the courtyard between the maltings and the old warehouses. The old drawings of the equipment still exist and one pair of stills with shell and tube condensers will be fabricated. There will also be a second, smaller pair with the intention of creating experimental whiskies. The plan is to have the distillery, with an 800,000 litre capacity up and running sometime in 2021. There will also be a visitor centre, or brand home as Diageo calls it.

The founder of the distillery, Alexander Mackay, went bankrupt a few months after the distillery had opened and instead it was a relative of his, John Ramsay, who would run the distillery until the late 1800s and that with great success. There was no intention of ever bottling the spirit as a single malt – all the production went to blends. In 1930 the distillery was mothballed and didn't reopen until 1967. The final era would last but 16 years and in 1983 the distillery was closed for good (or so it would seem). Ten years before the closure, a huge drum maltings was opened on the site and this continues to produce malted barley for several of the Islay distilleries. At its height, Port Ellen was equipped with four stills, producing 1.7 million litres of alcohol.

Port Ellen single malt has been released twice in the Rare Malts range (1998 and 2000). It wasn't until 2001, when the first Port Ellen Special Release turned up that things started to change and the malt became a cult whisky. There is still some stock of old Port Ellen left but going forward, this will not be launched in the Special Releases as it used to be. However, a new range from Diageo, Untold Stories, appeared in spring 2019 with a **39 year old** Port Ellen as the inaugural release.

History:

1825 Alexander Kerr Mackay assisted by Walter Campbell founds the distillery. Mackay runs into financial troubles after a few months and his three relatives John Morrison, Patrick Thomson and George Maclennan take over.

1833 John Ramsay, a cousin to John Morrison, comes from Glasgow to take over.

1836 Ramsay is granted a lease on the distillery from the Laird of Islay.

1892 Ramsay dies and the distillery is inherited by his widow, Lucy.

1906 Lucy Ramsay dies and her son Captain Iain Ramsay takes over.

1920 Iain Ramsay sells to Buchanan-Dewar who transfers the administration to the company Port Ellen Distillery Co. Ltd.

1925 Buchanan-Dewar joins Distillers Company Limited (DCL).

1930 The distillery is mothballed.

1967 In production again after reconstruction and doubling of the number of stills from two to four.

1973 A large drum maltings is installed.

1980 Queen Elisabeth visits the distillery and a commemorative special bottling is made.

1983 The distillery is mothballed.

1987 The distillery closes permanently but the maltings continue to deliver malt to all Islay distilleries.

2001 Port Ellen cask strength first edition is released.

2014 The 14th release of Port Ellen - a 35 year old from 1978.

2015 The 15th release of Port Ellen - a 32 year old from 1983.

2016 The 16th release of Port Ellen - a 37 year old from 1978.

2017 The 17th release of Port Ellen - a 37 year old from 1979. Diageo announces that the distillery will re-open in 2020.

2019 A Port Ellen 39 year old is released as the first in a new range – Untold Stories.

37 years old

Rosebank

[rows•bank]

Owner:
Ian Macleod Distillers

Region/district:
Lowlands

Founded: 1840 **Status:** Closed

Capacity: 6-800,000 litres

Address: Falkirk FK1 4DS

Website: rosebank.com

Tel: -

When Distillers Company Limited (later to become Diageo) launched the groundbreaking Classic Malts in 1988, Glenkinchie was selected as the distillery to represent the Lowlands. To be honest, there weren't that many to choose from.

More than a decade later, whisky aficionados started to question the choice, wondering why Rosebank hadn't been singled out. For many, this was a more interesting whisky. Favourable for Glenkinchie was its proximity to Edinburgh and the easy access for visitors to the distillery. Eventually, Rosebank stopped production in 1993. During the last decade, there have been plans (not by Diageo though) to resurrect the distillery but things have been complicated. Diageo sold the buildings to British Waterways in 2002 and in 2008, the stills and most of the equipment was stolen.

Eventually, in 2017, Ian Macleod bought the property from Scottish Canals, (British Waterways' successor) and the trademark and stock from Diageo. In early 2019, they received planning permission from Falkirk council and the plan is to have the distillery up and running by autumn 2020. The new distillery, with the iconic chimney being kept, will be equipped with three stills (for triple distillation) and wormtubs with a capacity to produce 1 million litres of pure alcohol per year. The cost for the project is estimated to be £12m.

Established in 1798, Rosebank single malt enjoyed a good reputation during most of its lifespan even though the distillery also produced its fair share of grain whisky which was common especially in the Lowlands at the time. Most of the production went into blends but in 1982, Rosebank 8 year old single malt became a part of the owners Ascot Malt Cellar range together with Lagavulin, Talisker and Linkwood. Six years later, The Classic Malts saw the light of day and when the owners were to decide which malt to represent the Lowlands, their choice was Glenkinchie.

The latest official bottling of Rosebank was a **21 year old** in the Special Releases autumn 2014.

History:

1840 James Rankine founds the distillery.

1845 The distillery is expanded.

1864 Rankine buys Camelon Distillery on the west bank of the Forth-Clyde canal.

1894 Rosebank Distillery Company is formed.

1914 Rosebank, togehter with Clydesdale, Glenkinchie, St. Magdalene and Grange form Scottish malt Distillers (SMD).

1919 SMD becomes a part of Distillers company Limited (DCL).

1982 DCL launches the series The Ascot Malt Cellar with Rosebank, Linkwood, Talisker, Lagavulin and two blendeed malts.

1993 The distillery closes in June.

2002 The buildings are bought by British Waterways.

2008 The stills and other equipment are stolen.

2014 A 21 year old is launched as part of the Special Releases.

2017 The site is bought from Scottish Canals by Ian Macleod Distillers and at the same time they acquire the trademark and stocks from Diageo.

21 years old

Banff

Owner:	Region:	Founded:	Status:
Diageo	Speyside	1824	Demolished

The distillery has a tragic history of numerous fires, explosions and bombings. The most spectacular incident was when a lone Junkers Ju-88 bombed one of the warehouses in 1941. The distillery was closed in 1983 and the buildings were destroyed in a fire in 1991.

Ben Wyvis

Owner:	Region:	Founded:	Status:
Whyte & Mackay	N Highlands	1965	Dismantled

Built on the same site as Invergordon grain distillery, the distillery was equipped with one mash tun, six washbacks and one pair of stills. The stills are in use today at Glengyle distillery. Production stopped in 1976 and in 1977 the distillery was closed and dismantled.

Caperdonich

Owner:	Region:	Founded:	Status:
Chivas Bros.	Speyside	1897	Demolished

Founded by the owners of Glen Grant. Five years after the opening, the distillery was shut down but was re-opened again in 1965 under the name Caperdonich. In 2002 it was mothballed yet again. Sold in 2010 to Forsyth´s in Rothes and the buildings were demolished.

Coleburn

Owner:	Region:	Founded:	Status:
Diageo	Speyside	1897	Dismantled

Coleburn was used as an experimental workshop where new production techniques were tested. In 1985 the distillery was mothballed and never opened again. Since 2014, the warehouses are used by Aceo Ltd, who owns the independent bottler Murray McDavid.

Convalmore

Owner:	Region:	Founded:	Status:
Diageo	Speyside	1894	Dismantled

This distillery is still intact and can be seen in Dufftown next to Balvenie distillery. The buildings are used by William Grant´s for storage while Diageo still holds the rights to the brand. In the early 20th century, distilling of malt whisky in continuous stills took place. Closed in 1985.

Dallas Dhu

Owner:	Region:	Founded:	Status:
Diageo	Speyside	1898	Closed

The distillery is still intact, equipment and all, but hasn´t produced since 1983. Today it is run by Historic Scotland as a museum which is open all year round. In 2013 a feasibility study was commissioned to look at the possibilities of re-starting production again.

Glen Albyn

Owner:	Region:	Founded:	Status:
Diageo	N Highlands	1844	Demolished

One of three Inverness distilleries surviving into the 1980s. In 1866 the buildings were transformed into a flour mill. but then converted back to a distillery in 1884 and continued producing whisky until 1983 when it was closed. Three years later the distillery was demolished.

Glenesk

Owner:	Region:	Founded:	Status:
Diageo	E Highlands	1897	Demolished

Operated under many names; Highland Esk, North Esk, Montrose and Hillside. In 1968 a large drum maltings was built adjacent to the distillery and the Glenesk maltings still operate today under the ownership of Boortmalt. The distillery building was demolished in 1996.

Glen Flagler

Owner:	Region:	Founded:	Status:
InverHouse	Lowlands	1965	Demolished

Glen Flagler was one of two malt distilleries (Killyloch being the other) that were built on the site of Garnheath grain distillery. Killyloch was closed in the early 1970s, while Glen Flagler continued to produce until 1985. A year later, Garnheath was closed only to be demolished in 1988.

Glenlochy

Owner:	Region:	Founded:	Status:
Diageo	W Highlands	1898	Demolished

Glenlochy was one of three distilleries in Fort William at the beginning of the 1900s. For a period of time, the distillery was owned by Joseph Hobbs who, after having sold the distillery to DCL, bought the second distillery in town, Ben Nevis. Glenlochy was closd in 1983.

Glen Mhor

Owner:	Region:	Founded:	Status:
Diageo	N Highlands	1892	Demolished

Glen Mhor was one of the last three Inverness distilleries and probably the one with the best reputation when it comes to the whisky that it produced. Glen Mhor was closed in 1983 and three years later the buildings were demolished. Today there is a supermarket on the site.

Glenugie

Owner:	Region:	Founded:	Status:
Chivas Bros	E Highlands	1831	Demolished

Glenugie produced whisky for six years before it was converted into a brewery. In 1875 whisky distillation started again, but production was very intermittent until 1937 when Seager Evans took over. Following several ownership changes, the distillery closed in 1983.

Glenury Royal

Owner:	Region:	Founded:	Status:
Diageo	E Highlands	1825	Demolished

The founder of Glenury was the eccentric Captain Robert Barclay Allardyce, the first to walk 1000 miles in 1000 hours in 1809. The distillery closed in 1983 and part of the building was demolished a decade later with the rest converted into flats.

Imperial

Owner:	Region:	Founded:	Status:
Chivas Bros	Speyside	1897	Demolished

In over a century, Imperial distillery was out of production for 60% of the time, but when it produced it had a capacity of 1,6 million. In 2012, the owners announced that a new distillery would be built. The old distillery was demolished and in 2015 Dalmunach distillery was commissioned.

Inverleven

Owner:	Region:	Founded:	Status:
Chivas Bros	Lowlands	1938	Demolished

Inverleven was built on the same site as Dumbarton grain distillery, equipped with one pair of traditional pot stills. In 1956 a Lomond still was added. Inverleven was mothballed in 1991 and finally closed. The Lomond still is now working again since 2010 at Bruichladdich.

Killyloch

Owner:	Region:	Founded:	Status:
InverHouse	Lowlands	1965	Demolished

Publicker Industries converted a paper mill in Airdrie into a grain distillery (Garnheath) and two malt distilleries (Glen Flagler and Killyloch). Killyloch (originally named Lilly-loch after the water source) was closed in the early 1970s, while Glen Flagler continued to produce until 1985.

Kinclaith

Owner:	Region:	Founded:	Status:
Chivas Bros	Lowlands	1957	Demolished

The last malt distillery to be built in Glasgow and constructed on the grounds of Strathclyde grain distillery by Seager Evans. In 1975 it was dismantled to make room for an extension of the grain distillery. It was later demolished in 1982.

Ladyburn

Owner:	Region:	Founded:	Status:
W Grant & Sons	Lowlands	1966	Dismantled

In 1963 William Grant & Sons built their huge grain distillery in Girvan in Ayrshire. Three years later they also decided to build a malt distillery on the site which was given the name Ladyburn. The distillery was closed in 1975 and finally dismantled during the 1980s.

Littlemill

Owner:	Region:	Founded:	Status:
Loch Lomond Co.	Lowlands	1772	Demolished

Scotland's oldest working distillery until production stopped in 1992. Triple distillation was practised until 1930. In 1996 the distillery was dismantled and part of the buildings demolished and in 2004 much of the remaining buildings were destroyed in a fire.

Lochside

Owner:	Region:	Founded:	Status:
Chivas Bros	E Highlands	1957	Demolished

Most of the output from the distillery was made for blended whisky. One of the owners combined grain and malt whisky production. In 1992 the distillery was mothballed and five years later all the equipment and stock were removed. The distillery buildings were demolished in 2005.

Millburn

Owner:	Region:	Founded:	Status:
Diageo	N Highlands	1807	Dismantled

The oldest of those Inverness distilleries that made it into modern times. With one pair of stills, the capacity was 300,000 litres. In 1985 it was closed and three years later all the equipment was removed. The buildings are now a hotel and restaurant owned by Premier Inn.

North Port

Owner:	Region:	Founded:	Status:
Diageo	E Highlands	1820	Demolished

The names North Port and Brechin are used interchangeably on the labels of this single malt. The distillery had one pair of stills and produced 500,000 litres per year. Closed in 1983, it was dismantled piece by piece and was finally demolished in 1994 to make room for a supermarket.

Pittyvaich

Owner:	Region:	Founded:	Status:
Diageo	Speyside	1974	Demolished

Built by Arthur Bell & Sons on the same ground as Dufftown distillery. For a few years in the 1990s, Pittyvaich was also a back up plant for gin distillation (Gordon's gin). The distillery was mothballed in 1993 and has now been demolished.

St Magdalene

Owner:	Region:	Founded:	Status:
Diageo	Lowlands	1795	Dismantled

The distillery came into ownership of DCL in 1912 and was at the time a large distillery with 14 washbacks, five stills and with the possibility of producing more than one million litres of alcohol. Ten years after the closure in 1983, the distillery was re-built into flats.

Japanese Alchemy

by Stefan Van Eycken

The Japanese whisky scene is marked by
an unbridled optimism and there are more bona fide whisky
producers than ever making whisky or getting ready to make whisky at
present. That side of the story is in the pages that follow.
However, there's another side to the story.

While consumers are waiting for all that spirit to turn into liquid gold, some 'creative' whisky producers are doing their best to fill the void using the sort of magic that wouldn't fly in other established whisky-making regions of the world. Last year, we looked at the laundering of whisky imported in bulk from abroad as 'Japanese' whisky. This year, we're turning the spotlight on another type of magic, with the semantic wizardry shifting from the adjective to the noun: i.e. the recategorization of spirits made in Japan that started life as shochu (and awamori) as Japanese 'whisky' in the U.S. A rose is a rose is a rose. Or is it?

Shochu is a traditional Japanese distilled spirit made from a wide range of base ingredients. The most common varieties are rice shochu, sweet potato shochu and (unmalted) barley shochu. Just like in sake production, a mold called koji-kin is used to break down starch molecules into sugar molecules, which can then be processed by yeast cells.

In terms of distillation method, there are two types: one is distilled in a continuous still to a very high ABV and then diluted for sale; the other, authentic, type is single distilled usually in a stainless steel still – wooden stills are also used, copper being the exception – to an ABV of no more than 45%. Most quality shochu is then 'aged' in large earthenware pots or stainless and enamel-lined tanks. However, some shochu, rice and barley shochu, for the most part, is matured in oak barrels.

To avoid confusion with whisky, the Japanese government put certain regulations in the shochu category in place. Most significantly, in this context, is the fact that bottled shochu cannot exceed a certain, very low, absorbance value – meaning it must be markedly paler in colour than whisky, by law.

Obviously, shochu maturing in a barrel doesn't care about the colour threshold and it will happen, depending on the cask types used and policies in place (or not) in the warehouse, that the liquid comes out darker than allowed. Up until recently, there were three options: heavily filter it (but that can take the 'soul' out of an otherwise excellent beverage), blend it with much lighter shochu before bottling to get it under the threshold or sell it as a liqueur (but for that you need a separate license). None of those options are really attractive to producers, but losing money is even less attractive so one did what was necessary. Until a fourth, more lucrative option presented itself like a deus ex machina coming from the U.S.

"In Japan, the shochu makers are in a tough position," shochu expert Stephen Lyman points out. "However, the U.S. has quite liberal whisk(e)y definitions, to which these barrel-aged shochu manage to comply despite their highly irregular production methods relative to standard whiskies."

All that's necessary for a distilled spirit to be legally sold as whisky in the U.S. is that it is made from a fermented mash of grain, stored in oak containers and bottled at over 40% abv. The word 'malt' does not appear in the regulations and the use of koji-kin is not a problem either (which it is in other markets). And that's how barrel-aged shochu and awamori – which is made in Okinawa with long grain rice and a different type of mold – is magically transformed into 'Japanese whisky' in the U.S. market.

Consumers are often surprised, and this is part of the attraction, that the flavour profile of these 'Japanese whiskies' is quite different from what they are used to associating with that category.

"Most if not all barrel-aged shochu currently sold as whisky in the U.S. is single distilled," Lyman explains. "This results in a much more flavorful new pot than would be expected from a double or triple distillation process, which is, I think, what makes these products so different, and sometimes wild/rough, compared to Japanese whiskies made in a Scottish tradition. There are no low wines in shochu production because the ferment itself is 15-18% abv prior to distillation resulting in a 42-45% abv first run."

While the repurposing of shochu as Japanese whisky in the U.S. can seem like a win-win situation for everyone involved – the producers, the distributors and the consumer looking for new taste sensations – there is potential for frustration. As Lyman points out,

"Shochu makers may be very good at making shochu but they are not necessarily adept at making whisky,

nor in the arts of barrel maintenance or blending. This certainly does not hold for all producers and some even have highly sophisticated barrel management programs and blending teams. However, one particular rice-shochu-as-whisky release that I tried a few months back smelled distinctly of acetone – not a favourable whisky profile. This risks diluting the reputation of Japanese whisky more broadly. I think, also, it risks damaging the reputation of the shochu producers if they're seen as opportunistic."

The pragmatist would say: where's the problem? The U.S. doesn't regulate the shochu category and Japanese whisky makers are not able to prevent shochu makers from selling barrel-aged spirits in the U.S. as 'whisk(e)y' unless they can get WTO geographical indication status, which seems unlikely for a 100-year old whisky tradition borrowed nearly whole cloth from Scotland. And yet, there's considerable confusion among consumers who are getting increasingly tired of what they see as 'Wild West skulduggery' in the Japanese whisky category.

One answer would be to give these interesting spirits a category of their own and call them 'koji whisky', or 'Takamine whisky', after Jokichi Takamine, the Japanese scientist who developed a method to make whisky using koji in the early 1890s in Peoria, Illinois, for The Whiskey Trust. This would allow these spirits to be judged on their own merits and remove the suspicion of switcheroo tactics on the part of producers and distributors.

Until that happens, get used to more weird and wonderful Japanese brand names on the whisky shelves appearing out of nowhere.

Stefan Van Eycken grew up in Belgium and Scotland and moved to Japan in 2000. Editor of Nonjatta, he is also the man behind the 'Ghost Series' bottlings and the charity event 'Spirits for Small Change'. He is regional editor (Japan) for Whisky Magazine UK, and a regular contributor to Whisky Magazine Japan and France. His book "Whisky Rising: The Definitive Guide to the Finest Whiskies and Distillers from Japan" is available in English, Chinese and Japanese.

Akkeshi

Owner:	Location:	Founded:	Capacity:
Kenten Jitsugyo	Hokkaido	2015	109,000 l

Malt whisky range:
Akkeshi New Born Foundations

Asaka

Owner:	Location:	Founded:	Capacity:
Sasanokawa Shuzo	Fukushima P.	2015	32,000 l

Malt whisky range:
none yet

Akkeshi distillery is located near the sea and surrounded by beautiful wetlands with an abundance of peat. This is no coincidence. Company president Keiichi Toita is an aficionado of Islay malts so that's where the inspiration came from.

The inspiration, equipment and methods used may be Scottish, but his goal is to create a whisky that is shaped by the Akkeshi environment: "an aroma and flavour like nothing found elsewhere". They've already got the perfect pairing sorted out as Akkeshi is the only town in Japan where oysters can be shipped out all year round.

Winters are particularly harsh in the Akkeshi area and temperatures can drop to -20°C, which makes whisky production all but impossible, so the maintenance season here is in the winter, rather than in the summer.

Production began in the fall of 2016. After a year of distilling, their warehouse was full, so they built a second one across from the distillery. In 2018, they built a third one close to the sea (above sea level). A fourth one, next to the third one, is currently in the planning stage. Construction should be completed by April 2020. A bottling hall is also being built. This should be completed by the end of 2019 and be operative in early 2020.

On the production, there have been a few interesting developments the last season. The fermentation time was extended to about 120 hours and an original yeast strain is being tested. The staff at the distillery is also pushing ahead with their "Akkeshi All-Stars" project, i.e. whisky made with all materials sourced from Akkeshi. After surveying several possible sites to source peat, a suitable candidate site has been found. In 2017 and 2018, local farmers grew a small amount of barley with satisfactory results, so in 2019, the first proper cultivation of local barley destined for Akkeshi whisky was carried out. In 2018, they managed to get some local mizunara for the first time. In 2019, they also managed to get some puncheons built from mizunara.

In 2018, the first two New Born Foundations were released, in small 200ml bottles. Meant to highlight the progress of the distillery, the quartet was completed in 2019 with the third release (unpeated spirit matured in a Hokkaido mizunara puncheon) and the fourth release was a blend of in-house distilled malt whisky and grain whisky imported from Scotland but matured in Akkeshi. Next up is the first proper single malt whisky release. Watch this space.

Sometimes a small gesture at the right time can have a big impact. If Sasanokawa Shuzo hadn't offered their warehouse space to Ichiro Akuto when he was desperately trying to save the old Hanyu stock from being poured down the drain, there may have been no Chichibu distillery and no post-Ichiro craft whisky boom in Japan.

For that alone, Sasanokawa deserves our respect, but there's more: they also produce fabulous whisky of their own. The company was founded in 1765 and turned its hand to whisky making straight after World War II. Sasanokawa applied for a license to make whisky in 1945 and the year after they got to work. Their focus was on the lowest grade of blended whisky.

As the economy recovered so did people's palates, so Sasanokawa – looking to up their game – started making whisky in makeshift stills (not made out of copper!) Sales weren't always great but the structure of the company kept their whisky business afloat. Sake making took up 2/3 of the year, so the remaining 1/3 the staff was kept busy making whisky.

To mark the 250th anniversary of the company in 2015, Sasanokawa decided to set up a proper malt whisky distillery. By the end of the year, two small pot stills (2,000 litre and 1,000 litre respectively) had been installed in a vacant warehouse, and by June 2016 the distillery was ready to start producing. All processes from milling to filling taking place under one roof. What's more, everything from carrying and feeding the barley to the mill to cleaning the equipment is done by hand.

Asaka distillery's 3rd season ran from September 2018 to mid-July 2019. According to head distiller Daisuke Taura, there were no major changes to the production parameters, but production was increased from 5 to 6 batches a week and the production staff doubled from 2 to 4. "Our goal this season", Taura explains, "was to keep producing the same quality of spirit, but make more of it." The final month of the season was dedicated to making heavily-peated spirit (50ppm). At the end of 2019, we can expect the highly-anticipated first Asaka single malt release, which will be at cask-strength and 100% matured in ex-bourbon wood. For the follow-up releases, the idea is to tap into the wide variety of other cask types in the warehouse.

Chichibu

Owner:
Venture Whisky

Location:
Saitama P.

Founded:
2007

Capacity:
60,000 l

Malt whisky range:
Occasional limited releases

Chichibu #2

Owner:
Venture Whisky

Location:
Saitama P.

Founded:
2019

Capacity:
240,000 l

Malt whisky range:
None yet

Another year, another batch of high profile awards for Ichiro Akuto, the rock-and-roll star of the Japanese whisky world, and his team.

Not only did he take one of the top awards at the World Whiskies Awards for the third year in a row ('World's Best Blended Limited Release', like last year), Ichiro was also crowned 'Master Distiller of the Year' at the 2019 International Spirits Challenge. Meanwhile, his brand ambassador Yumi Yoshikawa was chosen as 'World Whisky Brand Ambassador of the Year' at Whisky Magazine's Icons of Whisky awards. Not bad for a modest operation started up in the hills of Saitama a little over a decade ago.

Chichibu distillery may be a small distillery, but the staff dream big and work hard. There's a 2,400 litre mashtun (manually stirred with a wooden paddle!), 8 mizunara washbacks, 3,000 litre each and a pair of 2,000 litre pot stills. Every year, about 10% of production is dedicated to local barley so there is an area for floor malting. There are 5 warehouses and there's also a cooperage.

Since 2010, Ichiro and his team have been making regular trips to Hokkaido to buy mizunara wood and the two in-house coopers have been perfecting their mizunara-barrel-making skills since 2016. This year they managed to reach another milestone. "We've been making mizunara casks for years," Ichiro explains, "but the wood we used was always sourced at the hardwood log auctions in Hokkaido. This year, however, we started making casks made out of local mizunara – Chichibu mizunara." It's taken 3 years to get to this point and they managed to make 11 casks. Clearly, the effort and cost involved is gigantic, but that's beside the point. It's all R&D for Ichiro and that's what matters.

There's always room for meaningful experiments at Chichibu distillery, so it seems. This year, for example, rather than moving straight into maintenance season after the peated season, they decided to use some non-peated barley after the peated season.

Supply and demand is Ichiro's biggest headache, and in spite of continued production, the situation doesn't show any signs of improvement. Quite the contrary, in fact. At this year's Tokyo International Bar Show in early May 2019, riots broke out on the morning of the first day of the show among those hoping to score a Chichibu single cask bottling available on a first-come first-served basis.

The latest 'big' release (11,500 bottles) was The Peated 2018, released to mark the 10th anniversary of the distillery, but even that vanished into thin air pretty quickly.

Always ahead of the curve, Ichiro Akuto started thinking about setting up a second distillery around 2014.

By the time you are reading this, his new distillery will be up and running. Unlike Suntory, Nikka and even Hombo Shuzo, who built their second distilleries in locations that were distinctly different from the environment of their first distilleries, Ichiro wanted to stay in his hometown of Chichibu. As it turned out, his new distillery is just a two-minute drive away from the 'old' one.

Construction began in April of 2018 and the first spirit (test production) came off the stills on 9 July 2019. There are quite a few features of the new distillery that are the same as at Chichibu distillery. What's very different, however, is the scale: the new distillery is 5 times bigger than the first one. For the first couple of years, Ichiro is planning on working in just one shift.

At Chichibu #2, 2t of malted barley will be processed per batch. The water is the same as at Chichibu distillery. Mashing takes place in a semi-lauter tun – no more hand-stirring the mash like at Chichibu. For the fermentation process, Ichiro is sticking with wooden vessels, but unlike at Chichibu where the washbacks are made of mizunara, the washbacks at the new distillery are made of French oak. There are 5 washbacks at the moment (with 10,000 litres of wort going in), but there's room for a few more when they move to a 2-shift system. The yeast is the same as that used at Chichibu distillery.

The stills are the same shape as at Chichibu distillery and even the lyne-arm angle is the same (12° downward), but they are much bigger (10,000 litres and 6,500 litres respectively) and both stills are direct-fired, whereas those are Chichibu are indirect-heated. Ichiro expects this to have the biggest impact on the character of the spirit:

"I am expecting a more robust, more complex spirit." The distillation process is harder to control, but Ichiro and his team are ready. In 2018, he hired 6 new people to work at Chichibu. This allowed them to train under his experienced staff, who – once the new distillery was ready – could be moved to take on the new challenges there.

There's also a brand new warehouse (No.6) next to Chichibu #2, dunnage style, just like the others, but there's room for more warehouses – either bigger dunnage warehouses or big racked warehouses – at the new site. The future is looking bright and hopefully there'll be plenty of amber nectar to go around in a few years' time.

Fuji Gotemba

Owner: Kirin Holdings
Location: Shizuoka P.
Founded: 1973
Capacity: 2,000,000 l

Malt whisky range:
17 year old Small Batch and occasional limited releases

Kanosuke

Owner: Komasa Jozo
Location: Kagoshima P.
Founded: 2017
Capacity: 110,000 l

Malt whisky range:
none yet

Fuji-Gotemba distillery is nestled at the foot of Mt. Fuji, less than 12 kilometres from the peak.

The 'mother water' at Fuji Gotemba distillery is taken from three bores on site that top into underground streams 100 metres deep. Analysis has shown that the water used today fell on Mt Fuji as snow 50 years ago. That's how long it takes for the water to filter through the hardened lava.

The distillery was established in 1972 by Kirin Brewery, Seagram and Sons and Chivas Brothers as a comprehensive whisky manufacturing plant where all production processes – from malt and grain whisky distilling to blending and bottling – take place on site. Unlike most Japanese distilleries, which followed Scottish whisky-making practice, Fuji Gotemba adopted production techniques and methodologies from all over the world. After Seagram started selling off its beverage assets worldwide, Kirin became the sole owner of Fuji Gotemba Distillery.

The pot stills for malt whisky production at Fuji Gotemba distillery are said to be modeled after those at Strathisla distillery in Scotland, owned by Chivas Bros. The goal at Fuji Gotemba distillery is to produce a malt spirit that is 'clean and estery' so the pot stills were designed with that in mind.

In addition to malt whisky, three types of grain whisky are made at the distillery using a multi-column still, a kettle and a doubler in a modular way. Understandably, given this production set up, most new products coming out of Fuji Gotemba are blended whiskies. These are put together by Master Blender Jota Tanaka and his team and are well worth seeking out. The flagship Fuji-Sanroku Signature Blend was launched in 2017 and revolves around the concept of 'maturation peak', with all component whiskies in the blend at their peak of maturation.

Those keen to try a Fuji Gotemba malt whisky should look out for the 17 year old Small Batch or one of the recent 12 year old Red Wine Cask Finish single cask releases. With regards to the latter, Tanaka points out they don't simply buy casks from just any winery: "We get high quality French oak casks used for the ageing of red wine at Chateau Mercian, which is part of the Kirin group."

If the timing is right – and you don't mind a trip to the distillery – you may be able to pick up one of the Distiller's Select Single Malt bottlings (there is also a Single Grain) which is put together ever year in the spring.

Japan doesn't have any equivalent of the Scottish whisky producing regions and distilleries are pretty much scattered all over the place. The only exception is Kagoshima prefecture, where Tsunuki and Kanosuke Distillery are pushing the envelope.

Not only are they located on the same street, albeit 40km apart, but there's also friendly cooperation between the two distilleries, which is rare for a country where producers play their cards close to their chest and don't swap any stock.

Kanosuke distillery is the youngest of the two. It is owned by Komasa Jozo, one of the leading shochu makers in the area. Their claim to fame is 'Mellowed Kozuru', a barrel-aged shochu developed by the second president of the company Kanosuke Komasa and launched in 1957.

The idea to establish a whisky distillery was born in 2015 and the project was spearheaded by Yoshitsugu Komasa, who represents the fourth generation of the family to play a leading role in the company. As location for the new distillery, they picked some vacant land next to three warehouses where the company's shochu is matured. The location itself is stunning. The distillery overlooks the East China Sea and Fukiage beach, which is the longest beach in Japan.

Kanosuke Komasa had a vision to build a brand home for Mellowed Kozuru on that piece of land, so in recognition of his contribution to the company, the decision was made to name the distillery after him. The equipment was installed in the summer of 2017: a 6,000 litre mash tun, 5 stainless steel washbacks and 3 pot stills of 6,000 litres, 3,000 litres and 1,600 litres capacity respectively, all with wormtub condensers. This peculiar set up allows for various double distillation permutations as the middle one can function as either wash or spirit still. Production officially started on November 13 2017.

Limited edition releases of whisky-in-progress matured in various cask types have been presented at whisky fairs throughout Japan and well received. One of the plans for the future is to use their shochu-making know-how to produce a rice whisky. This would be done in the copper pot stills, using Kagoshima rice and their in-house shochu yeast.

Photo © Kirin

Kurayoshi

Owner:
Matsui Shuzo

Location:
Tottori P.

Founded:
2017

Capacity:
undisclosed

Malt whisky range:
The Matsui Sakura Cask, Mizunara Cask and Peated.

Nagahama

Owner:
Nagahama Roman Beer Co.

Location:
Shiga P.

Founded:
2016

Capacity:
undiscl.

Malt whisky range:
none yet

Kurayoshi is, without a shadow of a doubt, the most controversial distillery in Japan.

Kurayoshi first emerged on the Japanese whisky scene in 2015 as a brand rather than a distillery, with age-statement releases of 'Japanese pure malt whisky' at a time when age-statement Japanese whisky had become rare as hen's teeth. The fact that Japanese whisky-makers don't swap stock together with the extremely lax regulations governing Japanese whisky made it easy for savvy consumers to figure out that the liquid in the bottles was imported in bulk from abroad. However, everything about the presentation and wording on the labels screamed 'Japanese whisky' and therefore casual (especially non-Japanese) consumers snapped up these Kurayoshi whiskies in the belief that they had, against all odds, managed to find some rare, (long-)aged Japanese whisky. Needless to say, they didn't make many friends on the Japanese whisky scene.

In 2017, they set up an actual distillery, but all of this happened away from the public eye, unbeknownst even to industry-people in Japan. They started with three small 1,000 litre Hoga alembic-type stills of the kind used at Nagahama distillery. In 2018, they added two larger stills (3,000 and 5,000 litres respectively) and slowly started to let people into their distillery. Age statements were dropped from their various bottlings, which was a clever tactical move, as it was no longer possible to know whether the liquid inside the bottles was made in-house, imported in bulk from abroad or a mix of both.

At the time of writing, whisky was only produced using the large pair of pot stills, the plan being to use the three small alembic stills to make brandy from local wine. The single malt expressions mentioned above are all made in-house and about 1 1/2 year old (which is legally whisky in Japan).

Ironically, it's one of the company's dreams to "produce a whisky wholly sourced from the Kurayoshi area". Since 2018, they have been working with local farmers who are growing Shunrei, a two-row spring barley cultivar developed in Fukuoka in 2004, on their behalf. It will be interesting to see if Matsui Shuzo will be able to lure customers turned off by their deceptive tactics back and to get the whisky community to evaluate their whiskies made in-house with an open mind.

Nagahama Distillery's motto is "one distillation, one barrel" which suits them well, as they are the smallest distillery in Japan at the time of writing. It is located in the picturesque town of the same name in Shiga and was set up in a record time of 7 months.

The owners didn't have to start from scratch, because the distillery is, in fact, an extension of Nagahama Roman (with the emphasis on the second syllable, as in "romantic") Brewery, which was established in 1996 as a brewpub. The first half of the whisky-making process – mashing and fermentation – takes place in the equipment used for beer making. For the second half of the process, a small "still room" was created behind the bar counter.

Inspired by some of the new wave of craft distillers in Scotland (Strathearn and Eden Mill, in particular), the team at Nagahama decided to go for small stills with alembic heads of the type seen more often in calvados, cognac or pisco distillation than in whisky making. They started with a 1,000 litre wash still and a 500 litre spirit still, but in 2018, they took out the small spirit still, and put in 2 new 1,000 litre stills. Two of these identical stills are used as wash stills and one as a spirit still. If they want to produce more they will have to relocate as three is indeed a crowd as far as this stillroom is concerned.

Nagahama Distillery is all about variety. Using different types of barley and a wide plethora of casks, their goal is to have as many distinctly different types of whisky as they have casks maturing. In the spring and summer of 2019, they received Swedish oak octaves, Oloroso sherry quarter casks, ex-Koval casks and ex-bourbon quarter casks.

Up until last year, releases coming out of Nagahama distillery have all been new make, both unpeated and peated, but in the summer of 2019, the distillery started selling some limited quantities of aged "prototypes" in small 100ml bottles at various events in Japan. The first three of these were: Amontillado Cask Finish, Laphroaig Quarter Cask and French Oak. The distillery also set up a 2-day 'distillation experience' that is run a few times a month. The first of these mini-schools was held in April 2019.

Nukada

Owner: Kiuchi Shuzo

Location: Ibaraki P.

Founded: 2016

Capacity: 7,200 l

Malt whisky range: none yet

Okayama

Owner: Miyashita Shuzo

Location: Okayama P.

Founded: 2011

Capacity: 7,000 l

Malt whisky range: Okayama Single Malt Whisky

In terms of output, Nukada is the smallest whisky distillery in Japan, but as we all know, size isn't everything. It may be small in size, but the people here dream big.

Nukada disitllery was set up by Kiuchi Shuzo in a corner of their new Hitachino Nest brewhouse in 2016. Head distiller Isamu Yoneda's uses a 1,000 litre hybrid still to make whisky as well gin.

The production volume varies from year to year, but is usually no more than 7,200 litres of puree alcohol. Production is also limited by the fact that the staff is occupied with a multitude of other tasks beside making whisky. The big project that occupied them was setting up their second distillery near Mt. Tsukuba (see Yasato distillery, page 208), for which they sourced equipment in half a dozen countries around the world. Understandably, the output at Nukada distillery was down a bit the past year compared with previous years. Between April 2018 and March 2019, roughly 3,600 litres were produced.

"Since April 2019," head distiller Sam Yoneda reports, "Nukada distillery has only been distilling beer for spirits and liqueurs. The barrels allocated to whisky have all been filled."

Atypically for the new wave of Japanese craft distilleries, Nukada has resisted the temptation (or need) to release new-make or 'newborn' whisky in their first few years, so the general whisky-drinking public has no idea of the style(s) or flavor profile(s) involved. This year, however, the company offered a sneak preview, again in an atypical way, by launching a limited edition 9%abv canned highball on 1 April 2019 – and no, this wasn't an April Fool's joke. The whisky used was both malt and grain distilled at Nukada distillery, aged in Spanish sherry and wine casks for 3 years.

"It's quite rare," Yoneda points out, "for a craft distillery to have the equipment to carbonate their own water and have their own canning line, but we do, so it seemed like an interesting side project." And a side project it will remain, as they don't want to use too much of their Nukada stock.

Small batch whisky bottlings may start to appear towards the end of 2019, but this will be in limited quantities and to a limited clientele.

As far as whisky production is concerned, Miyashita Shuzo is undoubtedly the most under-the-radar distillery. That's by design. The bread and butter of the company is beer and sake, and whisky is a side-gig.

Miyashita Shuzo was founded as a sake brewery in 1915. In 1994, they became one of the pioneers of Japanese craft beer. In 2003, some folks at the company had the bright idea to distill some hoppy beer in their stainless steel shochu still and put the spirit in American white oak casks. Pleased with the way this was developing, they decided to have a go at producing malt whisky. They acquired their license in 2011 and started double-distilling batches in their shochu still – but stainless steel and whisky are awkward bedfellows so in the summer of 2015, they installed a copper hybrid still. That's been used for whisky making ever since.

With more sunny days and fewer rainy days than most other prefectures in Japan, Okayama is known as 'The Land of Sunshine' and is a traditional barley-growing region. The people at Okayama Distillery are keen on using as much local barley as they can, which comes at a price, but is worth the effort, they say.

Unlike other craft whisky producers in Japan, Okayama Distillery has resisted the temptation to sell new make or whisky-in-progress. In fact, they have sold very little whisky so far. Once in a while, they sell 30-odd bottles of their Okayama Single Malt (a vatting of 3-5 year old whisky, made using local barley and aged in a sherry and brandy casks) on their website by lottery, but this is all very un-hyped and goes by almost unnoticed among Japanese whisky aficionadoes. A recent limited edition Okayama Single Malt Triple Cask (not aged in 3 different cask types consecutively, as one might assume, but simply a vatting of malt matured for over 3 years in 3 different cask types, i.e. brandy, sherry and mizunara) was awarded gold at the 2019 Meiningers International Spirits Awards.

Production is very limited so it is unlikely that the Okayama Single Malt will ever have a big presence on the whisky scene, but it would be good to have some products more widely available. Even in Japan and even at bars, a bottle of Okayama Single Malt is like a unicorn… reputed to exist, but who has ever seen one in the wild?

Saburomaru

Owner:
Wakatsuru Shuzo

Location:
Toyama P.

Founded:
ca. 1990

Capacity:
15,000 l

Malt whisky range:
occasional releases

Sakurao

Owner:
Chugoku Jozo

Location:
Hiroshima P.

Founded:
2018

Capacity:
t.b.d.

Malt whisky range:
none yet

Wakatsuru Shuzo traces its history back to 1862, when they started making sake. Like so many sake breweries hit by the rice shortages following the end of the Pacific War, they turned their attention to whisky-making.

Throughout the second half of the 20th century, they dabbled in cheap whisky. It wasn't until September 2016 that they decided to start revitalizing their distillery with the aim of producing high-quality malt whisky. Up until 2017, malt whisky was produced at Saburomaru distillery using an alumite pot still of the type commonly used in shochu-making. After a successful crowdfunding campaign, the distillery building was refurbished, a brand new mill and mashtun was installed, which increased production efficiency, and the pot still got a copper swan neck.

Encouraged by the positive developments, distillery manager Takahiko Inagaki decided to keep pushing forward. In June 2019, the distillery unveiled its brand new pair of pot stills: the world's first cast pot stills. The nearby town of Takaoka has a 380-year history in copperware production so it wasn't a big a stretch to have the pot stills made locally. Oigo Works, a company specializing in temple bells, made the stills from an alloy consisting of 90% copper and 8% tin.

The stills have thicker walls so they're expected to have a longer working life. Both stills have lantern heads and downward lyne arms and are fitted with shell-and-tube condensers. The wash still (2,600 litres) is heated with a hybrid indirect and direct steam system, the spirit still (3,800 litres) with indirect steam only. The first distillation using the new so-called Zemon stills took place on 20 June 2019.

The whisky season at Saburomaru in 2019 was very short: from mid-June to the end of the August. Distillation took place 4 times a week, from Thursday to Sunday. All barley used was heavily peated (50ppm), and the yeast was a mix of ale and distillers yeast. The total production volume in 2019 was 15,000 litres of pure alcohol. This season saw an interesting new development on the maturation front: the use of mizunara cask heads made in the local woodcarving town of Inami. Watch this space!

Wakatsuro Shuzo has been releasing a blended limited edition called 'Moonglow' for the past few years, but we're still waiting for malt whisky bottlings made under the new regime. It's still early days, of course.

Sakurao may be a new distillery, but the liquor company behind it is not exactly a new-kid-on-the-block. Chugoku Jozo was established in October 1918 as a limited partnership and incorporated in 1938, when it was given its current name.

Their liquor portfolio comprises shochu, sake, mirin and various liqueurs. One of the company's claims to fame is that it was the first to release sake sold in cartons (in 1967).

Chugoku Jozo started 'producing' whisky in 1938, but exactly how is lost in the mists of time. Up until the liquor-tax change of 1989, their field was second-grade whisky, bottled at the lowest legally-allowed abv (37%), mostly in 1.8l bottles – in other words, utilitarian products meant to consumed without much attention to depth of flavour. In 2003, they launched the Togouchi brand, but the expressions in that range are all made up of whisky imported in bulk from abroad.

To mark the 100th anniversary of the company, a proper whisky distillery was set up at the company's main site. It represents a new challenge for the company, and visually that's made clear as well: a modern, raven-black distillery building standing out against the dull white and grey dilapidated surroundings. The new distillery is very compact and is used to make gin a well. It's hard to imagine one could carry out all steps of the whisky and gin-making process in a smaller space than the one occupied by the 'black box' that is Sakura distillery, but they seem to manage.

Half of the production is non-peated and the other half lightly-peated (20ppm). Everything is double distilled even though there is only one hybrid still at the distillery. For the second distillation, the column is used. The first casks filled were sherry butts. Casks are matured on site as well as in disused tunnels in the town of Togouchi, in the Nishi-Chugoku Sanshi Quasi-National Park.

The company has chosen to focus on selling gin while waiting for the whisky to come of age, so it will be a while before we have an idea of what the character of the Sakurao whisky will be like.

Shinshu

Owner: Hombo Shuzo
Location: Nagano P.
Founded: 1985
Capacity: 90,000 l

Malt whisky range:
Komagatake (various limited edition releases)

Shizuoka

Owner: Gaia Flow Distilling
Location: Shizuoka P.
Founded: 2015
Capacity: 60,000 l

Malt whisky range:
none yeat

"Strikes and gutters" is how you could summarize the colourful history of Mars Shinshu Distillery.

It was built at the peak of whisky consumption in Japan, but the trouble with peaks is that they are followed by a fall – in the case of Japan, a 25-year long decline. Parent company Hombo Shuzo had been making whisky since 1949. Initially, they simply blended sourced components with neutral spirits, but from 1960, they distilled proper malt whisky in various places. It was a history of starts and stops, at the mercy of nebulous market forces. In 1985, they moved their whisky operation to Shinshu but with a gradually shrinking domestic market, they were forced to mothball their distillery in 1992. In 2011, with whisky booming again, they fired up the stills once more.

Production used to be limited to the winter months, but now it's closer to a typical year-round production schedule with a silent summer season. The silent season seems to get shorter every year. The 2018-19 season ran from early September to late June. The first mashing of the current season took place on August 24, so production is at an all-time high.

Since 2014, the distillery has received some much-needed upgrades. In November 2014, the old pot stills were replaced with brand new ones, built following the original blueprints. At the end of August 2018, the 5 rusty cast-iron washbacks got new company in the form of 3 Douglas fir washbacks. The plan is to phase out the old cast-iron ones and replace them with new stainless steel washbacks. At the time of writing, a new maturation warehouse was under construction, which is part of a massive 1.2 billion yen investment, which will also see the construction of a new distillation facility and visitor centre, slated for completion in September 2020.

In terms of malted barley used, four types of distillate are made at Mars Shinshu: non-peated and peated at 3.5, 20 and 50ppm. Keen to explore the influence of climate on the maturation process, some Mars Shinshu spirit has been sent to Tsunuki in Kagoshima and Yakushima island annually since 2014. Eager to get the fans on board for this adventure, the company has been releasing limited editions the past two seasons showcasing how impressive the results are, even after just a few years: Komagatake Single Malt Limited Edition, Komagatake Yakushima Aging (self explanatory), Komagatake Double Cellars (Shinshu distillate matured at home and at Tsunuki) and a handful of single casks in the much-loved Papillon series.

Most craft distillers in Japan don't give much thought to the aesthetics of their production site. Shizuoka distillery is the proverbial exception to this rule.

It's the most ingeniously designed distillery in Japan. Inspired by Karuizawa distillery, everything from milling to filling (the barley and casks, resp.) takes place under one roof, but in different 'rooms'. Another thing that's carefully considered is the way in which the landscape – small green tea farms and forested mountains – is visible from various points in the distillery building. Also, the 'visitor experience' is integrated into the design of the distillery, which is the exception rather than the rule in Japan.

Inspired by a visit to Kilchoman distillery in 2012, Gaia Flow founder Taiko Nakamura started thinking about setting up a distillery of his own back home. He set up a liquor import company to get a foot in the door of the drinks business, and kept working on his distillery project. Shizuoka distillery was officially opened in February 2017.

Up until 2017, there were 5 washbacks at the distillery: four made from Oregon pine and one made from local, Shizuoka cedar. In February 2018, three more Shizuoka cedar washbacks were installed, bringing the total to 8. The stillhouse has 3 pot stills: one from the old Karuizawa distillery and a new pair made by Forsyths. Both of the new stills have a bulge (or boil ball). Interestingly, Nakamura opted for direct (wood-fired) heating for the wash still. The old Karuizawa still and the new spirit still are steam heated.

The most interesting development during the last production season is the 'All-Shizuoka Whisky' project: whisky 100% made with Shizuoka-sourced raw materials. This applies to the water and the barley, obviously, but also to the yeast (a new strain was developed at the nearby Numazu Industrial Support Center), the wood of the fermenters used and the fuel used to heat the wash still (local lumber). The first all-Shizuoka spirit came off the spirit still on November 23, 2018.

Since December 2018, the site is open to the public. There was also exciting news for enthusiasts abroad keen to follow the progress of Shizuoka distillery more closely, too. Up until now, the Shizuoka Private Cask program was limited to the Japanese market. In early 2019, it was made available in selected foreign markets, giving interested parties the chance to choose from new make distilled with the Karuizawa wash still or with the wood-fired still.

Tsunuki

Owner:
Hombo Shuzo

Location:
Kagoshima P.

Founded:
2016

Capacity:
90,000 l

Malt whisky range:
none yet

White Oak

Owner:
Eigashima Shuzo

Location:
Hyogo P.

Founded:
1984

Capacity:
60,000 l

Malt whisky range:
Akashi NAS and occasional, limited releases

Towards the end of 2015, Hombo Shuzo surprised friend and foe when they announced they were in the process of setting up a second distillery in their homebase of Tsunuki in Kagoshima. This was a first for a craft producer in Japan, and it seemed like a strange move to lots of people at the time.

Wouldn't it have been more efficient to make more spirit at Mars Shinshu distillery? Efficient, yes – but this wasn't about efficiency. It was about diversity and expanding the Hombo whisky palette.

The very first distillation was on October 27th (stripping run) and 28th (spirit run). The distillery is the playground of Tatsuro Kusano, the 31 year old head distiller. Kusano learned the ropes at Mars Shinshu under distillery manager Koki Takehira, but he has his own vision for Tsunuki and an inquisitive mind. Some things are the same: the season runs roughly parallel with Mars Shinshu, i.e. September to June, and the barley used is the same too: non-peated, lightly-peated, medium peated and heavily peated. But there are marked differences, too – not just in terms of the equipment in place, but the approach to making whisky.

In 2018, Kusano moved from a 1 ton batch to a 1.1 ton batch size, in preparation for the addition of 100g of specialty malts (caramelized and roasted) in some batches this season. In addition to various beer yeasts, he's also trialed some in-house shochu yeast – always in combination with their regular distillers yeast.

So far, there have only been occasional releases of various types of new make (which, confusingly, is the term they use for both 'white dog' and whisky matured for under 3 years) bottled for the visitor centre and the Tsunuki Distillery Matsuri, which takes place in the fall every year.

One thing is already clear, though: Tsunuki won't be a one-trick pony. Between the different types of spirit, the creativity of the team, the wide variety of casks used and the three different maturation locations (home, at Mars Shinshu and on Yakushima Island), there's bound to be a plethora of Mars Tsunuki whiskies coming at the discerning drinker very soon.

Eigashima is one of the few distilleries in Japan that seems to lack the 'ambition gene', which is endearing, in a way.

Whisky making used to be a marginal enterprise and they didn't participate in any whisky shows in Japan until recently, simply because they felt they didn't have the artillery to match their colleagues in the field. That has slowly been changing and now, if there is a whisky festival somewhere in Japan, chances are they'll have a little table with some interesting new releases. No crowds and no hype, but you'll always find something to stimulate the taste buds.

The supreme irony is that this humble producer from Hyogo prefecture has been part of the Japanese whisky scene longer than anyone else that is still active today. On paper, it's the oldest whisky producer in Japan – having acquired a distilling license in 1919, four years before Yamazaki. It took them four decades to get their act together, though, and another four decades to release their first single malt (an 8 year old in 2007). The current distillery was built in 1984.

Since two years ago, the company has expanded its whisky-making season to seven months. The old spirit and wash still were retired in February 2019 and replaced with brand new stills made by Miyake Industries in Japan. The first distillation in the new stills took place on March 28 (stripping run) and March 29 (spirit run). This also marked the start of the 2019 season. It'll be interesting to see how the character of the spirit will be affected by the new stills. Those nostalgic for the old dilapidated stills can see them out in the courtyard on display.

All production is matured on site, near the Akashi strait, in old single-story rickety warehouses. They mostly fill into ex-bourbon wood, but they also have sherry butts, cognac casks, wine, tequila and recharred ex-barley shochu casks, as well as domestically-made virgin oak.

In November 2018, a very interesting limited edition Akashi came out. A first for whisky as far as we know, this was matured for 3 years, in the case of this release, in casks that previously held sake. A limited edition, single cask expression of this was released in the Ghost Series. In the past, the distillery was not open to the public, but this has changed in recent months. In fact, now there is a 100ml new make bottle exclusively available at the distillery to tempt you into making the trek.

Yasato

Owner:
Kiuchi Shuzo

Location:
Ibaraki P.

Founded:
2019

Capacity:
not determined

Malt whisky range:
None yet.

Yuza

Owner:
Kinryu

Location:
Yamagata P.

Founded:
2018

Capacity:
not determined

Malt whisky range:
None yet.

Yasato distillery is Kiuchi Shuzo's second whisky distillery, located in the Yasato part of Ishioka city, very close to Mt. Tsukuba.

It is situated in a building that was owned by the city and previously used as a public hall, school lunch kitchen, restaurant and community center. The building has been completely renovated inside and outside while keeping the original shape for the locals.

The distillery will be making malt and grain whisky, like at Nukada distillery, but on a much bigger scale. It's equipped with a four-roller mill, a 5,000 litre cereal cooker for step mashing and rice/buckwheat/corn cooking and a 6,000 litre lauter tun, four 12,000 litre stainless steel fermenters and four 6,000 litre wooden fermenters (two acacia and two made of European oak). There's a 12,000 litre wash still and an 8,000 litre spirit still, both with a straight head and a downward lyne arm.

The idea is to display the character of the grain(s) used in the spirit produced. Ex-bourbon and ex-sherry casks will be used for maturation for the most part, but casks partially made of 'sakura' (cherry blossom tree) and Japanese chestnut wood will also be used.

The staff at Yasato distillery is aiming to start mashing at the end of September 2019. The plan is to process two 1.2 tonne batches a day. They will also try to work with farmers growing Japanese barley as much as possible. At the moment, it looks like they may even get a supply of corn from a farmer in Hokkaido.

"The goal," head distiller Sam Yoneda says, "like at our other distillery in Ibaraki, is to be the most innovative whisky producer in Japan, and to be creative with different grains and yeast strains. The ultimate goal is to create something that people would categorize as 'Ibaraki whisky' and to contribute to developing a whisky region with a particular identity, not unlike whisky regions in Scotland, in the same way that Kagoshima is starting to emerge as such a region in Japan."

Yasato will be a distillery to watch, not in the least because Kiuchi Shuzo is not a company known to rush product to market. Don't they say good things come to those who wait?

Kinryu, the company behind Yuza Distillery, was founded in 1950 in Sakata city in Yamagata prefecture.

It was a joint venture funded by 9 local sake producers, initially to make neutral spirit (which is added to most sake to improve the taste and/or inexpensively increase volume, depending on who you ask). Over time, they started making and selling shochu made in a continuous still, in the case of Kinryu, mostly from molasses. Kinryu is the only specialized shochu maker in Yamagata and the bulk of what they make is sold in their home prefecture. And therein lies the problem. Overall consumption of shochu (as well as sake) has been on the decline for decades. But there's an even more alarming downward curve in play here. Over the next three decades, the population of Yamagata is expected to fall by over 30%, twice as hard as the national average.

The folks at Kinryu knew the very survival of their company was at stake and they also realized the time to act was now rather than 20 years later. Unsurprisingly, they turned to whisky in search of a brighter future. The team spent a year looking for a suitable location and found it in Yuza city, at the foot of Mt Chokai. Mt Chokai has the highest precipitation of any mountain in Japan, so the abundance of good quality spring water was an important consideration.

For the equipment, Kinryu decided to go with the Scottish company Forsyth. The first official distillation took place on November 4th and the first cask was filled on November 6th. The distillery is run by a team of absolute beginners – a mashwoman and stillwoman fresh out of college and a warehouseman who left his job in banking to pursue his dream of making whisky – under the watchful eye of company veteran Masaharu Sasaki. "I didn't want to bring in an expert," Sasaki explains, "because then it becomes that person's distillery. I wanted to start from zero, with young, motivated people and put our own stamp on things."

The set up and the processes are textbook Scottish and the goal is to produce a high-quality single malt. They're aiming for a rich yet clean spirit. No gimmicks, no hype and not selling the bear's fur before it's shot. If they stick to their guns, they may turn out to become the Japanese equivalent of Daftmill Distillery in Scotland – if not in character of spirit, at least in terms of ethos. Time will tell.

Yamazaki

Owner: Beam Suntory   **Location:** Osaka P.   **Founded:** 1923   **Capacity:** 6,000,000 l

Malt whisky range:
NAS, 12, 18, 25 years old and occasional limited releases.

Hakushu

Owner: Beam Suntory   **Location:** Yamanashi P.   **Founded:** 1973   **Capacity:** 4,000,000 l

Malt whisky range:
NAS, 18, 25 year old plus occasional limited releases

The first proper malt distillery in Japan, Yamazaki Distillery was founded by Shinjiro Torii in 1923.

The distillery has been expanded many times over the years, first in 1957, and most recently in 2013, when four pot stills were added bringing the count to 16. There's plenty of variety in terms of heating method, shape, size, lyne-arm orientation and condenser type. Fans of the distillery were delighted with the release of the second 'Essence of Suntory' trilogy in February 2019, which were all Yamazaki single malts: a 9yo virgin Spanish oak, a 9yo ex-Montilla wine cask and a 10yo refill sherry cask expression.

Hakushu was built 50 years after the first Suntory malt whisky distillery and is nestled in a forest area at the foot of Mt Kaikomagatake.

The original distillery had 6 pairs of stills. In 1977, capacity was doubled and another 6 pairs added. In 1981, Suntory gave up on what was then the biggest malt whisky distillery in the world, and built a new distillery on the site, this time focusing on variety of spirit rather than volume. In 2010, a small grain whisky facility was established. Hakushu has been completely out of the spotlight this year with no new releases but production is in full swing so hopefully this is a case of no news is good news.

Yoichi

Owner: Nikka Whisky   **Location:** Hokkaido   **Founded:** 1934   **Capacity:** 2,000,000 l

Malt whisky range:
NAS and occasional limited releases.

Miyagikyo

Owner: Nikka Whisky   **Location:** Miyagi P.   **Founded:** 1969   **Capacity:** 3,000,000 l

Malt whisky range:
NAS and occasional limited releases

Yoichi distillery was founded by Masataka Taketsuru in 1934, after ten years as Shinjiro Torii's right-hand man.

The first spirit ran off the still in 1936. Initially equipped with a single still that doubled as spirit and wash still, the distillery now houses 6 stills, all coal-heated. The 'house style' is peaty and heavy, but Yoichi is set up to create a wide range of distillates. In September 2018, a Yoichi Manzanilla Wood Finish was released, limited to the domestic market and 4,000 bottles only. March 2019 saw the release of the very pricy 'Single Malt Yoichi Limited Edition 2019' (300k yen), which is a NAS release, but contains malt whisky spanning five decades (1960s to 2000s).

Originally known as 'Sendai', the distillery was renamed 'Miyagikyo' when Asahi took control of Nikka in 2001.

The distillery is equipped with 22 steel washbacks and 8 huge pot stills with boil balls and upward lyne arms which results in a lighter, cleaner spirit. The site also houses two enormous Coffey stills. To mark the 50th anniversary of the distillery, Nikka released a 'Single Malt Miyagikyo Limited Edition 2019', which contains malt spanning five decades, including some of the very first whisky made at the distillery. In September 2018, a limited release of Miyagikyo single malt finished for 18 months in 50+yo ex-Manzanilla casks was released for the domestic market.

Grant Stevely - founder and owner
of Dubh Glas Distillery in Canada

Distilleries
around the globe

Including the subsections:
Europe
North America | Australia & New Zealand
Asia | Africa | South America

To get a clear picture of how the world of malt whisky distilleries has expanded since I wrote the first Malt Whisky Yearbook back in 2005, I chose six non-Scottish countries. I counted the number of distilleries in MWY2006 and compared them to the number in this book, MWY2020.

This year there are 121 malt distilleries in the USA (there were only six in the first edition of the book). For the other countries the corresponding figures are Canada 14 (1), Australia 52 (5), England 16 (0), France 34 (2) and Germany 55 (2). A number of countries were not represented in MWY2006 but can boast several distilleries today: The Netherlands, Denmark, Norway, Iceland, Argentina, Taiwan, Israel, New Zealand, Liechtenstein and Italy.

All these distilleries produce an incredible plethora of whiskies where some of the producers aim to mimic the traditional, Scottish way of making whisky down to smallest detail,

while others see no limits in product development. Comments like "Scotch whisky no longer best in the world" and "world whisky could become a threat to Scotch" are often heard.

In my view, that's a narrow-minded and limited way of describing the state of the whisky world and in most cases, producers of Scotch agree. More countries getting involved in the production of malt whisky means that the interest for the entire category is growing among consumers and it benefits the industry as a whole.

Today, a large number of whisky drinkers can enjoy the possibility of "travelling the world" with a glass in hand. Sometimes looking for similarities or discrepancies but above all enjoying the wide offering of different styles of malt whisky that exist today compared to 15 years ago. We, the consumers are the winners of this expanded world of whisky.

Europe

Austria

Destillerie Haider

Roggenreith, founded in 1995

www.whiskyerlebniswelt.at

In the small village of Roggenreith in northern Austria, the Haider family has been distilling whisky since 1995 and three years later, the first Austrian whisky was released. In 2005, they opened up a Whisky Experience World with guided tours, a video show, whisky tasting and exhibitions. The wash is allowed to ferment for 72 hours before it reaches either of the two 450 litre Christian Carl copper stills. The desired strength is reached in one single distillation, thanks to the attached column. The main part (70%) of the whisky production is made from either 100% malted rye or a combination of rye and malted barley rye while the rest is from malted barley. Three expressions make up the core range and apart from them there are peated versions as well and limited releases are launched regularly. Some of the latest include a single malt matured for nine years in Austrian oak and then finished for a year in ex-Laphroaig casks.

Broger Privatbrennerei

Klaus, founded in 1976 (whisky since 2008)

www.broger.info

The production of whisky at this distillery owned by the Broger family is supplementing the distillation and production of eau de vie from apples and pears. The distillery is equipped with a 150 litre Christian Carl still. The current range of whiskies consists of Triple Cask, Medium Smoked (smoked using beech wood), Burn Out (heavily peated), Riebelmais (corn whisky), Dinkel and the limited Distiller's Edition which has been maturing in madeira casks.

Other distilleries in Austria

Reisetbauer

Kirchberg-Thening, founded in 1994 (whisky since 1995)

www.reisetbauer.at

Specialising in brandies and fruit schnapps, a range of malt whiskies is also produced. The distillery is equipped with five 350 litre stills. The 70 hour-long fermentation takes place in stainless steel washbacks. The current range of whiskies have all been matured in casks that have previously contained Chardonnay and Trockenbeerenauslese and include a 7, a 12 and a 15 year old.

Destillerie Rogner

Rappottenstein, founded in 1997

www.destillerie-rogner.at

Originally a producer of spirits from fruits and berries, whisky has recently been added to the range. The range consists of Rogner Waldviertel Whisky 3/3 (two versions, malted and unmalted). Rye Whisky No. 13 and a single malt, Whisky No. 2. A peated 12 year old named Old Power was released in June 2019.

Destillerie Weutz

St. Nikolai im Sausal, founded in 2002

www.weutz.at

The distillery added whisky to the range in 2004 when they started a cooperation with a local brewer. Some of the whiskies are produced in the traditional Scottish style while others are more unorthodox, for example based on elderflower.

Old Raven

Neustift, founded in 2004

www.oldraven.at

More than 250,000 litres of beer are produced yearly and the wash from the brewery is used for distillation of whisky. The triple distilled Old Raven comes in three expressions – Old Raven, Old Raven Smoky and the recently released limited edition Old Raven Black Edition.

Destillerie Hermann Pfanner

Lauterach, founded in 1854

www.pfanner-weine.com

In 2005, more than 150 years after the foundation, the owners expanded into whisky production. The two core expressions are Pfanner Single Malt Classic and Single Malt Red Wood with a maturation in red wine casks. There are also limited releases; the smoky Pfanner Whisky X-peated and the O-wood with a maturation in Grand Marnier casks!

Keckeis Destillerie

Rankweil, founded in 2003

www.destillerie-keckeis.at

Whisky production started in 2008 and today one expression, Keckeis Single Malt is for sale as well as the new make Keckeis

Jasmin Haider, CEO of Destilleri Haider

Baby Malt. Part of the barley has been smoked with beech and maturation takes place in small ex-sherry casks.

Dachstein Destillerie

Radstadt, founded in 2007

www.mandlberggut.com

Apart from production of various spirits from berries, malt whisky is also produced. Maturation takes place in a mix of casks – new Austrian oak, ex-sherry casks and red wine casks. Their only release so far is the five year old Rock-Whisky which is distilled 2,5 times.

Edelbrennerei Franz Kostenzer

Maurach/Achensee, founded in 1998, whisky since 2006

www.schnaps-achensee.at

A huge range of different spirits, mainly from fruits and berries, as well as whisky is produced. Several expressions under the name Whisky Alpin have been released including a 6 year old single malt with a sherry cask finish, a 6 year old 100% single malt rye and a rye with a finish in amarone casks.

Brennerei Ebner

Absam, founded in 1930

www.brennereiebner.at

Whisky production in this combination of a guesthouse, brewery and distillery, started in 2005. Whisky is just a small component of the business but, besides a single malt from barley, they have also released whiskies made from maize, dinkel and wheat.

Wieser Destillerie

Wösendorf in der Wachau, founded in 1996

www.wieserwachau.com

Traditionally a distillery producing schnaps and liqueur from fruits and berries, they have also launched a quadruple distilled whisky under the name Uuahouua where the oldest released so far is 10 years old.

Spezialitätenbrennerei Lagler

Kukmirn, founded in 2009

brennerei.lagler.cc

One of few distilleries that are using vacuum distillation. Two single malts and a blend are available under the name Pannonia.

Pfau Brennerei

Klagenfurt, founded in 1987

pfau.at

Focused on production of "Edelbrände" from apples, pears and berries, the owners also have whisky in the range. The core expression is a 7 year old single malt but recently a limited 15 year old has also been released

Belgium

The Owl Distillery

Grâce Hollogne, founded in1997

www.belgianwhisky.com

The first commercial bottling of Belgium's first single malt, 'The Belgian Owl', appeared in November 2008 and the core expression today is the un-chillfiltered Belgian Owl, a 3 year old bottled at 46% or at cask strength. Recent limited releases from the owner, Etienne Bouillon, include the 5 year old Eternity. The distillery is equipped with a 2.1 ton mash tun, four washbacks and two stills that had previously been used at Caperdonich distillery. All the barley used for production comes from farms close to the distillery.

De Molenberg Distillery

Blaasveld, founded in 1471 (whisky since 2003)

www.stokerijdemolenberg.be

Charles Leclef started out as a brewer and currently maintains this role at Brouwerij Het Anker. In 2010, he started a distillery of his own at the Leclef family estate, Molenberg, at Blaasveld. The wash still has a capacity of 3,000 litres and the spirit still 2,000 litres. The first bottles under the name Gouden Carolus Singe Malt, appeared on the market in 2008. The core expression is the 3 year old Gouden Carolus Single Malt. Limited expressions are released yearly with the 5 year old Victor, matured in first fill bourbon, as the latest.

Other distilleries in Belgium

Kempisch Vuur

Zandhoven, founded in 2011

www.kempisch-vuur.be

The distillery is equipped with a German continuous still and the spirit is matured for 18 months in ex-bourbon casks and then another 18 months in quarter casks from Laphroaig. The first batch was released in March 2016 and several releases have followed. The annual production is around 1.000 litres of pure alcohol.

Brouwerij Wilderen

Wilderen-St. Truiden, founded in 2011

www.brouwerijwilderen.be

A combination of a beer brewery with a history going back to 1642 and a distillery founded in 2011. There are currently two single malts in the range - Wild Weasel bottled at 46% or 62,4%.

Distillerie Radermacher

Raeren, founded in 1836

www.distillerie.biz

In a wide range of products from this classic distillery, there is also single malt whisky to be found. The 10 year old Lambertus has been aged in American oak casks that previously held tequila.

Czech Republic

Gold Cock Distillery

Founded in 1877

www.rjelinek.cz

The whisky is produced in three versions – a 3 year old blended whisky, a 12 year old single malt and different versions of Small Batch single malt. Production was stopped for a while but after the brand and distillery were acquired by R. Jelinek a.s., the leading Czech producer of plum brandy, the whisky began life anew. The malt whisky is double distilled in 500 litre traditional pot stills.

Denmark

Stauning Whisky

Stauning, founded in 2006

www.stauningwhisky.dk

The first Danish purpose-built malt whisky distillery entered a more adolescent phase in 2009, after having experimented with two small pilot stills bought from Spain. More stills were installed in 2012. The preconditions, however, were completely changed in December 2015 when it was announced that Diageo´s incubator fund project, Distil Ventures, would spend £10m to increase the capacity of Stauning. In August 2018, it became evident what the investment had meant to the distillery. A new distillery with no less than 24 copper stills, all directly fired, was opened. The floor malting were increased to 1,000 m² and the total production capacity is now 900,000 litres of pure alcohol. The most recent releases include a 5 year old rye matured in virgin American oak, a 3 year old rye with a rum cask finish, the 6 year old, lightly peated Heather and the 4 year old Kaos made from both malted rye and malted barley.

Braunstein

Köge, founded in 2005 (whisky since 2007)

www.braunstein.dk

Denmark's first micro-distillery, built in an already existing brewery in Køge, just south of Copenhagen. The wash comes from the own brewery. A Holstein type of still, with four plates in the rectification column, is used for distillation. Around 40% of the required barley is ecologically grown in Denmark. The lion's share of the whisky is stored in ex-bourbon (peated version) and first fill Oloroso casks (unpeated) from 190 up to 500 litres. The first release from the distillery and the first release of a malt whisky produced in Denmark was in 2010. The most recent releases include Library Collection 19:1, matured in a combination of oloroso sherry and port casks and Edition No: 10, matured in a cask that previously held sherry and Speyside whisky. Since 2010, the owners have also produced a bourbon-style whisky but this has yet to be released.

Other distilleries in Denmark

Fary Lochan Destilleri

Give, founded in 2009

www.farylochan.dk

The main part of the malted barley is imported from the UK but they also malt some of it themselves. The five day fermentation takes place in stainless steel washbacks and distillation is done in traditional copper pot stills. The first whisky was released in 2013 and a number of bottlings have been released since then. One of the latest was the 5 year old Distillery Edition in June 2019.

Braunstein Distillery

Trolden Distillery

Kolding, founded in 2011

www.trolden.com

The distillery is a part of the Trolden Brewery and the wash from the brewery is fermented for 4-5 days before a double distillation in a 325 litre alembic pot still. The first release of a single malt was Nimbus in 2014. In December 2018, the lightly peated Nimbus No. 5 was released, matured in a combination of bourbon and sherry.

Nordisk Brænderi/Thy Whisky

Fjerritslev, founded in 2009 (whisky since 2011)

www.nordiskbraenderi.dk, www.thy-whisky.dk

The ecological barley used for the production is grown in fields surrounding the distillery. In the first 7-8 years around ten whiskies have been released under the name Thy Whisky, the first in 2014. Bottlings during 2019 include a 3 year old Fjordboen matured in an ex-oloroso cask, the 4 year old Bøg, beech smoked and oloroso matured and Distillery Edition cask 64.

Nyborg Destilleri

Nyborg, founded in 1997 (whisky since 2009)

www.fioniawhisky.com

Originally opened in 2009 as an extension to an already existing brewery, the distillery moved in 2017 to new premises. The new distillery is equipped with washbacks made of oak and two copper pot stills with attached columns. The first release of Isle of Fionia single malt was in 2012. There are three ranges of single malt available; the aged Isle of Fionia, Ardor with a peated version and a Danish oak finish as the latest releases and Adventurous Spirit.

Braenderiet Limfjorden

Roslev, founded in 2013

www.braenderiet.dk

The distillery moved in spring 2018 to a new location and at the same time a brewery was added. Apart from peated and unpeated single malt and rye, the distillery also produces gin and rum. The first single malt was released in July 2016 and to celebrate the move to new premises, a limited range of Lindorm whisky was released.

Ærø Whisky

Ærøskøbing, founded in 2013

www.ærøwhisky.dk

This microdistillery has been working on the small island of Ærø since 2013 using stills from Portugal. In 2016, new and larger stills made in Germany were installed and the production increased. The first bottling (made from the pilot still) was a bourbonmatured single cask released in March 2017.

Mosgaard Whisky

Oure, founded in 2015

www.mosgaardwhisky.dk

Alambic stills made in Portugal are used for the distillation and the aim is to produce 20,000 bottles per year. The first single malts, matured in oloroso casks and PX sherry appeared in spring 2019.

Braenderiet Enghaven

Mellerup, founded in 2014

www.enghaven-whisky.dk

One of the newest distilleries in Denmark, producing rum, gin and whisky. The first whisky release, in autumn 2017, was a rye matured in both bourbon casks and port casks and in October 2018, the first single malt from a rum cask was released

England

St. George´s Distillery

Roudham, Norfolk, founded in 2006

www.englishwhisky.co.uk

St. George´s Distillery near Thetford in Norfolk was started by father and son, James and Andrew Nelstrop, and came on stream in December 2006. This made it the first English malt whisky distillery for over a hundred years. In December 2009, it was time for the release of the first legal whisky called Chapter 5 (the first four chapters had been young malt spirit). This has been followed by several more chapters with four of them still in the range; 14 (unpeated), 15 (heavily peated), 16 (peated sherry cask) and 17 (triple distilled). The core range however consists of The English Original and The English Smokey with a number of small batch, limited releases as well where Virgin Oak was the latest. In 2017 a new and innovative sub range was introduced – The Norfolk. Three whisky expressions have been bottled so far, Malt 'n' Rye made with malted barley and rye, matured in bourbon casks, Farmers where no less than eight different grains were used and Parched, a single grain. There are also a number of whisky liqueurs in the same range.

The distillery is equipped with a stainless steel semi-lauter mash tun with a copper top and three stainless steel washbacks with a fermentation time of 85 hours. There is one pair of stills, the wash still with a capacity of 2,800 litres and the spirit still of 1,800 litre capacity. First fill bourbon barrels are mainly used for maturation and all the whiskies from the distillery are un chill-filtered and without colouring. The distillery also has an excellent, newly expanded visitor centre, including a shop with more than 300 different whiskies. More than 50,000 people travel here every year.

Cotswolds Distillery

Stourton, founded in 2014

www.cotswoldsdistillery.com

The distillery is the brainchild of Dan Szor, who acquired an estate with two stone buildings and converted them into a distillery with a visitor centre. Production of both whisky and gin started in September 2014. There are three stills; one wash still (2,400 litres), one spirit still (1,600 litres) and a Holstein still (500 litres) for production of gin and other spirits. The rest of the equipment includes a 0.5 ton mash tun and eight stainless steel wash backs. The first product for sale was their Cotswolds Dry Gin in 2014 while the first single malt, the 3 year old Odyssey, was launched in 2017. A widely available bottling was released in November. In early 2018, the distillery started experimenting with rye whisky and in August the owners made their first trials producing rum. In December 2018, the distillery´s second release appeared. Founder´s Choice, with no age statement, had been matured in STR casks (red wine casks that had been shaved, toasted and re-charred). A limited release appeared in spring 2019 when the Lord Mayor´s Reserve Single Malt was launched and this was followed in August by a single malt that had matured in port casks, exclusively sold to consumers attending The Cotswolds Distillery Festival.

Lakes Distillery

Bassenthwaite Lake, founded in 2014

www.lakesdistillery.com

Headed by Paul Currie, who was the co-founder of Isle of Arran distillery, a consortium of private investors founded the distillery. Production started in autumn 2014 and the £2,5m distillery is equipped with two stills for the whisky production, each with both copper and stainless steel condensers, and a third still for the distillation of gin. Until recently, the capacity was 240,000 litres of pure alcohol. With eight new washbacks added in early 2019, the capacity has doubled to 450,000 litres. To help create a cash flow, the company launched a British Isles blended whisky called The One in 2013. Also gin and vodka are being distilled. The very first bottle of the distillery´s single malt, The Lakes Malt Genesis, was sold on 29th June 2018 at an auction fetching a staggering £7,900. More bottles were offered to members of the Founder´s Club in September and that same month saw the first installment in a four-year collection of single malts called The Quatrefoil Collection.

Spirit of Yorkshire Distillery

Hunmanby, founded in 2016

www.spiritofyorkshire.com

Plans for the distillery started in 2014 when Tom Mellor and David Thompson, the company directors, decided to make Yorkshire's

Spirit of Yorkshire Distillery

first single malt whisky. The distillery is actually situated in two separate locations with a one ton mash tun and two 10,000 litre washbacks standing at Tom's farm which also houses a brewery while the 5,000 litre wash still and a 3,500 litre spirit still are 2,5 miles down the road in Hunmanby. A four plate column are designed to run in tandem with the spirit still and currently 50% of the production is distilled using the column to achieve a lighter character of the new make. All the barley comes from the farm, is malted by Muntons and the fermented wash is tankered to the distillery every week. The distillery was commissioned in May 2016 and produces 80,000 litres of pure alcohol a year. There are plans to increase capacity, perhaps as soon as 2020. The distillery also has a visitor centre with daily tours. In December 2017, the owners released their first malt spirit called Distillery Projects Maturing Malt. The fifth and sixth editions were released in summer 2019 and the plan is to release the first mature single malt whisky in the autumn.

Other distilleries in England

The London Distillery Company

London, founded in 2012

www.londondistillery.com

Founded in 2012, the distillery started distilling gin at the beginning of 2013. The owners have one still designated for gin while a second still is used exclusively for whisky production. The first release in 2013 was Dodd's Gin. In December 2013 they got the licence to produce whisky and production started shortly thereafter. A rye whisky was launched in 2018 and early 2019, it was time for the company's first whisky made from malted barley – Cask 109. Matured for 3,5 years in ex-bourbon barrels, the whisky sold out instantly.

Adnams Copper House Distillery

Southwold, founded in 2010

www.adnams.co.uk

Adnams Brewery in Suffolk added distillation of spirits to their production in December 2010 and, apart from whisky – gin, vodka and absinthe are produced. The first two whiskies from the distillery were released in 2013 – Single Malt No. 1, and Triple Grain No. 2 from malted barley, oats and wheat and matured in new American oak. Several whiskies have been added to the range since then, one of the latest being one made from 75% rye and 25% malted barley.

Chase Distillery

Rosemaund Farm, Hereford, founded in 2008

www.chasedistillery.co.uk

The main product from the distillery is Chase Vodka made from potatoes and gin has also become part of their range. By the end of 2011, the first whisky was distilled and since then around 40 casks are filled every year but no bottling has yet been released. The distillery is equipped with a copper still with a five plate column and an attached rectification column with another 42 plates.

Bimber Distillery

London, founded in 2015

www.bimberdistillery.co.uk

The distillery buys its floor malted barley from Warminster Maltings and the spirit is distilled in two copper stills made by Hoga in Spain – a 1,000 litre wash still and a 600 litre spirit still. The owners use wooden washbacks and more were installed in 2019 to ensure that the distillery could continue with their long (seven days) fermentation. They also recently did trials with peated production using barley that they floor malted themselves. Distillation began in May 2016 and the first release from the distillery was a vodka followed by a blended 6-month-old malt spirit from bourbon, sherry and virgin oak casks. In September 2019, it was time for the distillery's inaugural single malt bottling – aptly named The First.

Copper Rivet Distillery

Chatham, founded in 2016

www.copperrivetdistillery.com

Situated in an old pump house in the Chatham Docks, this distillery is owned by the Russell family. There is one copper pot still with a column attached as well as a special gin still. Dockyard Gin and Vela Vodka were released early on and in April 2017, Son of a Gun, an 8 week old grain spirit made from rye, wheat and barley, was released. The first single malt is due for release in June 2020.

Dartmoor Distillery

Bovey Tracey, founded in 2016

www.dartmoorwhiskydistillery.co.uk

Starting the distillery, the founders, Greg Millar and Simon Crow acquired a 50 year old alembic still in Cognac which hadn't been used since 1994. The brought it to England, refurbished it and attached a copper "wash warmer" to pre warm the wash and increase the copper contact. The distillery is situated in Devon, just north of Torquay and the first distillation was in February 2017.

Isle of Wight Distillery

Newport, founded in 2015

www.isleofwightdistillery.com

The founders, Conrad Gauntlett and Xavier Baker, have years of experience in wine production and brewing but this is their first distillation venture. The fermented wash is bought from a local brewery and distilled in hybrid copper stills. Another 1,000 litre still was installed in summer 2019. The first whisky was distilled in December 2015 and the owners also produce vodka and Mermaid Gin.

Durham Whisky

Durham, founded in 2014 (whisky since 2018)

www.durhamwhisky.co.uk

The distillery was founded in 2014 with the aim to produce gin and vodka. In 2018, the owners decided to relocate to larger premises in Durham and whisky was added to the range. Using local malt, the distillery is equipped with a 1,200 litre wash still and a 1,000 litre spirit still.

Cooper King Distillery

Sutton-on-the-Forest, founded in 2018

www.cooperkingdistillery.co.uk

Inspired by a trip to whisky distilleries in Australia, Abbie Neilson and Chris Jaume decided to build a distillery of their own in Yorkshire. Equipped with a Tasmanian copper pot still, the distillery released its first gin in summer 2018 while whisky production started in June 2019. The distillery practises a combination of vacuum distillation and traditional distillation and use floor-malted barley from Warminster Maltings.

White Peak Distillery

Ambergate, Derbyshire, founded in 2017

www.whitepeakdistillery.co.uk

Founded by Max and Claire Vaughn and situated in the former Johnson & Nephew Wire Works in the Peak District, the distillery was commissioned in early 2018. The first distillation was in April that year. By that time, Shaun Smith from Cotswolds Distillery had been hired as the head distiller. Gin has already been launched and a rum release is imminent but the first bottlings of the lightly peated single malt whisky are not expected until at least 2021.

Henstone Distillery

Oswestry, Shropshire, founded in 2017

www.henstonedistillery.com

Commissioned in December 2017, the distillery released its

first gin in early 2018 and since then apple brandy has also been launched. The first single malt whisky, distilled in a Kothe copper hybrid still, was filled into an ex-bourbon cask in January 2018 and since then oloroso and PX casks have also been used. The first official whisky bottling is expected in early 2021.

Circumstance Distillery

Bristol, founded in 2018

www.circumstancedistillery.com

The owners idea is to make whisky and rum in a flexible distillery equipped with a pot still with attached columns. The very first bottling in March 2019 was Circumstantial Barley 1:1:1:1:6, a spirit from 100% malted barley that had matured for two months in charred oak spindles and then another four months in ex-bourbon casks. The wash had been fermented for 13 days!

Finland

Teerenpeli

Lahti, founded in 2002

www.teerenpeli.com

The original distillery, located in a restaurant in Lahti, is equipped with one wash still (1,500 litres) and one spirit still (900 litres). A completely new distillery, with one 3,000 litre wash still and two 900 litre spirit stills, was opened in 2015 in the same house as the brewery and today the old distillery serves as a "laboratory" for new spirits. The first single malt was launched as a 3 year old in 2005. The core range now consists of a 10 year old matured in bourbon casks, Kaski which is a 100% sherry maturation, Portti which is a 3 year old with another 1.5 years in port casks and the peated Savu matured in a combination of bourbon and PX sherry casks. Recent limited releases include the 12 year old Länki single bourbon cask and the 13 year old Juhlaviski.

Other distilleries in Finland

Helsinki Distilling Company

Helsinki, founded in 2014

www.hdco.fi

Whisky production started in September 2014. The distillery is equipped with one mash tun, three washbacks and two stills. The first gin was released in October 2014 and more gin and akvavit but also a one year old malt spirit has followed since. On the whisky side, the focus is on rye but also single malt made from barley. The first release was a 100% malted rye in autumn 2017.

Valamo Distillery

Heinävesi, founded in 2014

www.valamodistillery.com

The distillery is situated at the Valamo Monastery in eastern Finland. Experimental distillation started in 2011 in a small, 150 litre still. In 2015, production began in earnest when the distillery was equipped with a 5,000 litre mash tun, four stainless steel washbacks and a 1,000 litre Carl still. A 5 year old, matured in first fill bourbon casks, from the early days of the distillery has already been released.

France

Distillerie Warenghem

Lannion, Bretagne, founded in1900 (whisky since 1994)

www.distillerie-warenghem.com

Leon Warenghem founded the distillery at the beginning of the 20th century and in 1967, his grandson Paul-Henri Warenghem, together with his associate, Yves Leizour, took over the reins. They moved the distillery to its current location on the outskirts of Lannion in Brittany. Gilles Leizour, Yves' son, took over at the end of the 1970's and it was he who added whisky to the Warenghem range. WB (stands for Whisky Breton), a blend from malted barley and wheat both distilled in a pot still, saw the light in 1987 and Armorik – the first ever French single malt – was released in 1998. The distillery is equipped with a 6,000 litre semi-lauter mashtun, six stainless steel washbacks and two, traditional copper pot stills (a 6,000 litre wash still and a 3,500 litre spirit still). Around 180,000 litres of pure alcohol (including 20% grain whisky) are produced yearly. In 2019, a second warehouse was built as well as a stunning visiting centre with a beautiful tasting room.

The single malt core range consists of Armorik Édition Originale and Armorik Sherry Finish. Both are around 4 years old, bottled at 40%, have matured in ex-bourbon casks plus a few months in sherry butts and are sold in supermarkets in France. Armorik Classic, a mix of 4 to 8 year old whiskies from ex-bourbon and sherry casks and the 7 year old Armorik Double Maturation which has spent time in both new oak and sherry wood are earmarked for export as well as the Armorik Sherry Cask. Armorik Millésime (always a 2012 distillation) is a single cask bottling released every year while Maître De Chai is another annual, limited release. Warenghem has also distilled rye whisky which was first released in 2014 under the name Roof Rye and in 2018, the first peated expression, Triagoz, was released. At the end of 2018, to celebrate the 30th anniversary of their first single malt, Armorik released its first 10 year old.

The stunning, new visitor centre at Warenghem Distillery

Glann ar Mor

Pleubian, Bretagne, founded in 1999

www.glannarmor.com

The owner of Glann ar Mor Distillery in Brittany, Jean Donnay, already started his first trials back in 1999. He then made some changes to the distillery and the process and regular production commenced in 2005. The distillery is very much about celebrating the traditional way of distilling malt whisky. The two small stills are directly fired and Donnay uses worm tubs for condensing the spirit. He practises a long fermentation in wooden washbacks and the distillation is very slow. For maturation, a variety of casks are used and when the whisky is bottled, there is neither chill filtration nor caramel colouring. The full capacity is 50,000 bottles per year but the actual production is less than 10,000 bottles. There are two versions of the whisky – the unpeated Glann ar Mor and the peated Kornog. Core expressions are usually bottled at 46% but every year a number of limited releases are made including single casks and cask strength bottlings. Since 2016, Kornog Roc'hir (ex-bourbon casks) is the permanent Kornog bottling. Early 2019, Glann Ar Mor released its oldest bottling, a Kornog 12 year old and also the first ever French triple distilled single malt, Teir Gwech.

Distillerie des Menhirs

Plomelin, Bretagne, founded in 1986 (whisky since 1998)

www.distillerie.bzh

Originally a portable column still distillery, Guy Le Lay and his wife Anne-Marie decided in 1986 to settle down for good and the first lambig with the name Distillerie des Menhirs was released in 1989. Shortly after, Guy Le Lay came up with the idea of producing a 100% buckwheat whisky. Eddu Silver was launched in 2002, followed by Eddu Gold in 2006, Eddu Silver Brocéliande in 2013 and Eddu Diamant in 2015. Ed Gwenn (white cereal in English), aged for 4 years in ex-cognac barrels, was released for the first time in 2016 and in 2017, the third release in the Collector's Range (Eddu Dan Ar Braz) appeared. In June 2019, Les Menhirs released a vintage, 2004, on sale exclusively at its visiting center.

Distillerie Rozelieures

Rozelieures, Grand Est, founded in 1860 (whisky since 2003)

www.whiskyrozelieures.com

Hubert Grallet and his son-in-law, Christophe Dupic started with whisky production in 2003 and launched the Glen Rozelieures brand in 2007. Four versions are currently available: the first two are aged in ex-fino sherry casks, the third is lightly peated and aged in Sauternes casks and the fourth is peated. Fully automated in 2017 and with a production of 200,000 litres, Rozelieures distillery is now the largest distillery in France. Since 2015 Rozelieures is independent in energy with its own biogas plant a few hundred meters from the distillery. In 2018, Christophe Dupic opened his malting plant, with a capacity of 2,000 tons.

Distillerie Lehmann

Obernai, Grand Est, founded in 1850 (whisky since 2001)

www.distillerielehmann.com

The story of Lehmann distillery starts in 1850 when the family of the actual owner set up a still in Bischoffsheim. Yves Lehmann inherited the facility in 1982 but decided to move all the equipment to a new distillery in 1993. The first regular bottling from the distillery, aged for seven years in Bordeaux casks, was launched in 2008. The range now includes Elsass Origine (4-6 years) and Elsass Gold (6-8 years), both matured in ex-white wine casks, and Elsass Premium (8 years) matured in ex-Sauternes casks.

Miclo

Lapoutroie, Grand Est, founded in 1970 (whisky since 2012)

www.distillerie-miclo.com

Gilbert Miclo, the grandfather of Bertrand Lutt, the current manager, founded the distillery in 1970, specialising in fruit spirits. It is equipped with four Holstein waterbath pot stills and since 2012, wort from a local brewery is fermented and distilled into malt whisky. Under the brand name Welche's, three different whiskies were released in December 2016 - Welche, Welche Fine Tourbe, Welche Tourbé. In June 2018, Miclo released Welche Cherry Cask Finish, a small batch of four casks (ex-burgundy and sauternes) matured for 6 years and with a finish in an ex-cherry eau-de-vie cask.

Domaine des Hautes-Glaces

Saint Jean d'Hérans, Auvergne-Rhône Alpes, founded in 2009 (whisky since 2014)

www.hautesglaces.com

At an altitude of 900 metres in the middle of the French Alps, Jérémy Bricka and Frédéric Revol decided to produce whisky from barley to bottle. Apart from growing their own barley, all the parts of whisky production take place at the distillery – malting, brewing, distillation, maturation and bottling. Not only have they set out to create the first French single estate whisky, they are doing it organically. All of their cereals are harvested, malted, distilled and aged field by field and without any chemicals. Principium, the first whisky made at the distillery has been available since 2014. Domaine des Hautes Glaces was bought by Rémy Cointreau in 2015 and a second distillery, destined to open in 2020, is currently being built. The standard line-up includes two whiskies, Les Moissons Malt (100% malted barley) and Les Moissons Rye (100% malted rye). Single cask bottlings such as Ceros or Secale (rye), Flavis, Tekton, Ampelos or Obscuros (barley) are released from time to time.

Rouget de Lisle

Bletterans, Bourgogne-Franche Comté, founded in 1994 (whisky since 2006),

www.brasserie-rouget-lisle.com

Rouget de Lisle is a micro-brewery created by Bruno Mangin and his wife. In 2006, they commissioned the Brûlerie du Revermont to distil whisky for them. The first Rouget De Lisle single malt whisky was released in 2009 and in 2012, Bruno Mangin bought his own still. Current bottlings are from the numerous casks he filled during his association with the Tissot family and which lie maturing in his own warehouse. End of 2018, Bruno Mangin renamed the brand BM Signature with the launch of 4 bottlings: two NAS aged in ex-vin de paille and ex-Macvin casks and two ex-vin jaune bottlings, 9 and 12 years old respectively.

Domaine Mavela

Corsica, founded in 1991 (whisky since 2001)

www.domaine-mavela.com

Since 2001 whisky is produced in Corsica. The creators of P&M are the brewer Dominique Sialelli, also responsible for the creation of Pietra beer in 1996, and Jean-Claude Venturini who set up the Mavela distillery in 1991. Distilled in a Holstein still and aged in ex-Corsican muscat casks, the P&M single malt was sold for the first time in 2004 and its unique taste of the Corsican maquis surprised many whisky lovers. End of 2017, the distillery released its first 12 year old and in 2018, Fanu Venturini and his brother Lisandru, who now run the distillery, unveiled a completely new range of three expresions: P&M Signature, P&M Red Oak (aged in ex-red-wine casks) and P&M Tourbé (peated).

Distillerie Claeyssens de Wambrechies

Wambrechies, Hauts de France, founded in 1817 (whisky since 2000)

www.wambrechies.com

Owned by the Belgian company Grandes Distilleries de Charleroi, Claeyssens is one of the oldest in France. The distillery was originally famous for its genever. The first whisky, a 3 year old, was released in 2003 followed by an 8 year old in 2009. In 2013, two 12 year old bottlings were released: one aged in madeira casks and another in sherry casks, and in 2017 a limited 8 year old from sherry casks was launched to celebrate the distillery´s 200th anniversary. In June 2019, the distillery was bought by Saint-Germain brewery. The original installations will be kept in a museum and a new Holstein still is expected to make its appearance soon.

Distillerie Brunet

Cognac, Nouvelle Aquitaine, founded in 1920 (whisky since 2006)

www.drinkbrenne.com

In 2006, Stéphane Brunet made the bold move to start whisky production in the Poitou-Charentes region - famous for its cognac production. His whisky, Tradition Malt, was launched in 2009 and was launched in the USA by whisky enthusiast Allison Parc under the brand Brenne. In September 2015 a 10 year old version was released in small quantities in the USA. Since 2015 Brenne Cuvée Spéciale is also available in France. In 2017, the brand was acquired by Samson & Surrey with Allison Parc still being involved.

Other distilleries in France

Distillerie Meyer

Hohwarth, Grand Est, founded in 1958 (whisky since 2004)

www.distilleriemeyer.fr

Founded by Fridolin Meyer in 1958, and joined by his son Jean-Claude in 1975, Meyer soon became one of the most awarded distillers in France. At the beginning of the 2000's, Jean-Claude together with his two sons, Arnaud and Lionel, decided to start whisky production as well. They launched two no-age statement whiskies in 2007. There are currently three different versions: Meyer's Pur Malt (a single malt), Meyer's Blend Supérieur and Oncle Meyer Blend Supérieur. Meyer created a little buzz, in France in 2016 when he released the most expensive French whisky ever: a limited 12 year old – Hommage à JC Meyer – at the staggering price of 1 000 euros.

Distillerie Gilbert Holl

Ribeauvillé, Grand Est, founded in 1979 (whisky since 2000)

www.gilbertholl.com

In 1979, Gilbert Holl began to distill occasionally in the back of his wine and spirits shop but it wasn´t until the beginning of 2000, that he finally started producing also whisky. His first bottling, Lac'Holl, was put on sale in 2004 and was followed by Lac'Holl Junior in 2007 and Lac'Holl Vieil Or in 2009. In 2015, Lac'Holl Junior was replaced by Lac'Holl Or. The oldest whiskies, 10 and 12 year olds, can only be found at the distillery.

Distillerie Hepp

Uberach, Grand Est, founded in 1972 (whisky since 2005)

www.distillerie-hepp.com

A family-owned distillery with a no-age statement core expression by the name Tharcis Hepp. Two limited editions have also been released, the first one aged in ex-plum cask, the second one under the name Johnny Hepp. As well as producing their own whisky, Hepp also distils for Meteor, the brewery that supplies the wort to some Alsatian distilleries. In 2018, no less than three new bottlings were released from Distillerie Hepp; Ouisky, Tharcis Hepp Tourbé and French Flanker.

Brûlerie du Revermont

Nevy sur Seille, Bourgogne-Franche Comté, founded in 1991 (whisky since 2003)

www.marielouisetissot-levin.com

For many years, the Tissot family were travelling distillers offering their services to wine producers in the area. Relying upon a very unique distillation set-up, a Blavier still with three pots, designed and built for the perfume industry, they have been producing single malt whisky since 2003. Pascal and Joseph Tissot launched their

own whisky brand Prohibition in 2011. Aged in "feuillettes", the whisky is bottled without colouring.

Distillerie Bertrand

Uberach, Grand Est, founded in 1874 (whisky since 2002)

www.distillerie-bertrand.com

The distillery manager, Jean Metzger, gets the malt from a local brewer and then distils it in Holstein type stills. Two different types of whisky are produced. One is a non-chill filtered single malt with a maturation in both new barrels and ex Banyuls barrels. The other is a single cask matured only in Banyuls barrels. They have also experimented with maturation in a lot of other different ex-wine casks, and the new range became known as Cask Jaune.

Distillerie de Northmaen

La Chapelle Saint-Ouen, Normandie, founded in 1997 (whisky since 2002)

www.northmaen.com

Originally a craft brewery but later equipped with a distillery, Northmaen has every year since 2005 bottled and sold Thor Boyo, a 3 year old single malt, distilled in a small, mobile pot still. Several more releases have followed with the peated Fafnir from 2015 as one of the latest. The still is not portable anymore but now works in a real distillery.

Brasserie Michard

Limoges, Nouvelle Aquitaine, founded in 1987 (whisky since 2008)

www.bieres-michard.com

Started as a brewery, Jean Michard began to also produce whisky in 2008. Using their own unique yeast, the first batch of their whisky, released in 2011, was highly original and very fruity. It was followed by a second batch in 2013. Jean Michard has now retired and the still does not work very often but a brand new bottle and packaging design has been launched, announcing a new start.

Bercloux

Bercloux, Nouvelle Aquitaine, founded in 2000 (whisky since 2014)

www.distillerie-bercloux.fr

After many trials, Philippe Laclie opened his own brewery in 2000. In 2007 he decided to diversify by buying some Scotch whisky and finishing it for a few months in Pineau des Charentes barrels. In 2014, Philippe took the next step and invested around 100,000 euros, buying an 800 litre column still. After the first trial runs, regular whisky production started in September 2014. The first two bottlings of Bercloux Single Malt Whisky (peated and non peated) were released in September 2018.

Dreumont

Neuville-en-Avesnois, Hauts de France, founded in 2005 (whisky since 2011)

www.ladreum.com

Passionate about beer, Jérôme Dreumont decided to open a distillery as well in 2005. In 2011 he built his own 300 litre still and has since then been filling only one cask per year. His first whisky, distilled from a mix of peated and non-peated barley, was launched in March 2015 and was followed by more releases in 2016 and 2017..

Distillerie du Castor

Troisfontaines, Grand Est, founded in 1985 (whisky since 2011)

www.distillerie-du-castor.com

Founded by Patrick Bertin, the distillery produces both fruit and pomace brandies. It is equipped with two small stills which have been used since 2011 by Patrick's son to distil single malt whisky.

The malt is brewed by a local brewery and the distillate is aged in ex-white wine casks and finished in ex-sherry casks. The first release appeared in June 2015 under the name St Patrick.

La Roche Aux Fées

Sainte-Colombe, Bretagne, founded in 1996 (whisky since 2010)

www.distillerie-larocheauxfees.com

Gonny Keizer installed a micro-brewery in 1996 and became the first female master-brewer in France. In 2010, Gonny and her husband Henry bought a 400 litre portable automatic batch still. The still is wood-heated and equipped with a worm-tub condenser and the first spirit was put into cask in 2010. The first Roc'Elf bottling, distilled from three malted cereals (barley, wheat, oat), was released in January 2016 followed by a second in early 2017.

La Quintessence

Herrberg, Grand Est, founded in 2008 (whisky since 2013)

www.distillerie-quintessence.com

In 2008 Nicolas Schott took over the family distillery in Herrberg in order to continue the production of fruit spirits (raspberry, pear, plum, quetsche or quince) and also liqueurs (spices or asperule). End of 2016, the 34 year old distiller surprised everyone with the release of his first single malt whisky, Schott's, bottled at 42%.

Distillerie Castan

Villeneuve Sur Vère, Occitanie, founded in 1946 (whisky since 2010)

www.distillerie-castan.com

In 2010, Sébastien Castan decided to permanently house the portable still that had been in the family for three generations in a proper distillery. The same year, he distilled his first whisky and aged the spirit in ex-Gaillac wine casks. In 2016, a brewery was built to supply the beer. The range consists of five bottlings: Villanova Berbie (ex-white wine casks), Gost (new cask), Terrocita (peated), Roja (ex-red wine casks) and Segala (rye). In June 2019 the construction of a new distillery started.

La Piautre

Ménitré Sur Loire, Pays de Loire, founded in 2004 (whisky since 2014)

www.lapiautre.fr

Ten years after Yann Leroux and Vincent Lelièvre had founded a brewery, they decided to start malting their own barley and to start experiments with distillation. La Piautre is now equipped with a Charentais direct-fire heated still. The company released their first Loire Valley whiskies in January 2018: Malt, Tourbé and Seigle.

Leisen

Malling, Grand Est, founded in 1898 (whisky since 2012)

www.distillerie-leisen-petite-hettange.fr

Since 1898, the Leisen family has distilled fruit spirits but also spirits made from rye and barley. The distillery is equipped with two Carl stills (250 and 350 litres respectively). In 2018, Jean-Marie Leisen released his first bottles under the JML brand.

Distillerie Vercors

Saint-Jean En Vercors, Auvergne-Rhône Alpes, founded in 2015 (whisky since 2019)

www.distillerie-vercors.com

The distillery is an atypical installation equipped with a Charentais copper still and a steel boiler which works under vacuum for the first run. First distillation is carried out at low temperature (around 50°C). All operations are conducted on site, from barley to ageing. Sequoia Première Impression and Sequoïa Tourbé, two malt spirits, were launched in September 2018. The first proper whisky will be released in October 2019..

Ouche Nanon

Ourouer Les Bourdelins, Centre-Val de Loire, founded in 2015 (whisky since 2018)

www.ouche-nanon.fr

Originally a micro-brewery, the owner Thomas Mousseau acquired an old Guillaume still from 1930s. With a capacity of 500 litres,

The Vercors distillery in stunning surroundings west of Grenoble

the still is heated with wood. Thomas Mousseau started distilling in May 2015. The first whisky, aged in ex-Sauternes casks, was released in November 2018.

Ninkasi

Tarare, Auvergne-Rhône Alpes, founded in 2015 (whisky since 2018)

www.ninkasi.fr

Production at the distillery began in December 2015 and the whisky is distilled in a 2,500-litre Prulho Chalvignac still and aged in different types of barrels of the most famous wines of Bourgogne and Côte du Rhône areas. Track01, a limited edition of 1,000 bottles, was launched end of 2018. The second one, slightly peated, appeared in early in 2019.

Domaine Laurens

Clairvaux d'Aveyron, Occitanie, founded in 1983 (whisky since 2014)

www.domaine-laurens.com

The wine estate Domaine Laurens was founded in 1983 by two brothers Gilbert and Michel Laurens, assisted by their wives, Martine and Maryse. In 2014, Vincent started a whisky production with the help of a neighbor brewery. The first versions. Red Léon, aged in white dry aveyron wine ex-casks, and Blue Léon, red ratafia ex-casks, were launched in 2017.

Saint-Palais

Saint-Palais de Negrignac, Nouvelle Aquitaine, founded in 2016

www.alfredgiraud.com

The families Gautriaud and Naud have been settled for generations in Saint-Palais de Negrignac. In 1896, Louis Gautriaud built a distillery, Chevanceaux, which was taken over by his son and grandson. In 1963, a new distillery was built, with ten stills to be enlarged in 1990 when the number was doubled. In 2016, Julien Naud built a brewery dedicated to whisky production fitted with a filter press.

In 1995, Philippe Giraud who came from a famous cognac house, joined the company to form the Alfred Giraud French Malt Whisky in association with Julien Nau. Two versions of their blended malt have been released, Harmonie and Héritage

Distillerie Nalin

La Chana, Auvergne-Rhône Alpes, founded in 2015

www.distillerie-nalin.fr

A family distillery for three generations installed in Meys in the heart of the Monts du Lyonnais since 1919. For a long time, Nalin family distilled for other families or companies, mainly pear eaux-de-vie. In 2015, they started whisky production and the first bottling has been sold under the name of NP since November 2018.

Distillerie Ergaster

Passel, Hauts de Frances, founded in 2015

www.distillerie-ergaster.com

Founded by two friends, one of them being Éric Trossat who founded the Uberach brewery in Alsace. Peated and unpeated spirits are slowly aging in ex-Cognac, ex-Pineau des Charentes, ex-Vin Jaune and ex-Banyuls casks. The first whisky, the organic Ergaster Nature was launched in December 2018 and was followed by Ergaster Tourbé in October 2019.

Distillerie D'Hautefeuille

BeaucourtEnSanterre, Hauts de France, founded in 2015

www.distilleriedhautefeuille.com

Founded by Étienne d'Hautefeuille who took over the farming business of his parents in 2013, and Gaël Mordac, a recognized drink specialist retailer in Amiens. Eventually, the two men intend to carry out all production operations on site, including malting. The first whiskies will be released in 2020 but a bottling made in a similar still at another distillery was launched already in 2018

Harmonie - one of the two releases from Saint-Palais distillery

Germany

Whisky-Destillerie Blaue Maus

Eggolsheim-Neuses, 1980

www.fleischmann-whisky.de

The oldest malt whisky distillery in Germany distilling their first whisky in 1983. It took, however, 15 years before the first whisky, Glen Mouse 1986, appeared. A completely new distillery became operational in April 2013. All whisky from Blaue Maus are single cask and there are around ten single malts in the range. Some of them are released at cask strength while others are reduced to 40%. An unusual experiment started in 2016 when the owners transported some casks to the island of Sylt in the North Sea. They were lowered into the sea to continue the maturation process. Because of the tide, every six hours the casks were exposed to the air. When the whisky was released it was named Sylter Tide.

Slyrs Destillerie

Schliersee, founded in 1928 (whisky since 1999)

www.slyrs.de

The malt, smoked with beech, comes from locally grown grain and the spirit is distilled in 1,500 litre stills. The non chill-filtered whisky is called Slyrs after the original name of the surrounding area, Schliers. The core expressions are a 3 year old bottled at 43% and matured in new American oak and the 51 which has matured in casks that previously held sherry, port or sauternes and is bottled at 51%. In 2015, the distillery´s first 12 year old whisky was released and other limited editions occur from time to time, including the five year old Mountain Edition and a 12 year old finished in Islay casks.

Hercynian Distilling Co (formerly known as Hammerschmiede)

Zorge, founded in 1984 (whisky since 2002)

www.hercynian-distilling.de

Hammerschmiede´s main products used to be spirits from fruit, berries and herbs but whisky distilling was embarked on in 2002. In 2014, Alexander Buchholz, started the construction of a new still house with additional stills making it a total of five. The first bottles were released in 2006. The core range consists of three expressions; Glen Els Journey with a blend of different maturations, Ember, which is woodsmoked and the X Series. The latter range, introduced in 2016, represents the best whiskies from the distillery, ten years or older. One subrange from the distillery is called Alrik represented by experimental and smoky whiskies while the Willowburn range consists of whiskies finished in different types of casks.

Other distilleries in Germany

Spreewood Distillers

Schlepzig, founded in 2004 (whisky production)

www.stork-club-whisky.com

Founded by Torsten Römer, the distillery changed hands in 2016 when Spreewood Distillers took over. The distillery had a history of producing a wide range of spirits but the new owners decided to focus entirely on whisky and rum. The new range is called Stork Club with a Single Malt bottled at 47% and a Straight Rye bottled at 55%. Limited releases include a single Bordeaux red wine cask.

Brennerei Heinrich

Kriftel, founded in 1983 (whisky since 2009)

www.brennerei-henrich.de, www.gilors.de

The first whisky release from the distillery was the 3 year old single malt Gilors in 2012. The two core expressions, Gilors fino sherry matured and Gilors port matured, are both 3 years old. Recent limited editions include Gilors Peated (made from peated

malt and matured for three years in bourbon casks) and a 7 year old, finished in PX casks.

Bayerwald-Bärwurzerei und Spezialitäten-Brennerei Liebl

Kötzting, founded in 1970 (whisky since 2006)

www.coillmor.com

Around 30,000 litres of whisky are produced annually and in 2009 the first bottles bearing the name Coillmór were released. There is a wide range aged between 4 and 8 years currently available. Recent limited editions include a 12 year old oloroso finish and an 8.5 year with a one year finish in red wine casks from Tuscany.

Brennerei Höhler

Aarbergen, founded in 1895 (whisky since 2001)

www.brennerei-hoehler.de

The first whisky from the distillery was released in 2004 as a 3 year old. A couple of the more recent releases of their Whesskey (so called since it is from the province Hessen) include versions made from rye, oat, triticale and pilsner.

Stickum Brennerei (Uerige)

Düsseldorf, founded in 2007

www.stickum.de

The wash comes from their own brewery and the distillation takes place in a 250 litre column still. The single malt is called BAAS and the first bottling (a 3 year old) was released in 2010. In 2014 the owners released their first 5 year old whiskies.

Preussische Whiskydestillerie

Mark Landin, founded in 2009

www.preussischerwhisky.de

The spirit is distilled very slowly five to six times in a 550 litre copper still with a rectification column and is then matured in casks made of new, heavily toasted American white oak, German fine oak or German Spessart oak. Since 2013 only organic barley is used. The first whisky was launched as a 3 year old in December 2012. From 2015, all the whiskies have been at least 5 years old.

Kleinbrennerei Fitzke

Herbolzheim-Broggingen, founded in 1874
(whisky since 2004)

www.kleinbrennerei-fitzke.de

The first release of the Derrina single malt was in 2007 and new batches have been launched ever since including a lightly peated. The different varieties of Derrina are either made from malted grains (barley, rye, wheat, oats etc.) or unmalted (barley, oats, buckwheat, rice, triticale, sorghum or maize).

Glina Whiskydestillerie

Werder a.d. Havel, founded in 2004

www.glina-whisky.de

After 12 years of production, the distillery moved to larger premises in 2016, thereby ten-folding the capacity. Around 1,000 casks are now filled yearly. The first Glina Single Malt was released in 2008. Most of the whiskies are between 3 and 5 years old and have matured in a variety of casks. The oldest whisky so far is a 7 year old single malt matured in German Spessart oak.

Rieger & Hofmeister

Fellbach, founded in 1994 (whisky since 2006)

www.rieger-hofmeister.de

The first release was in 2009 and currently there are four expressions in the range – a single malt matured in pinot noir casks, a malt & grain (50% wheat, 40% barley and 10% smoked barley) from chardonnay casks, a malted rye and a single grain. A peated single malt is due for release in 2023.

Kinzigbrennerei

Biberach, founded in 1937 (whisky since 2004)

www.biberacher-whisky.de

The first release was Badische Whisky in 2008, a blend made from wheat and barley. Two years later came the 4 year old Biberacher Whisky, the first single malt and in 2012, the range was expanded with Schwarzwälder Rye Whisky and the smoky single malt Kinzigtäler Whisky. The oldest whisky so far is an 8 year old.

Destillerie Kammer-Kirsch

Karlsruhe, founded in1961 (whisky since 2006)

www.kammer-kirsch.de

The distillery is working together with the brewery Landesbrauerei Rothaus, where the brewery delivers a fermented wash to the distillery and they continue distilling a whisky called Black Forest Rothaus Single Malt Whisky. The whisky was launched for the first time in 2009 and every year, a new batch is released. Recent limited releases include a 2015 Banyuls cask finish.

Alt Enderle Brennerei

Rosenberg/Sindolsheim, founded in 1991 (whisky since 1999)

www.alt-enderle-brennerei.de

The first whisky distillation was in 2000 and the owners now have a wide range of Neccarus Single Malt for sale from a 4 year old to an 18 year old! The latest limited release was the 7 year old, smoky Terrador with a finish in rum casks.

AV Brennerei

Wincheringen, founded in 1824 (whisky since 2006)

www.avadisdistillery.de

Around 2,000 bottles are released yearly and the oak casks from France have previously been used for maturing white Mosel wine. Threeland Whisky is between 3 and 6 years old and the range also consists of finishes in oloroso and port casks..

Birkenhof-Brennerei

Nistertal, founded in 1848 (whisky since 2002)

www.birkenhof-brennerei.de

The first release from the distillery in 2008 was the 5 year old rye Fading Hill. This was followed a year later by a single malt. The most recent bottling was a single malt that had been finished in a cream sherry cask. Since 2015, peated whisky is also produced.

Brennerei Ziegler

Freudenberg, founded in 1865

www.brennerei-ziegler.de

One characteristic that distinguishes Ziegler from most other distilleries is that the maturation takes place not only in oak casks, but also in casks made of chestnut! Their current core bottling is a 5 year old called Aureum 1865 Single Malt and there is also a cask strength version. Limited releases occur regularly and in 2019, their first peated expression (a 5 year old) was released.

Brennerei Faber

Ferschweiler, founded in 1949

www.faber-eifelbrand.de

Established as a producer of eau-de vie from fruits and berries, whisky has been included in the poduction during the last few years. The only whisky so far is a single malt that has matured for 6 years in barrels made of American white oak.

Steinhauser Destillerie

Kressbronn, founded in 1828 (whisky since 2008)

www.weinkellerei-steinhauser.de

The main products are spirits which are derived from fruits, but whisky also has its own niche. The first release was the single malt Brigantia which appeared in 2011. It was triple distilled and an 8 year old has followed since.

Weingut Simons

Alzenau-Michelbach, founded in 1879 (whisky since 1998)

www.feinbrenner.eu

All the whisky was produced in a 150 litre still until 2013 when a new Holstein still was installed, raising the whisky production to 3-5,000 litres per year. A single pot still whisky has since been released and the first whisky from the new still, a 100% rye, was launched in 2016 followed by whisky made from rice and emmer.

Cornelia Bohn – founder and owner of Preussische Whiskydistillerie

Nordpfälzer Edelobst & Whiskydestille

Winnweiler, founded in 2008

www.nordpfalz-brennerei.de

The first release was in 2011, a 3 year old single malt by the name Taranis with a full maturation in a Sauternes cask and in 2013 a 4 year old Amarone finish was launched. The latest release, in 2019, was a 6 year old matured in a Sauternes cask.

Dürr Edelbranntweine

Neubulach, founded in 2002

www.blackforest-whiskey.com

The first release from the distillery, the 4 year old Doinich Daal, reached the market in 2012. The latest release, batch 4 in December 2017, with two expressions, Eichenacker and Alte Hau, had matured in a combination of casks (wine, bourbon, cognac).

Tecker Whisky-Destillerie

Owen, founded in 1979 (whisky since 1989)

www.tecker.eu

Apart from a variety of eau de vie and other spirits, around 1,500 litres of whisky is produced annually. The core expression is the 10 year old Tecker Single Malt matured for five years in ex-bourbon barrels, followed by five yers in oloroso casks. There is also a single grain aged for 18 years in bourbon, cognac and sherry barrels.

Märkische Spezialitäten Brennerei

Hagen, whisky since 2010

www.msb-hagen.de

The spirit is distilled four times, matured in ex-bourbon barrels for 12 months and then brought to a cave, with low temperature and high humidity, for further maturation. The first whisky, Tronje van Hagen, was released in 2013 and has now been renamed DeCavo.

Sperbers Destillerie

Rentweinsdorf, founded in 1923 (whisky since 2002)

www.salmsdorf.de

At the moment, four different expressions have been released, all of them 7 years old – single malt matured in bourbon casks, single malt matured in a mix of sherry and bourbon casks, a sherrymatured single malt bottled at cask strength, as well as a single grain whisky from a mix of bourbon, sherry and Spessart oak casks.

Brennerei Feller

Dietenheim-Regglisweiler, founded in 1820 (whisky since 2008)

www.brennerei-feller.de

In 2012 the 3 year old single malt Valerie matured in bourbon casks, was released. It was followed by two 5 year olds (sherry- and amarone-finish respectively) and a 4 year old finished in a madeira cask. A range called Augustus is reserved for single grain.

Marder Edelbrände

Albbruck-Unteralpfen, founded in 1953 (whisky since 2009)

www.marder-edelbraende.de

The first release in 2013 was the 3 year old Marder Single Malt matured in a combination of new American oak and sherry casks. The latest edition, a 5 year old matured in a combination of bourbon and port, was launched in 2015.

Destillerie Drexler

Arrach, whisky since 2007

www.drexlers-whisky.de

Apart from spirits made from herbs, fruits and berries, malt whisky has been produced since 2013. The first release was Bayerwoid in 2011 which was followed up by No. 1 Single Cask

Malt Whisky and a 100% malted rye whisky. The latest edition of the No. 1 single cask was 4 years old.

Edelbrände Senft

Salem-Rickenbach, founded in 1988 (whisky since 2009)

www.edelbraende-senft.de

The first 2,000 bottles of 3.5 year old Senft Bodensee Whisky were released in 2012 and they were later followed by a cask strength version (55%).

Schwarzwaldbrennerei Walter Seger

Calw-Holzbronn, founded in 1952 (whisky since 1990)

www.krabba-nescht.de

The first single malt was launched in 2009 and at the moment, the owner has two expressions in the range; the 6 year old Black-Wood single malt matured in amontillado sherry casks and an 8 year old wheat whisky.

Landgasthof Gemmer

Rettert, founded in 1908 (whisky since 2008)

www.landgasthof-gemmer.de

The only single malt released from the distillery is the 3 year old Georg IV which has matured for two years in toasted Spessart oak casks and finished for one year in casks that have contained Banyuls wine. Around 800 litres are produced per year.

St Kilian Distillers

Rüdenau, founded in 2015

stkiliandistillers.com

St Kilian is one of few German distilleries designated to make whisky and nothing else. The first single malt, Signature Edition One released in 2019, was a 3 year old matured in bourbon casks and it was followed by another matured in ex-Amarone casks. The distillery has a capacity of 200,000 litres of alcohol.

Hausbrauerei Altstadthof

Nürnberg, founded in 1984

www.hausbrauerei-altstadthof.de

The first German distillery to produce organic single malt. The current range consists of the 4 year old Ayrer´s Red matured in new American oak, Ayrer´s PX, finished in PX sherry casks and Ayrer´s Bourbon, matured in bourbon barrels. A limited expression called Louis XVI, released in 2018, had been matured in a combination of casks that had previously held Bordeaux wine and cognac.

Destillerie Mösslein

Zeilitzheim, founded in 1984 (whisky since 1999)

www.frankenwhisky.de

Originally a winery, whisky production was brought on board in 1999. The first whisky was released in 2003 and the core range consists of a single malt and a grain whisky, both 5 years old. In 2017, the first 12 year old, Ernest 25, was released.

Brennerei Josef Druffel

Oelde-Stromberg, founded in 1792 (whisky since 2010)

www.brennerei-druffel.de

The first single malt, Prum, was released in 2013 and had matured in a mix of different casks (bourbon, sherry, red wine and new Spessart oak) and was finished in small casks made of plum tree! In 2015, a 5 year old version was released.

Brauhaus am Lohberg

Wismar, whisky since 2010

brauhaus-wismar.de, hinricusnoyte.de

The first release of Baltach single malt was in December 2013. It was a 3 year old with a finish in sherry casks. The latest edition of Baltach, finished in PX sherry casks was released in May 2018 and a peated version followed in October 2018.

Wild Brennerei

Gengenbach, founded in 1855 (whisky since 2002)

www.wild-brennerei.de

Two 5 year old whiskies have been released so far – Wild Whisky Single Malt which has matured for three years in American white oak and another two in either sherry or port casks and Blackforest Wild Whisky, made from unmalted barley.

Brennerei Volker Theurer

Tübingen, founded in 1991

www.schwaebischer-whisky.de

Located in a guesthouse, the released its first whisky as a 7 year old in 2003 and since then they have released Sankt Johann, an 8 year old single malt and the 9 year old Tammer. Theurer is also selling a blended whisky called Original Ammertal Whisky.

Lübbehusen Malt Distillery

Emstek, founded in 2014

theluebbehusen.com

The distillery has one of the largest pot stills in the country, where whisky made from peated Scottish malt is distilled. The first release, a 3 year old, appeared in autumn 2017. Rye whisky has also been released and attached to the distillery is a visitor centre.

Burger Hofbrennerei

Burg, founded in 2007 (whisky since 2012)

sagengeister.de

Apart from distillates from fruits and berries, the distillery also produces whisky made from malted barley. Maturation is in small (100 litres) casks made of American white oak. The first release of Der Kolonist single malt was in spring 2015.

Number Nine Spirituosen-Manufaktur

Leinefelde-Worbis, founded in 1999 (whisky since 2013)

ninesprings.de

The production of liqueurs was expanded in 2013 to include rum, gin and whisky. The first single malt was launched in 2016 while one of the latest, Peated Breeze Edition, appeared in spring 2019.

Edelbrennerei Schloss Neuenburg

Freyburg, founded in 2012

schlossbrennerei.eu

Small volumes of single malt whisky are produced with the spirit maturing for two years in new, German oak and then for another year in pinot noir casks. The first release was in August 2016.

Gutsbrennerei Joh. B. Geuting

Bocholt Spork, founded in 1837 (whisky since 2010)

muensterland-whisky.de

The first releases from this distillery, two single malts and two single grain, appeared in September 2013. More releases of the J.B.G. Münsterländer Single Malt have followed, the latest a 3.5 year old matured in sherry casks.

Sauerländer Edelbrennerei

Kallenhardt, founded in 2000 (whisky since 2004)

sauerlaender-edelbrennerei.de

The first release of the Thousand Mountains McRaven appeared in 2007 as a 3 year old. The recipe was changed in 2011 from peated to unpeated and the first bottlings from the new era were launched

in 2014. In 2016, the distillery was expanded when they moved the production to an old sawmill.

Destillerie Ralf Hauer

Bad Dürkheim, founded in 1989 (whisky since 2012)

sailltmor.de

The first release from the distillery appeared in 2015 with the 3 year old Saillt Mor single malt. Recent bottlings include a 4 year old matured in ex-bourbon casks, a PX sherry cask finish bottled at 59.3% and their first peated single malt (45ppm).

Destillerie Thomas Sippel

Weisenheim am Berg, founded in 1992 (whisky since 2011)

destillerie-sippel.de

Wines as well as distillates of all kinds are on the menu with whisky being introduced in 2011. The first release of the Palatinatus Single Malt came in 2014 and there are several expressions available, including a 6 year old peated version.

Steinwälder Hausbrennerei

Erbendorf, founded in 1818 (whisky since 1920)

brennerei-schraml.de

A kind of whisky was made here already in the early 1900s but was then sold as "Kornbrand". When the current owner took over, the spirit was relaunched as a 10 year old single grain whisky under the name Stonewood 1880. Other releases include a wheat whisky as well as two 3 year old single malts, Dra and Smokey Monk.

Finch Whiskydestillerie

Heroldstatt, founded in 2001

finch-whisky.de

The distillery is one of Germany´s biggest with a yearly production of 250,000 litres and it is also equipped with one of the biggest pot stills in Germany - 3,000 litres. The range of whiskies is large and they are made from a variety of different grains. The age is between 5 and 8 years and included is a 5 year old single malt.

Mönchguter Hofbrennerei

Middelhagen, founded in 2006

ruegen-whisky.de

Located on the island of Rügen the distillery produces spirits from fruits and berries as well as whisky. The main product is the blended whisky Pommerscher Greif but occasional single malts have been released.

Schaubrennerei Am Hartmannsberg

Freital, founded in 2011

hartmannsberger.de

Working on a range of various spirits, the owner also produces small volumes of whisky. The first whisky was released in 2015 and the latest was a 5 year old single malt, matured in ex-sherry casks..

Old Sandhill Whisky

Bad Belzig, founded in 2012

sandhill-whisky.com

The first whisky was released as a 3 year old single malt in 2015. Since then a wide range of bottlings have been launched. The most recent are two single malts matured in port pipes and Bordeaux barriques respectively.

Bellerhof Brennerei

Owen, founded in 1925 (whisky since 1990)

bellerhof-brennerei.com

The production of whisky made from barley, wheat and rye started in 1990 and today there is one single malt in the range - the 5 year old Danne´s Gärschda Malt, available at cask strength or 43%.

Eifel Destillate

Koblenz, founded in 2009

eifel-destillate.de

Even though other spirits are produced, the focus is on whisky of all sorts. The core range consists of Single Rye, Malty Blend and Smoky Blend but several single malts can be found amongst the limited releases. No whiskies are chill-filtered or coloured.

Iceland

Eimverk Distillery

Reykjavik, founded in 2012

www.flokiwhisky.is

The country´s first whisky distillery emanated from an idea in 2008 when the three Thorkelsson brothers discussed the possibility of producing whisky in Iceland. In 2011 a company was formed, the first distillation was made in the ensuing year and full scale production started in August 2013. Only organic barley grown in Iceland is used for the production and everything is malted on site. Both peat and sheep dung is used to dry the malted barley. The distillery has a capacity of 100,000 litres where 50% is reserved for gin and aquavite and the rest for whisky. The first, limited release of a 3 year old whisky was in November 2017 and more bottlings have followed since then.

Republic of Ireland

Midleton Distillery

Midleton, Co. Cork, founded in 1975

www.irishdistillers.ie

Midleton is by far the biggest distillery in Ireland and the home of Jameson´s Irish Whiskey. The production at Midleton comprises of two sections – grain whiskey and single pot still whiskey. The grain whiskey is needed for the blends, where Jameson´s is the biggest seller. Single pot still whiskey, on the other hand, is unique to Ireland. This part of the production is also used for the blends but is being bottled more and more on its own.

Until recently, Midleton distillery was equipped with mash tuns both for the barley side and the grain side. After considerable

research and trials however, these have now been replaced by mash filters with an astonishing increase in spirit yield from 385 litres per ton of barley to 415 litres. Two major upgrades of the distillery (in 2013 and 2017) means that Midleton is now equipped with 48 washbacks, 6 column stills and 10 pot stills. A new maturation facility with 40 warehouses has also been built in Dungourney, not far from Midleton. More invetsments weree announced in autumn 2018 when the owners revealed that 150 million euros would be spent on more warehouses and an expansion of the bottling and packing facilities. In autumn 2015, a new micro distillery adjacent to the existing distillery, was opened. With a production capacity of 400 casks per year, it will be used for experiments and innovation. However, in spring 2019, a commercial release of Method an Madness Gin made at the micro distillery was introduced.

Of all the brands produced at Midleton, Jameson´s blended Irish whiskey is by far the biggest. In 2018 the brand sold 90 million bottles – an astonishing increase by 60% in the last five years! Apart from the core expression with no age statement, there are 12 and 18 year olds, Black Barrel, Gold Reserve and a Vintage. Since 2015, a number of special series have been launched; Deconstructed with three bottlings – Bold, Lively and Round, The Whiskey maker´s Series with The Cooper´s Croze and The Blender´s Dog and Jameson Caskmates with two whiskies aged in stout barrels and IPA barrels. In spring 2018, Jameson Bow Street 18 years old, the first cask strength release of a Jameson, appeared and a second batch was unveiled in August 2019. In July 2019, the triple-cask, triple-distilled Jameson Triple Triple was released as an exclusive to duty-free. Other blended whiskey brands include Powers and the exclusive Midleton Very Rare. In recent years, Midleton has invested increasingly in their second category of whiskies, single pot still, and that range now includes Redbreast 12, 12 cask strength, 15, 21 year old, the sherrymatured Lustau Edition and the Redbreast Dream Cask 32 years old. In May 2019, the limited 20 year old Redbreast Dream Cask Pedro Ximénez Edition was launched. Furthermore, there is Green Spot without age statement, the 12 year old Leoville Barton bordeaux finish and the Chateau Montelena finish as well as Yellow Spot 12 years old and the latest addition, Red Spot, released in late 2018 and matured in a combination of ex-bourbon, oloroso sherry and marsala casks. Powers (John´s Lane, Signature and Three Swallow) and Barry Crocket Legacy are other eaxmples of their single pot still whiskies. The first release of an Irish whiskey finished in virgin Irish oak in 2015, the Dair Ghaelach, was followed up by a seond edition in autumn 2017. More innovation followed in 2017 when a range of experimental whiskeys were released under the name Method and Madness. Included in the range were four expressions with onee of them finished in French chestnut casks. Two more expressions were added in summer 2018; a single pot still finished in virgin Hungarian oak and a 28 year old single pot still with six years in ex-bourbon barrels and a further 22 years in ruby port pipes.

Irish Distillers and their Midleton Distillery were instrumental in the re-birth of the Irish single pot still whiskey category

Tullamore Dew Distillery

Clonminch, Co. Offaly, founded in 2014

www.tullamoredew.com

Until 1954, Tullamore D.E.W. was distilled at Daly´s Distillery in Tullamore. When it closed, production was temporarily moved to Power´s Distillery in Dublin, and was later moved to Midleton Distillery and Bushmill´s Distillery. William Grant & Sons acquired Tullamore D.E.W. in 2010 and in May 2013, they started to build a new distillery at Clonminch, situated on the outskirts of Tullamore. The four stills produce both malt whiskey and single pot still whiskey and the capacity is 3.6 million litres of pure alcohol. In autumn 2017, a bottling hall and a grain distillery with a capacity of doing 8 million litres of grain spirit was opened on the same site. All whiskies at Tullamore are triple distilled. Tullamore D.E.W. is the second biggest selling Irish whiskey in the world after Jameson with 13 million bottles sold in 2018. The core range consists of Original (without age statement), 12 year old Special Reserve and 14 and 18 year old Single Malts. Recent limited releases include Trilogy (a triple blend whiskey matured in three types of wood), Phoenix and Old Bonded Warehouse. As an exclusive to duty free, the Tullamore D.E.W Cider Cask Finish was launched in summer 2015 and this was followed in autumn 2017 by a Carribean Rum Cask Finish.

Cooley Distillery

Cooley, Co. Louth, founded in 1987

www.kilbeggandistillingcompany.com

In 1987, the entrepreneur John Teeling bought the disused Ceimici Teo distillery and renamed it Cooley distillery. Two years later he installed two pot stills and in 1992 he released the first single malt from the distillery, called Locke´s Single Malt. A number of brands were launched over the years. In December 2011 it was announced that Beam Inc. had acquired the distillery for $95m. In 2014, Suntory took over Beam and the new company was renamed Beam Suntory. Cooley distillery is equipped with one mash tun, four malt and six grain washbacks all made of stainless steel, two copper pot stills and two column stills. There is a production capacity of 650,000 litres of malt spirit and 2,6 million litres of grain spirit. The range of whiskies is made up of several brands. Connemara single malts, which are all more or less peated, consist of a no age, a 12 year old and a cask strength. Another brand is Tyrconnell with a core expression bottled without age statement. Other Tyrconnell varieties include three 10 year old wood finishes, a 15 year old madeira finish and the recently launched (June 2019) Tyrconnell 16 year old with a finish in both oloroso and moscatel casks.

Teeling Distillery

Dublin, founded in 2015

www.teelingwhiskey.com

After the Teeling family had sold Cooley and Kilbeggan distilleries to Beam in 2011, the family started a new company, Teeling Whiskey. John´s two sons, Jack and Stephen, then opened a new distillery in Newmarket, Dublin in June 2015. This was the first new distillery in Dublin in 125 years. One year after the opening, an amazing 60,000 people had been welcomed to the distillery. In summer 2017, Bacardi acquired a minority stake in Teeling Whiskey for an undisclosed sum. This is the first time Bacardi gets involved with Irish whiskey.

The distillery is equipped with two wooden washbacks, four made of stainless steel and three stills made in Italy; wash still (15,000 litres), intermediate still (10,000 litres) and spirit still (9,000 litres) and the capacity is 500,000 litres of alcohol. Both pot still and malt whisky is produced. The core range from the distillery consists of the blend Small Batch which has been finished in rum casks, Single Grain which has been fully matured in Californian red wine barrels and Single Malt - a vatting of five different whiskies that have been finished in five different types of wine casks. All these had been distilled at Cooley. The first release from their own production appeared in August 2018, limited to 250 bottles. The first general

release came two months later when Teeling Single Pot Still (50% malted and 50% unmalted barley) was launched with batch 3 being released in August 2019. Recent limited bottlings include The Revival Volume V (a 12 year old finished in cognac and brandy barrels), the Brabazon Bottling where the second release was finished in port pipes, a 24 year old, a 30 year old and a 33 year old.

Roayl Oak Distillery (fomerly known as Walsh Whiskey Distillery)

Carlow, Co. Carlow, founded in 2016

www.walshwhiskey.com, www.roayloakdistillery.com

With succesful brands such as The Irishman and Writer´s Tears (both produced at Midleton), Bernard Walsh decided to open his own distillery at Royal Oak, Carlow. With a back-up from the major Italian drinks company, Illva Saronno, construction began in late 2014 and the distillery was commissioned in March 2016. The capacity is 2.5 million litres of alcohol and all types of whiskey is produced including grain- malt- and pot still whiskey. The equipment consists of a 3 ton semi-lauter mash tun, six washbacks, a 15,000 litre wash still, a 7,500 litre intermediate still and a 10,000 litre spirit still. There is also a column still for grain whiskey production. Apart from producing whiskey for its own brands, the distillery has allocated 15% of the output for a number of international partners. In January 2019, it was announced that Illva Saronno would take full control of the distillery while Bernard Walsh would continue with the brands Writer´s Tears and The Irishman, trading under the name Walsh Whiskey and in the future relying on whiskey from Irish Distillers (Midleton) for his needs. His latest releases (both in spring 2019) were Writer´s Tears Copper Pot – Deau XO Cognac Cask Finish and Double Oak (a blend of single pot still and single malt whiskey).

Great Northern Distillery

Dundalk, Co. Louth, founded in 2015

www.gndireland.com

In 2013, the Irish Whiskey Company (IWC), with the Teeling family as the majority owners, took over the Great Northern Brewery in Dundalk and turned it into a distillery. When it became operational in August 2015, it was the second biggest distillery in Ireland, with the capacity to produce 3.6 million litres of pot still whiskey and 8 million litres of grain spirit. The distillery is equipped with three columns for the grain spirit production and three pot stills for producing malt and single pot still whiskey. In summer 2019, another two washbacks were added to increase production. The main part of the business is supplying whiskey to private label brands but in 2017, the owners released their first own brand - a 14 year old single malt by the name Burke´s Irish Whiskey, which had been distilled during the family´s Cooley days.

Waterford Distillery

Waterford, Co. Waterford, founded in 2015

www.waterforddistillery.ie

Founded by the former co-owner of Bruichladdich distillery on Islay, Mark Reynier. In 2014 he bought the Diageo-owned Waterford Brewery and 16 months later, in December 2015, the first spirit was distilled. The distillery is equipped with two pot stills and one column still and, even though grain spirit will be produced, malt whiskey is the number one priority. The distillery also has a mash filter instead of a mash tun. There is a focus on local barley and Reynier is sourcing the barley from 72 farms on 19 different soil types. The distillery has a capacity of 1 million litres but the owners have plans to go up to 3 million litres in the future. After having distilled Ireland´s first organic whiskey in 2016, Reynier recently decided to also produce the first biodynamic whiskey where the barley comes from three self-sufficient farms. They also created a new role at the distillery in autumn 2018 – a "terroir agronomist" where Grace O´Reilly will oversee the distillery´s relationships with its barley growers.

Other distilleries in Ireland

Kilbeggan Distillery

Kilbeggan, Co. Westmeath, founded in 1757

www.kilbegganwhiskey.com

Brough back to life in 2007 by John Teeling, Kilbeggan is the oldest producing whiskey distillery in the world. The distillery, currently owned by Beam Suntory, is equipped with a wooden mash tun, four Oregon pine washbacks and two stills with one of them being 180 years old. The first single malt whiskey release from the new production came in 2010 and limited batches have been released thereafter. The core blended expression of Kilbeggan is a no age statement bottling but limited releases of aged Kilbeggan blend have occurred. There is also a Kilbeggan grain which was produced at Cooley distillery and in autumn 2018, Kilbeggan Small Batch Rye (30% rye), 100% distilled and matured at the Kilbeggan distillery, was released.

Pearse Lyons Distillery

Dublin, founded in 2017

www.pearselyonsdistillery.com

The founder, Dr Pearse Lyons, who passed away in March 2018, was a native of Ireland and used to work for Irish Distillers in the 1970s. In 1980 he changed direction and founded a company specializing in animal nutrition and feed supplements. In 2008, he opened a whiskey distillery in Lexington, Kentucky. Four years later, in a joint venture, he started a distillery in Carlow, Ireland. After a few years, the stills were moved to Dublin where Dr Lyons restored the old St James' church and converted it to a distillery. The first distillation was in September 2017 but the owners already have a range of aged whiskies; a 5 year old single malt made from whisky produced in the stills when they were situated at Carlow, a 12 year old, sourced single malt and three blends – 5 year old, 7 year old and one without age statement.

West Cork Distillers

Skibbereen, Co. Cork, founded in 2004

www.westcorkdistillers.com

The distillery, equipped with four stills, produces both malt whiskey and grain whiskey (from barley and wheat) and some of the malting is done on site. Apart from a range of vodka, gin and liqueurs, several single malts, including a 12 year old, and blends are sold under the name West Cork. In autumn 2018, a new expression was launched - Export Stout Cask Finished Blend where casks that had matured Irish whiskey were sent to Castle Island Brewing in Massachusetts. After having aged an imperial stout, the casks were sent back to Ireland to finish a West Cork Irish blend.

Connacht Whiskey Company

Ballina, Co. Mayo, founded in 2016

www.connachtwhiskey.com

The distillery is equipped with three pot stills made in Canada and has the capacity to produce 300,000 litres of pure alcohol per year. The first distillation of whiskey (double-distilled) was made in April 2016 and in 2017, triple distillation started as well. Apart from malt whiskey and single pot still whiskey, the owners also produce vodka, gin and poitin. Releases so far have been made from sourced whiskey including the blend Brothership – a vatting of 10 year old Irish pot still and 10 year old American whiskey.

The Dingle Whiskey Distillery

Milltown, Dingle, Co. Kerry, founded in 2012

www.dingledistillery.ie

The old Fitzgerald sawmills has been transformed into a distillery with three pot stills and a combined gin/vodka still and the first production of gin and vodka was in October 2012 with whiskey production commencing in December. The first whiskey, the limited Dingle Cask No. 2, was released in December 2015 and a general release was made in autumn 2016. Batch No. 3, a marriage of bourbon and port casks, was launched in July 2018, their second single pot still appeared in November 2018 and in April 2019, 30,000 bottles of Dingle Batch No. 4 Single Malt were released.

The Shed Distillery

Drumshanbo, Co. Leitrim, founded in 2014

www.thesheddistillery.com

Founded by entrepreneur and drinks veteran P J Rigney, the distillery cost €2m to build and is equipped with five pot stills, three column stills and six washbacks. The focus for the owners is triple distilled single pot still whiskey but Drumshanbo gin has been on the market for several years. The first whiskey distillation was in late 2014 and an un-official release from the first cask took place in December 2017. A general release isn't expected until 2019.

Boann Distillery

Drogheda, Co. Meath, founded in 2016

boanndistillery.ie

Assisted by the well-known whisky consultant, John McDougall, Pat Cooney built the distillery which is equipped with three Italian-made copper pot stills and a gin still. The owners have used a new technology where the copper contact is enhanced using nano-crystal coating in the lyne arms. The distillery was commissioned in autumn 2019 but in common with most Irish distillery start-ups, they have already launched a sourced whiskey called The Whistler.

Glendalough Distillery

Newtown Mount Kennedy, Co. Wicklow, founded in 2012

www.glendaloughdistillery.com

For the first three years, the company acted as an independent bottler. In 2015 Holstein stills were installed and gin production began. Apart from gin, the company regularly releases sourced whiskies. In 2016, the Canadian drinks distribution group Mark Anthony Brands invested €5.5m in the distillery.

Slane Distillery

Slane, Co. Meath, founded in 2017

www.slaneirishwhiskey.com

The Conyngham family, owners of the Slane Castle and Estate since 1703, established a whiskey brand a few years ago which became popular not least in the USA. The whiskey was produced at Cooleys but the family decided to start a distillery of their own. After an unsuccesful partnership with Camus Wine & Spirits, Brown-Forman stepped in and took over the entire project in 2015. Equipped with three copper pot stills, six column stills and washbacks made of wood, the distillery started production in summer 2018. Three types of whiskey will be produced; single malt, single pot still and grain whiskey.

Ballykeefe Distillery

Ballykeefe, Co. Kilkenny, founded in 2017

www.ballykeefedistillery.ie

A classic farm distillery and the first to operate in Kilkenny for over 200 years. The distillery is equipped with two copper pot stills and the first distillation was in spring 2018. Gin, vodka and poitin have already been released and the first Ballykeefe whiskey is expected in August 2020.

Powerscourt Distillery

Enniskerry, Co. Wicklow, founded in 2017

www.powerscourtdistillery.com

The distillery is situated at the Powerscourt Estate, owned by the

Slazenger family, south of Dublin. Three pot stills were ordered from Forsyths in Scotland, distillation started in autumn 2018 and a visitor centre was opened in summer 2019. Noel Sweeney from Cooley Distillery is in charge of the distillery which has a capacity of producing 1 million bottles per year. A range of sourced whiskies have been released under the name Fercullen.

Lough Mask Distillery

Tourmakeady, Co. Mayo, founded in 2017

www.loughmaskdistillery.com

Equipped with two alambic stills, the distillery started production in early 2018. The whiskey is double distilled and both peated and unpeated spirit will be produced. The first whiskey will be released in 2021 but both gin and vodka are already for sale.

Blackwater Distillery

Ballyduff, Co. Kerry, founded in 2014

www.blackwaterdistillery.ie

Originally located in Cappoquin, Co. Waterford, the distillery moved in 2018 to Ballyduff. At that time the company was already famous for their gin. In their new distillery, equipped with three stills from Frilli in Italy, whisky will now also be produced.

Roe & Co. Distillery

Dublin, founded in 2019

www.roeandcowhiskey.com

The opening of Roe & Co Distillery marks Diageo´s return to the Irish whiskey scene which they left in 2014 when they sold Bushmills. A blend named Roe & Co made from sourced whiskies was launched already in 2017 and at the same time plans to build a distillery in Dublin were revealed. The distillery is situated in The Liberties district in the former Guinness Power Station. Three stills made by Abercrombie in Alloa and wooden washbacks make up part of the equipment and the production capacity of double- and tripledistilled whiskey is 500,000 litres. The distillery is named after George Roe, a prominent whiskey maker in the 19th century.

Killowen Distillery

Newry, Co. Down, founded in 2017

www.killowendistillery.com

Founded by Brendan Cart, the distillery started producing gin in 2017. Whiskey was always on the charts however and in early 2019 the first batch of pot still whiskey was distilled. Carty's approach is a bit different compared to many other Irish distillers as the spirit is double distilled and peated and the mash bill contains not only malted and unmalted barley but also other grains (like oats). Furthermore, the two Portuguese stills are direct heated and worm tub condensers are being used. The current GI (geographical indication) for pot still whiskey allows for up to 5% of other cereal grains than barley but Cart is hoping for a change that would allow at least up to 30% of non-barley grains in line with how Irish pot still whiskey was produced in the old days.

The Dublin Liberties Distillery

Dublin founded in 2018

www.thedld.com

The birth of this, the third whiskey distillery in Dublin in modern days, and its location in the historical Liberties district, has been a struggle spanning at least eight years. Finally starting the production in early 2019, it is now owned by Quintessential Brands (75%) and East European drinks company Stock Spirits (25%). The master distiller is Darryl McNally who spent 17 years working for Bushmills. The three coper pot stills built by Firma Carl in Germany will produce both double and triple distilled whiskey as well as peated expressions. The mash tun charge is two tonnes of barley, the fermentation time is 60-72 hours and the total capacity is 700,000 litres of pure alcohol per year. The whiskeys currently available in the visitor centre have all been produced at either Bushmills or Cooleys.

Clonakilty Distillery

Clonakilty, Co. Cork, founded in 2016

www.clonakiltydistillery.ie

For eight successive generations, the Scully family have farmed the coastal lands near the resort town Clonakilty in West Cork.

The new Roe & Co Distillery opened in Dublin in 2019

Michael and Helen Scully, together with other members of the family, started production in early 2019. With Paul Corbett (former Teeling Whiskey Company) as the head distiller, the main product will be a triple distilled pot still whiskey. The first release is planned for late 2021

Lough Gill Distillery

Hazelwood House, Co. Sligo, founded in 2019

www.loughgilldistillery.com

A group of investors led by David Raethorne came up with the idea in 2014 to build a distillery adjacent to Hazelwood House, a country house built in the 18th century. Planning permission was secured in 2017 and in autumn 2019 the distillery commenced producing. It is unusually large with a capacity of 1 million litres. Several releases of sourced whiskey have already appeared under the brand name Athrú.

Italy

Puni Destillerie

Glurns, South Tyrol, founded in 2012

www.puni.com

There are at least two things that distinguish this distillery from most others. One is the design of the distillery – a 13-metre tall cube made of red brick. The other is the raw material that they are using. Malt whisky is produced but malted barley is only one of three cereals in the recipe. The other two are malted rye and malted wheat. In 2016, however, they also started distilling 100% malted barley and anticipate to release their first whisky from that production in 2019. The distillery is equipped with five washbacks made of local larch and the fermentation time is 96 hours. There is one wash still (3,000 litres) and one spirit still (2,000 litres) and the capacity is 80,000 litres of alcohol per year. The first single malt was released in October 2015 and the current core range consists of Nova (American oak), Alba (marsala casks with a finish in Islay casks) and Sole (two years in ex-bourbon and two years in PX casks). Recent limited editions include two 5 year olds – Gold (ex-bourbon) and Vina (marsala casks).

Psenner Destillerie

Tramin, founded in 1947 (whisky since 2013)

www.psenner.com

A producer of spirits from apples pears and later also a grappa, tried their hands at whisky for the first time six years ago. Their inaugural release of the 3 year old single malt eRètico appeared in October 2016 and had been matured in a combination of ex-grappa and ex-oloroso casks

Liechtenstein

Brennerei Telser

Triesen, founded in 1880 (whisky since 2006)

www.telserdistillery.com

Telser was probably the only distillery in Europe that still used a wood fire to heat their stills. The distillery has been closed since 2018 but new releases from the stock still occur. One of the latest expressions was L´Ultimo, matured in an ex-grappa casks.

The Netherlands

Zuidam Distillers

Baarle Nassau, founded in 1974 (whisky since 1998)

www.zuidam.eu

Zuidam Distillers was started in 1974 as a traditional, family distillery producing liqueurs, genever, gin and vodka and is today managed by Patrick van Zuidam. The first release of a single malt whisky, which goes by the name Millstone, was from the 2002 production and it was bottled in 2007 as a 5 year old. The current range is a 5 year old which comes in both peated and unpeated versions, American oak 10 years, French oak 10 years, Sherry oak 12 years and PX Cask 1999. Apart from single malts there is also a Millstone 100% Rye which is bottled at 50%. Limited expressions include 1998 PX Single Cask, the 8 year old Double Sherry Cask (oloroso and PX), American Oak Peated Moscatel, the 4 year old 92 Rye and the 7 year old Peated PX. In spring 2019, Zuidam collaborated with BrewDog brewery together with Compass Box and Duncan Taylor. The task was to design a whisky that could be paired with one of the brewery´s beers. The contribution from Millstone was Torpedo Tulip - a 100% rye whisky matured in ex-oloroso casks. The distillery has been expanded continuously over the years and two more stills were installed in 2019. The distillery has also started to grow their own barley and rye at a nearby farm.

Other distilleries in The Netherlands

Us Heit Distillery

Bolsward, founded in 2002

www.usheit.com

Frysk Hynder was the first Dutch whisky and made its debut in 2005 at 3 years of age. The barley is grown in surrounding Friesland and malted at the distillery. Some 10,000 bottles are produced annually and the whisky (3 to 5 years old) is matured in a variety of casks. A cask strength version has also been released.

Kalkwijck Distillers

Vroomshoop, founded in 2009

www.kalkwijckdistillers.nl

The distillery is equipped with a 300 litre pot still still with a column attached. The main part of the production is jenever, korenwijn and liqueurs but whisky has been distilled since 2010. In spring 2015, the first single malt was released. Eastmoor was 3 years old, made from barley grown on the estate and bottled at 40%. More batches have followed since.

Stokerij Sculte

Ootmarsum, founded in 2004 (whisky since 2011)

www.stokerijsculte.nl

The distillery is equipped with a 500 litre stainless steel mashtun, 4 stainless steel washbacks with a fermentation time of 4-5 days and two stills. The first Sculte Twentse Whisky was released in 2014 and this was followed by a 4 year old in 2016. The third release was heavily peated (30ppm) and this was followed by a fourth, unpeated batch in spring 2019.

Lepelaar Distillery

Texel, founded in 2009 (whisky since 2014)

www.landgoeddebontebelevenis.nl

Joscha and Inge Schoots started a brewery and shop in 2009 and continued five years later by adding distilling equipment. The business is a part of a larger crafts centre. Single malt (both peated and unpeated) as well as grain whisky and genever is produced. A first, limited release of the whisky was made for the crowd funders in 2018 and will be followed later by a general release.

Eaglesburn Distillery

Ede, founded in 2015

www.eaglesburndistillery.com

One of the latest whisky distilleries to be opened in The Netherlands. The owner, Bart Joosten, is a firm believer in long fermentation and at Eaglesburn, the wort is fermented for at least 10-12 days. The inaugural release of a 3 year old single malt whisky was in October 2018 from an ex-bourbon barrel followed by a second release in 2019 from virgin American oak. Besides whisky they produce gin, vodka and rum.

Den Hool Distillery

Holsloot, founded in 2015

www.denhoolwhisky.nl

Originally a farm, the owners also started brewing beer in the late 1990s. In 2009 some of the wash was sent to Zuidam Distillers for distillation into whisky. The first releases from those early batches appeared as a 6 year old Veenhaar single malt in 2016. Since 2015, distillation is carried out on the farm and they are also taking care of their own malting.

Northern Ireland

Bushmill´s Distillery

Bushmills, Co. Antrim, founded in 1784

www.bushmills.com

Diageo took the market by surprise when they announced in 2014 that they were selling the distillery. This was Diageo´s only part of the increasing Irish whiskey segment and commentators struggled to see the reason for the sale. The buyer was the tequila maker Casa Cuervo, producer of José Cuervo. Diageo already owned 50% of the company´s other, upscale tequila brand, Don Julio and with the deal, they got the remaining 50% as well as $408m.

Bushmills is the second biggest of the Irish distilleries after Midleton, with a capacity to produce 4,5 million litres of alcohol a year. In 1972 the distillery became a part of Irish Distillers Group which thereby gained control over the entire whiskey production in Ireland. Irish Distillers were later (1988) purchased by Pernod Ricard who, in turn, resold Bushmill´s to Diageo in 2005 at a price tag of €295.5 million. Since the take-over, Diageo invested heavily into the distillery and it now has ten stills with a production running seven days a week, which means 4,5 million litres a year. Two kinds of malt are used at Bushmills, one unpeated and one slightly peated. The new owners, Casa Cuervo, applied for a planning permission to expand the capacity and also to build another 29 warehouses on adjacent farmland. The approval process proved to be quite difficult due to environmental objections but in April 2019, the local council gave the green light. Apart from the new warehouses, an additional mash tun, eight new washbacks and ten new stills will be installed. The capacity after the expansion will be 9 million litres of alcohol. The total investment will be £30m.

Bushmill's core range of single malts consists of a 10 year old, a 16 year old Triple Wood with a finish in Port pipes for 6-9 months and a 21 year old finished in Madeira casks for two years. There is also a 12 year old Distillery Reserve which is sold exclusively at the distillery and the 1608 Anniversary Edition. Black Bush and Bushmill's Original are the two main blended whiskeys in the range but in December 2017 a third expression was added to the range – Bushmills Red Bush. The first new expression from the distillery for the domestic market in five years, Red Bush is a blend of triple distilled single malt and grain whiskey. In spring 2016, Bushmill´s launched their first whiskey exclusive for duty free, The Steamship Collection, with three special cask matured whiskies plus a number of limited releases. In December 2018, Bushmills became one of the first distilleries in the world to release a whiskey matured in acacia wood. The bottling, only available at the distillery, had been finished in acacia casks for more than a year. Bushmills is the third most sold Irish whiskey after Jameson and Tullamore D.E.W.

Other distilleries in Northern Ireland

Echlinville Distillery

Kircubbin, Co. Down, founded in 2013

www.echlinville.com

After having relied on Cooley Distillery for his mature whiskey, Shane Braniff decided in 2012 to build his own distillery. Located near Kircubbin on the Ards Peninsula he started production in August 2013. The distillery was further expanded with more equipment in 2015 and in April 2016, a visitor centre opened. The distillery also has its own floor maltings. Apart from single pot still and single malt whiskey, vodka and gin is also produced. Braniff recently revived the old Dunville´s brand of blended whiskey and released a 10 year old single malt with a finish in PX sherry casks. The latest expression was Dunville´s Three Crowns Peated. All these have been made from sourced whiskey. Their own produce has not yet been released.

Rademon Estate Distillery

Downpatrick, Co. Down, founded in 2012

shortcrossgin.com

Fiona and David Boyd-Armstrong opened their distillery on the Rademon estate in 2012. Since its inception, their main product has been Short Cross gin which quickly became a success story. In summer 2015 the production was expanded into whiskey and during

Bushmills Distillery is about to be expanded

the first year, around 100 barrels were filled. Through a £2.5m investment, the capacity of the distillery was further increased in 2018 with a new gin still as well as a new still for the whiskey production. The goal for the couple was to release their first single malt whiskey in 2019 but the uncertainty of the Brexit deal has made them put the plans on hold.

Norway

Det Norske Brenneri
Grimstad, founded in 1952 (whisky since 2009)
www.detnorskebrenneri.no

Founded in 1952 the company mainly produced wine from apples and other fruits. Whisky production started in 2009 and two Holstein stills are used for the distillation. In 2012, Audny, the first single malt produced in Norway was launched. Recent bottlings include two Eiktyrne editions, one finished in brandy casks and the other with an extra maturation in sauternes casks.

Other distilleries in Norway

Myken Distillery
Myken, founded in 2014
www.mykendestilleri.no

This distillery was built in Myken, a group of islands in the Atlantic ocean, 32 kilometres from mainland Norway. They distilled their first spirit in 2014 and the equipment consists of one wash still (1,000 litres), one spirit still (700 litres) and one gin still (300 litres). Both peated and unpeated whisky is produced. The first release of Myken Single Malt was in September 2018. This was followed in November by a version finished in ex-Pineau des Charentes casks and in July 2019 by Hungarian Touch matured in a combination of ex-bourbon and Hungarian oak.

Arcus
Gjelleråsen, founded in 1996 (whisky since 2009)
www.arcus.no

Arcus is the biggest supplier and producer of wine and spirits in Norway with subsidaries in Denmark, Finland and Sweden. The first whisky produced by the distillery was launched in 2013. Under the name Gjoleid, two whiskies made from malted barley and malted wheat were released. More recent bottlings, some up to 5 years old, include Blindpassasjeren and Praksis 1.1 and 1.2.

Aurora Spirit
Tromsö, founded in 2016
www.bivrost.com

At 69.39°N, Aurora is the northernmost distillery in the world. The mash is bought from a brewery, fermented at the distillery and distilled in the 1,200 litre Kothe pot still with an attached column. Both non-peated and peated whisky is produced. The plan is to release the first whisky in spring 2020. Apart from single malt whisky, the owners also produce gin, vodka and aquavit. All their products are sold under the name Bivrost.

Oss Craft Distillery
Flesland, founded in 2016
www.osscraft.no

Specialising in gin and other spirits made from herbs and botanicals, the distillery has already launched a range of spirits under the name Bareksten (from the founder and owner Stig Bareksten). In 2017, production of malt whisky began as well.

Feddie Distillery
Island of Fedje, founded in 2019
www.feddiedistillery.no

An already existing brewery was complemented by a distillery in 2019 and production of organic whisky started in the summer. The goal for 2020 is to produce 50,000 litres of pure alcohol and the owners have a plan to open a visitor centre in summer 2020. Fedje is the most westerly, populated island in Norway

Berentsen Distillery
Egersund, founded in 1895 (whisky since 2018)
www.berentsens.no

A producer of mineral water for more than 100 years, the company started brewing beer 15 years ago and just recently expanded into distillation of spirits (vodka, aquavit and whisky). With stills from Arnold Holstein in Germany and with the aid of Frank McHardy (ex Springbank) as a consultant, the owners hope to be one of the biggest whisky producers in Norway.

Spain

Distilerio Molino del Arco
Segovia, foundd in 1959
www.dyc.es

The distillery has a capacity for producing eight million litres of grain whisky and two million litres of malt whisky per year. In addition to that, vodka and rum are produced and there are also in-house maltings. The distillery is equipped with six copper pot stills and there are 250,000 casks maturing on site. The big seller when it comes to whiskies is a blend simply called DYC which is around 4 years old. It is supplemented by an 8 year old blend and, since 2007, also by DYC Pure Malt, a blend of malt from the distillery and from Scottish distilleries. To commemorate the distillery's 60th anniversary in 2019, they released a 15 year old single malt in a limited range called Colección Maestros Destiladores.

Other distilleries in Spain

Destilerias Liber
Padul, Granada, founded in 2001
www.destileriasliber.com

Apart from whisky, the distillery produces rum, marc and vodka. For the whisky production, the spirit is double distilled after a fermentation of 48-72 hours. Maturation takes place in sherry casks. The only available whisky on the market is a 5 year old single malt called Embrujo de Granada.

Sweden

High Coast Distillery (former BOX Distillery)
Bjärtrå, founded in 2010
www.highcoastwhisky.se

Set in buildings from the 19th century, the distillery started production in November 2010. Sales of their whisky, first released in 2014, has exceeded the owners´ expectations and in 2018 the distillery was expanded. The equipment now consists of a semilauter mash tun with a capacity of 1,5 tonnes, ten stainless steel washbacks, two wash stills (3,800 litres) and two spirit stills (2,500 litres). The expansion has increased capacity from 100,000 litres to 300,000. In 2014 an excellent visitor centre was opened which today attracts more than 10,000 visitors yearly.

The distillery makes two types of whisky – fruity/unpeated and peated. The distillery manager, Roger Melander, wants to create a new make which is as clean as possible by using a very slow distillation process with lots of copper contact in the still. The flavour of the spirit is also impacted by the effective condensation using what might be the coldest cooling water in the whisky world, namely 2-6°C, which is obtained from a nearby river. A fermentation time of 72-96 hours also affects the character.

The first whisky, The Pioneer, was released in June 2014. Between 3 and 4 years old, it was a vatting of unpeated and lightly peated whisky and the first in a range of four called Early Days Collection. In 2017, the first core expression, Dàlvve, was launched and in autumn that year, a new range, Quercus, was introduced where four expressions highlighting the influence of the oak were released. A new range, The Origin Series, appeared in 2019 with the bourbon matured, heavily peated Timmer, the medium peated Hav with some of the whisky matured in new oak from Sweden and Hungary, the unpeated Älv matured in first fill bourbon and unpeated Berg matured in PX sherry casks.

Mackmyra Svensk Whisky

Valbo, founded in 1999

www.mackmyra.se

Mackmyra´s first distillery was built in 1999 and, ten years later, the company revealed plans to build a brand new facility in Gävle, a few miles from the present distillery. In 2012, the distillery was ready and the first distillation took place in spring of that year. The construction of the new distillery is quite extraordinary and with its 37 metre structure, it is perhaps one of the tallest distilleries in the world. Since April 2013, all the distillation takes place at this new gravitation distillery. In 2017 however, the old distillery was re-opened as the Lab Distillery where the company aim to develop innovative spirits in collaboration with craft distillers.

Mackmyra whisky is based on two basic recipes, one which produces a fruity and elegant whisky, while the other is smokier. The first release was in 2006 and the distillery now has four core expressions; Svensk Ek, Brukswhisky, the peated Svensk Rök and, new since 2015, MACK by Mackmyra which competes in the lower price segment. A range of limited editions called Moment was introduced in 2010 and consists of exceptional casks selected by the Master Blender, Angela D´Orazio, who was inducted into the Whisky Hall of Fame in 2019. Two of the latest editions are Skogshallon where part of the whisky had been matured in casks that had previously held raspberry wine and Karibien where rum casks from Barbados and Jamaica had been part of the maturation. Seasonal expressions are also released regularly with Äppelblom being one

of the latest. The final part of the maturation was in ex-calvados casks. In autumn 2019, the distillery launched the first whisky in the world where AI (artificial intelligence) had been involved in the creation process. The distillery´s existing recipes, sales data and customer preferences generated more than 70 million recipes. The owners, however, are eager to stress that the blender had the final word what to bottle.

Spirit of Hven

Hven, founded in 2007

www.hven.com

The second Swedish distillery to come on stream, situated on the island of Hven right between Sweden and Denmark. The first distillation took place in May 2008. Henric Molin, founder and owner, is a trained chemist and very concerned about what type of yeast and grain he uses not to mention the right oak for his casks. The distillery is equipped with a 0,5 ton mash tun, six washbacks made of stainless steel, one wash still, one spirit still and a designated gin still. Apart from that, a unique wooden Coffey still was recently installed. Part of the barley is malted on site using Swedish peat, sometimes mixed with seaweed and sea-grass, for drying. Apart from whisky, other products include rum made from sugar beet, vodka, gin and aquavit.

Their first whisky was the lightly peated Urania, released in 2012. The second launch was the start of a new series of limited releases called The Seven Stars. The first expression was the 5 year old, lightly peated Dubhe, which was followed by Merak, Phecda, Megrez, Alioth, Mizar, Alcor and Alkaid. The first and so far only core expression, Tycho´s Star, was released in 2015. New and innovative bottlings include Sweden´s first rye whisky, Hvenus Rye, and Mercurious, the first whisky in Sweden made predominantly (88%) from corn which had been grown on the distillery grounds. The latter was released in summer 2019.

Smögen Whisky

Hunnebostrand, founded in 2010

www.smogenwhisky.se

Pär Caldenby – a lawyer, whisky enthusiast and the author of Enjoying Malt Whisky – is the founder and owner of Smögen Whisky on the west coast of Sweden. The distillery is equipped with three washbacks (1,600 litres each), a wash still (900 litres) and a spirit still (600 litres) and the capacity is 35,000 litres of alcohol a year. An interesting addition to the equipment setup was made in summer 2018 when Pär installed worm tubs to cool the

The High Coast Distillery recently doubled their capacity

spirits. Heavily peated malt is imported from Scotland and the aim is to produce an Islay-type of whisky. The first release from the distillery was the 3 year old Primör in 2014. This has over the years been followed by many limited releases and in spring 2019, an 8 year old, heavily peated, matured in Sauternes barriques was released.

Other distilleries in Sweden

Norrtelje Brenneri

Norrtälje, founded in 2002 (whisky since 2009)

www.norrteljebrenneri.se

The production consists mainly of spirits from ecologically grown fruits and berries. Since 2009, a single malt whisky from ecologically grown barley is also produced. The first bottling was released in summer 2015 and several limited editions have followed.

Gammelstilla Whisky

Torsåker, founded in 2005

www.gammelstilla.se

Unlike most of the other Swedish whisky distilleries, the owners chose to design and build their pot stills themselves. The wash still has a capacity of 600 litres and the spirit still 300 litres and the annual capacity is 20,000 litres per year. The first, limited release for shareholders was in May 2017 with a general release in January 2018 of the 4 year old Jern. The latest release was the bourbonmatured Anna Christina in spring 2019.

Gotland Whisky

Romakloster, founded in 2011

www.gotlandwhisky.se

The distillery is equipped with one wash still (1,600 litres) and one spirit still (900 litres). The local barley is ecologically grown and part of it is malted on site. Both unpeated and peated whisky is produced and the capacity is 60,000 litres per year. The first release of Isle of Lime single malt was in early 2017 with a general launch in August. The latest bottling, released in June 2019, was the 4 year old Tjaukle matured in American oak.

Uppsala Destilleri

Uppsala, founded in 2015

www.uppsaladestilleri.se

With a yearly production of 1,500 litres (but with a goal to increase production in the future) this is currently one of the smallest distilleries in the country. Production started in early 2016 with a 100 litre alambic still from Portugal but yet another still has already been installed. Apart from whisky, gin and rum are also produced.

Tevsjö Destilleri

Järvsö, founded in 2012

www.tevsjodestilleri.se

The owners of this combination of distillery and restaurang are primarily focused on distillation of aquavit and other white spirits and malt whisky production did not start until spring 2017. However, whisky has been produced earlier in the way of a "bourbon" with a mash bill of 70% corn, 10% malted barley, 10% unmalted barley, 5% wheat and 5% rye.

Agitator Whiskymakare

Arboga, founded in 2017

www.agitatorwhisky.se

The owners of this new distillery have chosen some rather unusual techniques in the production. Water is added during the milling

in order to make the mashing more efficient. The same fermented wash is split in half and distributed to the two pairs of stills in order to achieve different characters. The stills, by the way, operate under vacuum which is extremely rare in pot still whisky making. The distillery is experimenting with different kinds of grain apart from barley - oat, wheat and rye. Finally, the maturation takes place in casks where extra staves have been inserted, some of them made from chestnut. The distillery has a capacity of 500,000 litres of pure alcohol and the first distillation was made in February 2018.

Nordmarkens Destilleri

Årjäng, founded in 2018

www.nordmarkensdestilleri.se

The first products to come from this new distillery were aquavit, vodka and limoncello. Sourced whisky has also been released. The first whisky distillation was in December 2018 and both peated and unpeated spirit is produced.

Switzerland

Käsers Schloss (a.k.a Whisky Castle)

Elfingen, Aargau, founded in 2002

www.kaesers-schloss.ch

The first whisky from this distillery, founded by Ruedi Käser, reached the market in 2004. It was a single malt under the name Castle Hill. Since then the range of malt whiskies has been expanded and today include Castle Hill Doublewood (3 years old matured both in casks made of chestnut and oak), Smoke Barley (at least 3 years old matured in new oak), the portmatured Family Reserve and the 8 year old Edition Käser, the distillery´s premium expression. In the last couple of years, Rudi´s two sons Michael and Raphael have taken over the running of business and have also changed the brand name to Käser´s Schloss.

Brauerei Locher

Appenzell, founded in 1886 (whisky since 1999)

www.saentismalt.com

Brauerei Locher is unique in using old beer casks for the maturation. The core range consists of three expressions; Himmelberg, bottled at 43%, Dreifaltigkeit which is slightly peated having matured in toasted casks and bottled at 52% and, finally, Sigel which has matured in very small casks and is bottled at 40%. A new bottle was also introduced in 2019, Föhnsturm, bottled at 46%. A range of limited bottlings under the name Alpstein is available. The most recent, Edition XV, was released in 2019 and had matured in beer casks for five years and then another two years in casks that had held a pinot noir wine. Finally, a new range of limited releases, Edition Genesis, was introduced in spring 2019.

Other distilleries in Switzerland

Langatun Distillery

Langenthal, Bern, founded in 2007

www.langatun.ch

The distillery was built in 2005 and under the same roof as the brewery Brau AG Langenthal. The casks used for maturation are all 225 litres and Swiss oak (Chardonnay), French oak (Chardonnay and red wine) and ex sherry casks are used. The two 5 year old core expressions are Old Deer and the peated Old Bear. Other bottlings include the single cask rye Old Eagle, a single cask "bourbon" Old Mustang and the organic Old Woodpecker. Recent limited bottlings include Winter Wedding, Avo Jazz, a Nero d´Avola cask finish and the distillery´s first 10 year old.

Bauernhofbrennerei Lüthy

Muhen, Aargau, founded in 1997 (whisky since 2005)

www.brennerei-luethy.ch

The first single malt from the distillery was Insel-Whisky, matured in Chardonnay casks and released in 2008. Several releases have since followed. Starting in 2010, the yearly bottling was given the name Herr Lüthy and the 12th release from these had been matured in a combination of chardonnay casks, ex-oloroso casks and ex.bourbon barrels. Since 2016, whisky from malted rice is also produced!

Brennerei Stadelmann

Altbüron, Luzern, founded in 1932 (whisky since 2003)

www.schnapsbrennen.ch

The distillery is equipped with three Holstein-type stills and the first generally available bottling (a 3 year old) appeared in 2010. In autumn 2014, the sixth release was made, matured in a Bordeaux cask. The first whisky from smoked barley was distilled in 2012.

Etter Distillerie

Zug, founded in 1870 (whisky since 2007)

www.etter-distillerie.ch

The main produce from this distillery is eau de vie from various fruits and berries. A sidetrack to the business was entered in 2007 when they decided to distil their first malt whisky. The first release was made in 2010 under the name Johnett Single Malt Whisky and this is currently sold as a 7 year old. In 2016, a limited Johnett with a 12 months finish in Caroni rum casks was released.

Spezialitätenbrennerei Zürcher

Port, Bern, founded in 1954 (whisky from 2000)

www.lakeland-whisky.ch

The main focus of the distillery is specialising in various distillates of fruit, absinth and liqueur but a Lakeland single malt is also in the range. A limited version was released in June 2018 - an 8 year old which had matured the whole time in a Chateau d`Yquem cask.

Whisky Brennerei Hollen

Lauwil, Baselland, founded in 1999

www.single-malt.ch

The first Swiss whisky was distilled at Hollen in July 1999. In the beginning most bottlings were 4-5 years old but in 2009 the first 10 year old was released and there has also been a 12 year old, the oldest expression from the distillery so far.

Z´Graggen Distillerie

Lauerz, Schwyz, founded in 1948

www.zgraggen.ch

Focusing mainly on spirits distilled from fruits and berries, the owners also produce gin, vodka and whisky. The distillery is quite large, with a combined production of 400,000 litres per year. There are three single malts in the range – 3, 8 and the 10 year old Bergsturz..

Macardo Distillery

Strohwilen, Thurgau, founded in 2007

www.macardo.ch

Built on a former cheese factory, the distillery produces fruit brandy, gin, rum and whisky. The range consists of a core expression without age statement, two 10th anniversary bottlings (bottled at 42% and at cask strength) and special bottlings under the label Distillers Selection

Wales

Penderyn Distillery

Penderyn, founded in 2000

www.penderyn.wales

When Penderyn began producing in 2000, it was the first Welsh distillery in more than a hundred years. A new type of still, developed by David Faraday for Penderyn, differs from the Scottish and Irish procedures in that the whole process from wash to new make takes place in one single still. In 2013, a second still (almost a replica of the first still) was commissioned and in 2014, two traditional pot stills, as well as their own mashing equipment was installed. In February 2019, planning permission was secured for yet another distillery, this time in Swansea. The plan is to have the second unit up and running in 2022.

The first single malt was launched in 2004. The core range today is divided into two groups. Dragon consists of the Madeira finished Legend, Myth which is fully bourbon matured and Celt with a peated finish. The other range is Gold with Madeira, Peated, Portwood, Sherrywood and, most recently, Rich Oak which has matured in bourbon casks and then finished in rejuvenated ex-wine casks. In 2019, the distillery´s first travel retail exclusive, Penderyn Faraday, was launched. Over the years, the company has released several single casks and limited releases and a new range of whiskies called Icons of Wales was introduced in 2012 with the sixth edition, the peated and portfinished Royal Welsh Whisky, being released in 2019. An excellent visitor centre opened in 2008.

Other distilleries in Wales

Dà Mhìle Distillery

Llandyssul, founded in 2013

www.damhile.co.uk

Focusing on gin and grain whisky but there is also a single malt in the pipeline. Meanwhile, they have been offering aged, organic single malts that were distilled by Springbank back in the 1990s, when John Savage-Onstwedder, one of the founders, commissioned Springbank to produce the world´s first organic whisky

Aber Falls Distillery

Abergwyngregyn, founded in 2017

www.aberfallsdistillery.com

In common with so many other distilleries, Aber Falls started producing gin and the first product was released in late 2017. Malt whisky, however, is also produced and in spring 2019, they distilled rye for the first time. A visitor centre was opened in May 2018.

Penderyn´s Faraday still and safe

North America

USA

Westland Distillery

Seattle, Washington, founded in 2011

westlanddistillery.com

Until November 2012, Westland was a medium sized craft distillery where they brought in the wash from a nearby brewery and had the capacity of doing 60,000 litres of whiskey per year. During the summer of 2013 the owners, the Lamb family, moved to another location which is equipped with a 6,000 litre brewhouse, five 10,000 litre fermenters and two Vendome stills. The capacity is now 260,000 litres per year. In 2017, global spirits giant Remy Cointreau bought Westland Distillery. According to Westland master distiller and co-founder Matt Hoffman, the buyer´s views, not least on the influence of terroir, were very much in line with Westland´s way of working. The distillery has been focusing on local barley varieties and also local peat.

The first 5,500 bottles of their core expression, Westland American Single Malt Whiskey, were released in 2013. The current core range consists of American Oak and Peated Malt, both matured in a combination of new American oak and first fill bourbon casks and Sherry Wood which has matured in new American oak as well as in casks that previously held Oloroso and PX sherry. All three varieties have been mashed with a 5-malt grain bill and the wash has been fermented for 6 days. Over the years there have also been a large number of different releases of single casks. Apart from experimenting with a multitude of different barley varieties, Westland is also exploring different types of oak for maturation. One of them, Garryana oak, is native to the Pacific Northwest and was used for the first time for whisky maturation by Westland. The fourth release of a single malt partly matured in Garry oak appeared in September 2019. November 2018 saw the release of the first expression in a new range named Reverie where the blender has been given a free hand to create new and innovative bottlings. In January 2019, the 5th edition of the limited Peat Week was released and in July/August three new expressions made from three different barley types (Maris Otter, Pilsen and Golden Promise) were released in order to highlight the flavours that the grains impart.

Balcones Distillery

Waco, Texas, founded in 2008

balconesdistilling.com

Originally founded by Chip Tate who left the company in 2014, Balcones celebrated it´s 10th anniversary last year. All of Balcones´

whisky is mashed, fermented and distilled on site and they were the first to use Hopi blue corn for distillation. The core range currently consists of five expressions; Texas Single Malt, two corn whiskies made from blue corn, Baby Blue and True Blue 100 Proof, Texas 100 Rye and Pot Still Bourbon.

Recent limited expressions include a rum cask finished single malt, a new addition of Fr. Oak Single Malt and High Plains Texas Single Malt. All whiskies are un chill-filtered and without colouring. The demand for Balcones whiskies grew rapidly and in January 2014, another four, small stills were installed. The big step though, was a completely new distillery which was built 5 blocks from the old site. Distillation started in February 2016 and the official opening was in April. The new distillery is equipped with two pairs of stills and five fermenters and they now distill approximately 350,000 litres per year.

Stranahans Whiskey Distillery

Denver, Colorado, founded in 2003

stranahans.com

Founded by Jess Graber and George Stranahan, the distillery was bought by New York based Proximo Spirits (makers of Hangar 1 Vodka and Kraken Rum among others) in 2010. Rob Dietrich took over as master distiller in 2011 but left the company in 2019 to work with Blackened whiskey, introduced by the heavy metal band Metallica together with the late Dave Pickerell in 2018. Except for the core Stranahans Colorado Whiskey, the range is made up of Diamond Peak, a vatting of casks that are around 4 years old, a Single Barrel and Sherry cask. The latter is a version where the classic single malt has received a finish in oloroso sherry butts. Every year in December there is the a release of the limited Snowflake edition. In December 2018 it was Mount Elbert which was a vatting Stranahan´s Original that had been finished in a variety of casks; syrah, muscat, port, merlot, zinfandel, madeira, rum and chocolate stout.

Hood River Distillers

Hood River, Oregon, 1934

hrdspirits.com

Since the foundation, the company acts as importer, distiller, producer and bottler of all kinds of spirits. Some of the products are distilled in-house while others are sourced. The role in the single malt segment came through buying Clear Creek Distillery in 2014. Founded by Steve McCarthy the distillery was one of the first to produce malt whiskey in the USA but was most famous for their eau-de-vie made from pears. The only single malt whiskey

Westland´s Master Distiller, Matt Hoffman

produced by the company is the peated McCarthy´s Oregon Single Malt. In December 2017, Clear Creek closed their distillery in Portland and moved to Hood River. At the same time they sold their Canadian whisky brand Pendleton to Becle (parent company to José Cuervo tequila) for $205m.

Tuthilltown Spirits

Gardiner, New York, founded in 2003

tuthilltown.com

The distillery, 80 miles north of New York City, was founded in 2003 by Ralph Erenzo and Brian Lee. In 2010, William Grant & Sons aquired the Hudson Whiskey brand while the founders still owned the distillery. In spring 2017, William Grant followed up the deal by buying the entire company. The first products came onto the shelves in 2006 in New York and the whiskey range now consists of Hudson Baby Bourbon, a 2-4 year old bourbon made from 100% New York corn and the company´s biggest seller by far, Four Grain Bourbon (corn, rye, wheat and malted barley), Single Malt Whiskey (aged in small, charred American oak casks), Manhattan Rye, Maple Cask Rye and New York Corn Whiskey. The most recent limited release (January 2019) was the single barrel Port Wine Cask Manhattan Rye with a 6 months port finish. There is also gin, vodka and liqueur in the range.

RoughStock Distillery

Bozeman, Montana, founded in 2005

montanawhiskey.com

The owners, Kari and Bryan Schultz, buy the 100% Montana grown and malted barley and then mill and mash it themselves. It is then fermented on the grain in two 1,000 gallon open top wooden fermenters for a 72 hour fermentation before distillation in two Vendome copper pot stills. In 2009, the first bottles of RoughStock Montana Pure Malt Whiskey were released. Since then a single barrel bottled at cask strength has been added (Black Label Montana Whiskey) and apart from whiskey made from 100% malted barley, the product range also includes whiskey made from wheat and rye and bourbon.

Copper Fox Distillery

Sperryville, Virginia, founded in 2000

www.copperfoxdistillery.com

Founded in 2000 by Rick Wasmund, the distillery moved to another site in 2006 and in November 2016, he opened up a second distillery in Williamsburg. Wasmund does his own floor malting of barley and it is dried using smoke from selected fruitwood. After mashing, fermentation and distillation, the spirit is filled into oak barrels, together with plenty of hand chipped and toasted chips of apple and cherry trees, as well as oak wood. The first single malts (known as Red Top) were just four months old but the current batches are more around 12-16 months. An older version, Blue Top, has matured for up to 42 months. Other expressions include Peachwood Single Malt and Copper Fox Rye Whiskey. In 2019, Rick Wasmund rebranded the entire range which now goes by the name Copper Fox and at the same time he introduced a 100% malted Sassy Rye which had been made using rye that had been smoked with sassafras wood.

St. George Distillery

Alameda, California, founded in 1982

stgeorgespirits.com

The distillery is situated in a hangar at Alameda Point, the old naval air station at San Fransisco Bay. It was founded by Jörg Rupf, who came to California in 1979 and who was to become one of the forerunners when it came to craft distilling in America. In 1996, Lance Winters joined him and today he is Distiller, as well as co-owner. In 2005, the two were joined by Dave Smith who now has the sole responsibility for the whisky production. The main produce is based on eau-de-vie and a vodka named Hangar One. Whiskey production was picked up in 1996 and the first single malt appeared on the market in 1999. St. George Single Malt used to be sold as a three year old but, nowadays, comes to the market as a blend of whiskeys aged from 4 to 19 years. The latest release was Lot 19 in October 2019 and every lot is around 3-4,000 bottles. A new addition to the range was released in 2018 - Breaking & Entering - where the distillery sources bourbon and rye from other producers and blend it themselves. The latest edition also included single malt from St George.

Corsair Distillery

Bowling Green, Kentucky and Nashville, Tennessee, founded in 2008

corsairdistillery.com

The two founders of Corsair, Darek Bell and Andrew Webber, first opened up a distillery in Bowling Green, Kentucky and two years later, another one in Nashville, Tennessee (followed by a second one in Nashville a few years later). In March 2018 a third site in Nashville was acquired for $6,8m. This will be used for warehousing, distilling, offices etc. It will not only increase the production capacity hugely but will also create more than 50 new jobs. Apart from producing around 20 different types of beer, the brewery is also where the wash for all the whisky production takes place. Corsair Distillery has a wide range of spirits – gin, vodka, absinthe, rum and whiskey. The number of different whiskies released is growing constantly and the owners are experimenting with different types of grain. The big sellers are Triple Smoke Single Malt Whiskey (made from three different types of smoked malt) and Ryemaggedon (made from malted rye and chocolate rye). Included in the current range of whiskis are also Quinoa Whiskey, Oatrage and the hickory smoked single malt Wildfire.

House Spirits

Portland, Oregon, founded in 2004

westwardwhiskey.com

In 2015, Christian Krogstad and Matt Mount moved their distillery a few blocks to bigger premises. The main products for House Spirits used to be Aviation Gin and Krogstad Aquavit but with their new equipment they drastically increased whiskey capacity from 150 barrels per year to 4,000 barrels. But it didn´t stop at that. In September 2018, Diageo´s "spirits accelerator" Distill Ventures acquired a minority stake in the distillery and the brand which will help expanding the capacity in 2019 by nearly 40%. The first three whiskies were released in 2009 and in 2012 it was time for the first, widely available single malt under the name of Westward Whiskey. It was a 2 year old, double pot distilled and matured in new American oak. Recent releases have been up to 5 years old and now each release is a single barrel.

Kings County Distillery

Brooklyn, New York, founded in 2010

kingscountydistillery.com

This distillery is the oldest and largest in Brooklyn. The founders, Colin Spoelman and David Haskell, have made a name for themselves as being both experimental and yet at the same time true to traditional Scottish methods of distilling whiskey. The wash is fermented for four days in open-top, wooden fermenters and they practise a double distillation in two copper pot stills. A third pot still from Vendome was also recently installed. In June 2019, they went from small bottles to full-sized 750ml bottles with the ambition to increase their distribution beyond the 20 states and six countries where their products are currently available. The first single malt (60% unpeated and 40% peated), matured in ex-bourbon barrels was distilled in 2012 but didn´t hit the market until 2016. It has then been released in batches aged between 1.5 and 4 years. In the product range is also bourbon with the unusual mash bill of 60% corn and 40% malted barley. Even more unusual is their peated bourbon!

Virginia Distillery

Lovingston, Virginia, 2008 (production started 2015)

vadistillery.com

The whole idea for this distillery was conceived in 2007. The copper pot stills arrived from Turkey in 2008 but following several changes in ownership and struggling with the financing, the first distillation didn´t take place until November 2015. The distillery has the capacity of making 1.1 million litres of alcohol and is equipped with a 3.75 ton mash tun, 8 washbacks, a 10,000 litre wash still and a 7,000 litre spirit still. Only sourced whisky (from Scotland) is currently available under their label. Their first single malt from own production will be a very limited release of Prelude: Courage & Conviction in autumn 2019 followed by a general launch in April 2020. The first bottling of Courage & Conviction is a vatting of whiskies matured in bourbon, sherry and wine casks. In autumn 2020, individual bottlings from each cask type will be available.

Long Island Spirits

Baiting Hollow, New York, founded in 2007

www.lispirits.com

Long Island Spirits, founded by Rich Stabile, is the first distillery on the island since the 1800s. The starting point for The Pine Barrens Whisky, the first single malt from the distillery, is a finished ale with hops and all. The beer is distilled twice in a potstill and matures for one year in a 10 gallon, new, American, white oak barrel. The whisky was first released in 2012 and was followed in 2018 by a bottle-in-bond version (at least four years old) and later by an expression that is cherrywood smoked. The whiskey range also includes Rough Rider bourbon and rye.

Great Wagon Road Distilling Co.

Charlotte, North Carolina, founded in 2014

gwrdistilling.com

The distillery, founded by Ollie Mulligan, started with a 15 litre still but is now equipped with a 3,000 litres Kothe still in the 15,000 sq foot facility. The mash comes from a neighbouring brewery and the fermentation is made in-house in four tanks. The first batch of his Rua Single Malt was launched at Christmas 2015 and several batches have since followed, including vodka and Drumlish poteen. New releases in 2018 included a straight Rua single malt, two finishes - port and sherry - and a rye whiskey. They were followed in 2019 by a Rua fully matured in port casks.

Hamilton Distillers

Tucson, Arizona, founded in 2011

www.whiskeydelbac.com

Stephen Paul came up with the idea of drying barley over mesquite, instead of peat. He started his distillery using a 40 gallon still but since 2014, a 500 gallon still is in place. In 2015, new malting equipment was installed which made it possible to malt the barley in 5,000 lbs batches, instead of the previous 70 lbs! The first bottlings of Del Bac single malt appeared in 2013 and they now have three expressions – aged Mesquite smoked (Dorado), aged unsmoked (Classic), unaged Mesquite smoked (Old Pueblo) and aged unsmoked, bottled at cask strength (Distiller´s Cut).

Deerhammer Distilling Company

Buena Vista, Colorado, founded in 2010

www.deerhammer.com

The location of the distillery at an altitude of 2,500 metres with drastic temperature fluctuations and virtually no humidity, have a huge impact on the maturation of the spirit. Owners Lenny and Amy Eckstein released their first single malt, aged for only 9 months, in 2012. More and older batches (2 to 3 years) of their Deerhammer Single Malt have followed including a port cask finish. The distillery´s range also includes a straight bourbon, a hickory smoked corn whiskey, a limited rye, finished in porter barrels and a bourbon finished in casks that had held Apis IV (a dark, Belgian-style ale).

Hillrock Estate Distillery

Ancram, New York, founded in 2011

www.hillrockdistillery.com

What makes this distillery unusual, at least in the USA, is that they are not just malting their own barley – they are floor malting it. When Jeff Baker founded the distillery he equipped it with a 250 gallon Vendome pot still and five fermentation tanks. In spring

Gareth Moore, CEO and Maggie Moore, Chief Experience Officer at Virginia Distillery

2019, the distillery was substantially expanded with a new pot still, a lauter mash tun and more fermentation tanks. This tripled the capacity to 20,000 cases of whiskey per year. The first release from the distillery was in 2012, the Solera Aged Bourbon. Today, the range has been expanded with a Single Malt and a Double Cask Rye. Over the years, limited bottlings have appeared such as the peated Single Malt and a Napa cabernet cask finished bourbon

Santa Fe Spirits

Santa Fe, New Mexico, founded in 2010

www.santafespirits.com

Colin Keegan, the owner of Santa Fe Spirits, is collaborating with Santa Fe Brewing Company which supplies the un-hopped beer that is fermented and distilled in a 1,000 litre copper still from Christian Carl in Germany. The whiskey gets a hint of smokiness from mesquite. The first product, Silver Coyote released in 2011, was an unaged malt whiskey. The first release of an aged (2 years) single malt whiskey, Colkegan, was in October 2013. Since then, the range has been expanded to include also a version finished in apple brandy casks and one bottled at cask strength.

Copperworks Distilling Company

Seattle, Washington, founded in 2013

www.copperworksdistilling.com

Jason Parker and Micah Nutt obtain their wash from a local brewery and then ferment it on site. The distillery is equipped with two, large copper pot stills for the whiskey production, one smaller pot still for the gin and one column still. The whiskey is matured in 53-gallon charred, American oak barrels. The first distillation was in 2014 and the first batch of the single malt was released in 2016. In October 2018, to celebrate the distillery´s fifth anniversary, they released their first single malt to be fully matured for 38 months in French oak instead of new American oak. Previously in 2018, Copperworks had been named Distillery of the Year by the American Distilling Institute.

Rogue Ales & Spirits

Newport, Oregon, founded in 2009

www.rogue.com

The company consists of one brewery, two combined brewery/pubs, two distillery pubs and five pubs scattered over Oregon, Washington and California. The main business is still producing Rogue Ales, but apart from whiskey, rum and gin are also distilled. The first malt whiskey, Dead Guy Whiskey, was launched in 2009. In 2016, it was time for the first straight malt whiskey - the two year old Oregon Single Malt Whiskey. Spring 2018, saw the launch of the company´s first 5 year old single malt as well as a 3 year old rye malt whiskey. The latest addition to the range was in February 2019 when Rolling Thunder Stouted Whiskey was released - a single malt aged for one year in new garryana oak and a further two years in casks that had held imperial stout.

FEW Spirits

Evanston, Illinois, founded in 2010

fewspirits.com

Former attorney (and founder of a rock and roll band) Paul Hletko started this distillery in Evanston, a suburb in Chicago in 2010. It is equipped with three stills; a Vendome column still and two Kothe hybrid stills. Bourbon and rye had been on the market for a couple of years when the first single malt, with some of the malt being smoked with cherry wood, was released in 2015. In autumn 2018 a new core bottling appeared when American Whiskey, a vatting of bourbon, rye and smoked single malt, was released. Limited releases include bourbons finished in Italian red wine casks and casks that had previously held American brandy.

Sons of Liberty Spirits Co.

South Kingstown, Rhode Island, founded in 2010

www.solspirits.com

Founded by Michael Reppucci the distillery is equipped with a stainless steel mash tun, stainless steel, open top fermenters and one 950 litre combined pot and column still from Vendome. Sons of Liberty is first and foremost a whiskey distillery, but the first product launched was Loyal 9 Vodka. In 2011 the double distilled Uprising American Whiskey was launched, made from a stout beer and it was followed in 2014 by Battle Cry made from a Belgian style ale. Both Uprising and Battle Cry have also been released as PX and oloroso finishes respectively. The latest, limited releases include Wheated Single Malt (70% malted barley and 30% malted wheat) and Heavily Peated Single Malt.

Do Good Distillery

Modesto, California, founded in 2013

dogooddistillery.com

Founded by six friends and family members, the goal was to make whiskey and, in particular, single malt. First production was in 2014 and since autumn 2015 a number of different releases have been made; Beechwood Smoked, Peat Smoked, Cherrywood Smoked - all of them single malts - and The Nighthawk bourbon. A couple of the latest additions include The Benevolent Czar, a single malt made from three different types of malt, California Wheat Bourbon and a hop flavoured single malt.

Other distilleries in USA

Dry Fly Distilling

Spokane, Washington, founded in 2007

www.dryflydistilling.com

The first batch of malted barley was distilled in 2008 but a 100% single malt has yet to be launched. However, several other types of whisky have been released – Bourbon 101, Straight Cask Strength Wheat Whiskey, Port Finish Wheat Whiskey, Peated Wheat Whiskey and Straight Triticale Whiskey. A new limited bottling, first released in 2015, is the triple distilled O´Danaghers which is a mix of barley, wheat and oats. A later edition was a single potstill made from malted and unmalted barley.

Triple Eight Distillery

Nantucket, Massachusetts, founded in 2000

www.ciscobrewers.com

Apart from whiskey, Triple Eight also produces vodka, rum and gin. Whiskey production was moved to a new distillery in 2007. The first 888 bottles of single malt whiskey were released on 8th August 2008 as an 8 year old. To keep in line with its theme, the price of these first bottles was also $888. More releases of Notch (as in "not Scotch") have followed, aged up to 12 years.

Cedar Ridge Distillery

Swisher, Iowa, founded in 2003

www.crwine.com

Malt whiskey production started in 2005 and in 2013 the first single malt was launched. More releases of the single malt have been made since then. Other spirits in the range include both bourbon, malted rye and malted wheat. Their first bottled-in-bond bourbon was launched in June 2019.

Nashoba Valley Winery

Bolton, Massachusetts, founded in 1978
(whiskey since 2003)

www.nashobawinery.com

Mainly about wines, the business has been expanded with a

brewery and a distillery as well. In autumn 2009, Stimulus, the first single malt was released. The second release of a 5 year old came in 2010 and an 11 year old was launched in 2016.

Woodstone Creek Distillery

Cincinnati, Ohio, founded in 1999

www.woodstonecreek.com

Opened a farm winery 1999, a distillery was added to the business in 2003. The first whiskey, a five grain bourbon, was released in 2008 followed by a 10 year old single malt. Whiskey production is very small and just a handfull of releases have appeared since.

Cutwater Spirits (former spirit division of Ballast Point)

San Diego, California, founded in 2016

www.cutwaterspirits.com

In December 2015, Ballast Point Brewing was bought by Constellation Brands for the staggering sum of $1bn! The distilling side of Ballast Point, which started in 2008, was never a part of the deal and during 2016, a handful of executives and co-founders started a new company and distillery called Cutwater Spirits. That company in turn was sold to brewing giant Anheuser-Busch InBev in February 2019. The whiskyside of the business consists of Devil's Share Whiskey which comes in two versions - single malt and bourbon. Lately the range has been expanded with Black Skimmer rye and bourbon.

Edgefield Distillery

Troutdale, Oregon, 1998

mcmenamins.com

The distillery is a part of the McMenamin chain of more than 60 pubs and hotels in Oregon and Washington. More than 20 of the pubs have adjoining microbreweries and the chain's first distillery opened in 1998 at their huge Edgefield property in Troutdale with the first whiskey, Hogshead Whiskey, being bottled in 2002. Hogshead is still their number one seller. Limited releases occur every year on St Patrick's Day under the name The Devil's Bit. A second distillery was opened in 2011 at the company's Cornelius Pass Roadhouse location in Hillsboro.

Town Branch Distillery

Lexington, Kentucky, 1999

lexingtonbrewingco.com

The founder Dr Pearse Lyons, who passed away in March 2018, had an interesting background. A native of Ireland, he used to work for Irish Distillers in the 1970s. In 1980 he changed direction and founded Alltech Inc, a biotechnology company specializing in animal nutrition and feed supplements. Alltech purchased Lexington Brewing Company in 1999 and in 2008, two traditional copper pot stills were installed with the aim to produce Kentucky's first malt whiskey. The first single malt whiskey was released in 2010 under the name Pearse Lyons Reserve and it was later followed by Town Branch bourbon Town Branch malt and Town Branch Rye. In June 2018, Alltech opened yet another distillery in Pikeville - Dueling Barrels Brewery and Distillery.

Prichard's Distillery

Kelso, Tennessee, 1999

prichardsdistillery.com

Phil Prichard started the distillery in 1999 and in 2012 the capacity was tripled with the installation of a new mash cooker and three additional fermenters. In 2014, a second distillery equipped with a new 400-gallon alembic copper still was opened at Fontanel in Nashville. The main track of the production is rum. The first single malt was launched in 2010 and later releases usually have been vattings from barrels of different age (some up to 10 years old). The whiskey range also includes rye, bourbon and a Tennessee whiskey.

Charbay Winery & Distillery

St. Helena, California, founded in 1983

www.charbay.com

With a wide range of products such as wine, vodka, grappa, pastis, rum and port, the owners decided in 1999 to also enter in to whiskey making. They were pioneers distilling whiskey from hopped beer and over the years several releases have been made including Double-Barrel Release I, II, III and IV and Charbay R5 (five releases so far). The spirit distillation takes place in alambic pot stills with sometimes extraordinary long distillation time (up to ten days!). In spring 2017, the company was split in two with Marko and his wife Jenni focusing on the spirit side while Marko's father Miles continues with the wine production.

High West Distillery

Park City, Utah, founded in 2007

highwest.com

The founder, David Perkins, has made a name for himself mainly as a blender of sourced rye whiskies. None of these have been distilled at High West distillery. In 2015, they opened another distillery at Blue Sky Ranch in Wanship, Utah. It started off with two 6,000 litre pot stills but the plan is to eventually have 18 washbacks and four pot stills with the possibility of producing 1,4 million litres. Even though they consider themselves blenders first and foremost, the 2018 versions of Rendezvous Rye and Double Rye, were the first expressions which included whiskey from their own production. In 2016 Constellation Brands (makers of Corona beer and Svedka vodka) bought High West Distillery for a sum of $160 million.

New Holland Brewing Co.

Holland, Michigan, founded in 1996 (whiskey since 2005)

www.newhollandbrew.com

After ten years, this beer brewery opened up a micro-distillery as well. The first cases of New Holland Artisan Spirits were released in 2008 and among them were Zeppelin Bend, a 3 year old (minimum) straight malt whiskey which is now their flagship brand. Included in the range are also Zeppelin Bend Reserve, matured for four years and then finished for an additional 9 months in sherry casks, Beer Barrel Bourbon and Beer Barrel Rye.

DownSlope Distilling

Centennial, Colorado, founded in 2008

www.downslopedistilling.com

The first whiskey, Double-Diamond Whiskey, was released in 2010. It was made from 65% malted barley and 35% rye and is still the core whiskey. It was followed by a number of varieties of bourbon, rye and 3 year old single malt. All malt whiskies are made from floor malted Maris Otter barley.

Bull Run Distillery

Portland, Oregon, founded in 2011

www.bullrundistillery.com

The distillery is equipped with two pot stills (800 gallons each) and the main focus is on 100% Oregon single malt whiskey. First release was the sourced bourbon Temperance Trader. The first release of a single malt under the name Bull Run was a 4 year old in 2016 (now 5 years old). Shortly after that the Oregon Single Malt Whiskey was also released at cask strength (56%).

Cut Spike Distillery (formerly Solas Distillery)

La Vista, Nebraska, founded in 2009

www.cutspikedistillery.com

Originally opened as Solas distillery in 2009, it was later renamed Cut Spike distillery. In 2010 single malt whiskey was distilled and the first bottles were launched in August 2013. New batches of the 2 year old whiskey have then appeared regularly and in autumn 2017, the first single barrel version was launched.

Journeyman Distillery

Three Oaks, Michigan, founded in 2010

www.journeymandistillery.com

The first release from the distillery (Ravenswood Rye) was sourced from Koval Distillery in Ravenswood. The range of whiskies distilled at their own premises now include bourbon, rye, wheat and single malt. The first release of Three Oaks Single Malt Whiskey was in 2013. All products are 100% organic and kosher.

Wood´s High Mountain Distillery

Salida, Colorado, founded in 2011

www.woodsdistillery.com

Whiskey is the main product at this distillery. The first expression (and current big seller), Tenderfoot Whiskey was a triple malt and so is the latest release, Sawatch, made from malted barley (a mix of chocolate malt and cherrywood smoked malt), malted rye and malted wheat. There is also an Alpine Rye Whiskey in the range.

Door County Distillery

Sturgeon Bay, Wisconsin, founded in 2011

doorcountydistillery.com

A winery founded in 1974 was complemnted by a distillery in 2011. Gin, vodka and brandy are the main products but they also make single malt whiskey. The first Door County Single Malt was released in 2013 and there are also bourbon and rye in the range.

Immortal Spirits

Medford, Oregon, founded in 2008

www.immortalspirits.com

A wide range of spirits are produced including gin, rum, vodka and limoncello. The only whiskey made from barley (unmalted) is the 3 year old Single Grain. The Single Barrel range of selected casks has sometimes ben represented by a single malt but currently it´s a bourbon.

Painted Stave Distilling

Smyrna, Delaware, founded in 2013

paintedstave.com

Whiskey production started in 2014, first with bourbon and rye, then followed by whiskey from malted barley. Most of the whiskey production is centered on bourbon and rye but they have also released Ye Old Barley made from 100% malted barley and Diamond State Pot Still Whiskey made from malted barley and rye.

Van Brunt Stillhouse

Brooklyn, New York, founded in 2012

www.vanbruntstillhouse.com

Part of the Brooklyn Spirits Trail in New York, the distillery made their first release of Van Brunts American Whiskey in 2012, a mix of malted barley, wheat and a hint of corn and rye. This has been followed by a malt whiskey from 100% malted barley, a wheated bourbon, a rye and a smoked corn whiskey.

Civilized Spirits

Traverse City, Michigan, founded in 2009

www.civilizedspirits.com

The spirits are produced at a distillery on Old Mission Peninsula, just outside Traverse City in a 1,000 litre pot still with a 24-plate column attached. The whiskey side of the business includes Civilized Single Malt (at least 3 years old), Civilized Whiskey (made from locally grown rye), Civilized White Dog Whiskey (an unoaked wheat whiskey) and Civilized Bourbon.

Spirits of St Louis Distillery (formerly known as Square One)

St. Louis, Missouri, founded in 2006

spiritsofstlouisdistillery.com

A combined brewery and restaurant in St. Louis. Apart from rum, gin, vodka and absinthe, the owners also produce J.J. Neukomm Whiskey, a malt whiskey made from toasted malt and cherry wood smoked malt.

Maine Craft Distilling

Portland, Maine, founded in 2013

www.mainecraftdistilling.com

The distillery offers vodka, gin, rum and Chesuncook, which is a botanical spirit using barley and carrot distillates as well as the Fifty Stone single malt in limited batches. The barley is floor malted on site and both peat moss and seaweed is used to dry the barley.

3 Howls Distillery

Seattle, Washington, founded in 2013

www.3howls.com

The malted barley is imported from Scotland including a small amount of peated malt. For the distillation they use a 300 gallon hybrid still with a stainless steel belly and a copper column. Their first whiskies were released in 2013, a single malt and a hopped rye, and these were followed in 2014 by a rye whiskey and a bourbon.

Montgomery Distillery

Missoula, Montana, founded in 2012

www.montgomerydistillery.com

The owners mill the barley and rye to a fine flour using a hammer mill and the wash is then fermented on the grain. Distillation takes place in a Christian Carl pot still with a 21 plate column attached. The first whiskey was a rye in 2015, followed up by the 3 year old Montgomery Single Malt in 2016 and a 4 year old a year later. In summer 2019 a collaboration with Draught Works Brewery resulted in the limited, 5 year old Gwin Du whiskey.

Two James Spirits

Detroit, Michigan, founded in 2013

www.twojames.com

Equipped with a 500 gallon pot still with a rectification column attached, the distillery started production in 2013. Vodka, gin, bourbon (even a peated version) and rye have been released while a single malt is still maturing in the warehouse. Aged in ex-sherry casks the whiskey has been made from peated Scottish barley.

Seven Stills Distillery

San Francisco, California, founded in 2013

www.sevenstillsofsf.com

The first releases were made at Stillwater distillery in Petaluma. Since 2016, the owners have been producing their own spirit from a distillery in Bayview, an area in the San Francisco environs. Soon, the concept was expanded with three taprooms and in autumn 2019, the whole production (distillery and brewery) was moved to Mission Bay together with a restaurant, cocktail bar and taproom. The idea is to make whiskey from craft beers and the range is made up of the Core Series (in-house beer), Collaboration working with other breweries and Experimental where experimentation without boundaries is the key word.

Ranger Creek Brewing & Distilling

San Antonio, Texas, founded in 2010

www.drinkrangercreek.com

The owners focus on beer brewing and whiskey production. They have their own brewhouse where they mash and ferment all their beers, as well as the beer going for distillation. The first release was Ranger Creek .36 Texas Bourbon in 2011. Their first single malt, Rimfire, was launched early in 2013.

Vikre Distillery

Duluth, Minnesota, founded in 2012

www.vikredistillery.com

Together with whisky - gin, vodka and aquavit are produced at the distillery. Whiskies include Iron Range American Single Malt, Northern Courage Smokey Rye, Sugarbush Whiskey and Honor Brand Hay & Sunshine. The single malt was released in March 2017. The distillery was expanded in 2016 with six new fermentation tanks.

Rennaisance Artisan Distillers

Akron, Ohio, founded in 2013

renartisan.com

Apart from whiskey, the distillery produces gin, brandy, grappa and limoncello. The first whiskey release, The King's Cut single malt, was made from a grain bill including toasted and caramel malts and new batches appear every 6 months. In autumn 2019 an Islay Style Single Malt and a Blue Corn Bourbon were released.

John Emerald Distilling Company

Opelika, Alabama, founded in 2014

www.johnemeralddistilling.com

With the wash being fermented on the grain, the main product is

John's Alabama Single Malt which gets its character from barley smoked with a blend of southern pecan and peach wood. The first release was made in 2015. In spring 2017, the owners also started trial distillations using triticale.

Eleven Wells Distillery

St. Paul, Minnesota, founded in 2013

11wells.com

The distillery is equipped with a 650 gallon mash tun, stainless steel open-top fermentation tanks and two stills. Whiskey is the main product and the first two releases, aged bourbon and rye, were released in 2014 followed by a wheat whiskey in 2015 and finally a single malt made from malted barley.

Blaum Bros. Distilling

Galena, Illinois, founded in 2012

blaumbros.com

The distillery equipment consists of a 2,000 litre mash tun, five 2,000 litre wash backs and a 2,000 litre Kothe hybrid still. Apart from gin and vodka, the first two releases were the sourced Knotter Bourbon and Knotter Rye. The first whiskey from their own production was a rye in 2015 followed by a straight bourbon in 2018. It will be a few years before the first single malt is released.

Sugar House Distillery

Salt Lake City, Utah, founded in 2014

sugarhousedistillery.net

The first release from the distillery in 2014 was a vodka, followed later that year by a single malt whisky. More releases of the single malt have followed and bourbon, rye and rum have also been added to the range.

Venus Spirits

Santa Cruz, California, founded in 2014

venusspirits.com

Production is focused on whiskey, but gin and spirits from blue agave have also been released. The first single malt was Wayward Whiskey, made from crystal malt and released in 2015. This was followed up by a rye and later a bourbon.

Blue Ridge Distilling Co.

Bostic, North Carolina, founded in 2010

www.blueridgedistilling.com

The first distillation at the distillery was in June 2012 and in December the first bottles of Defiant Single Malt Whisky were released. The maturation part is very unorthodox. The spirit is matured for 60 days in stainless steel tanks with oak spirals inserted. According to the owners, this ensures a greater contact between the whisky and the wood which speeds up the maturation process. In autumn 2017, a 100% rye was added to the range.

Bent Brewstillery

Roseville, Minnesota, founded in 2014

bentbrewstillery.com

This combined brewery and distillery produces, apart from a range of beers, also gin and whiskey. A rye whiskey named Punish95 has been released and a single malt is currently aging in a combination of charred oak and charred apple wood.

Orange County Distillery

Goshen, New York, founded in 2013

orangecountydistillery.com

Every ingredient needed for the production is grown on the farm,

Ranger Creek Distillery

including sugar beet, corn, rye, barley and even the botanicals needed for their gin. They malt their own barley and even use their own peat when needed. Since 2014, they have launched a wide range of whiskies, including corn, bourbon, rye and peated single malt. The first aged single malt was launched in summer 2015.

Key West Distilling

Key West, Florida, founded in 2013

kwdistilling.com

The main track is to produce rum but they are also distilling whiskey. The mash is brought in from Bone Island Brewing, fermented, distilled and filled into new barrels or used rum barrels. The first release of Whiskey Tango Foxtrot was in 2015.

Thumb Butte Distillery

Prescott, Arizona, founded in 2013

thumbbuttedistillery.com

A variety of gin, dark rum and vodka, as well as whiskey are produced by the owners, Dana Murdock, James Bacigalupi and Scott Holderness. Rodeo Rye, Bloody Basin Bourbon, Crown King Single Malt and a limited grain whiskey have all now been released. Maris Otter barley is being used for the malt whiskies.

Seattle Distilling

Vashon, Washington, founded in 2013

seattledistilling.com

The distillery produces gin, vodka, coffee liqueur, as well as a malt whiskey. The latter, named Idle Hour was first launched in 2013 followed by more batches. The style is Irish with both malted and unmalted barley being used in the mashbill.

Hewn Spirits

Pipersville, Pennsylvania, founded in 2013

hewnspirits.com

Apart from rum, gin and vodka the distillerty produces bourbon, rye and the Reclamation American Single Malt Whiskey. After maturing the malt whiskey in barrels for 1-4 months, it receives a second maturation in stainless steel vats where charred staves of chestnut and hickory wood add to the profile.

Wright & Brown Distilling Co.

Oakland, California, founded in 2015

wbdistilling.com

The distillery is focused on barrel aged spirits, i. e. whiskey, rum and brandy. The first whiskey was distilled in 2015 and the first product, a rye whiskey, was launched in autumn of 2016 followed by a bourbon in autumn 2017. So far, no single malt made from barley has been released.

Stark Spirits

Pasadena, California, founded in 2013

starkspirits.com

The first single malt whiskey was distilled in July 2015 and the first release was a barrel of peated single malt in February 2016. The first official distillery release of single malt (both peated and un-peated) came in February 2017. They have two stills with one reserved for all the peated production.

Cotherman Distilling

Dunedin, Florida, founded in 2015

cothermandistilling.com

All the whiskies are made from 100% malted barley. The mash is

brought in from local breweries, fermented at the distillery and then distilled in a pot still and a 3-plate bubble-cap still. First launched in July 2016, several batches have followed since. Apart from whiskey – gin and vodka are also produced.

Quincy Street Distillery

Riverside, Illinois, founded in 2011

quincystreetdistillery.com

The distillery produces an impressive range of spirits including gin, vodka, absinth, bourbon, corn whiskey and rye. So far, single malt whiskey made from barley only forms a small part. The only single malt released so far is a 2 year old Golden Prairie which was launched in December 2015.

Boston Harbor Distillery

Boston, Massachusetts, founded in 2015

bostonharbordistillery.com

The distillery started production in summer 2015 and while it concentrates mainly on whiskey, it is also making a variety of spirits based on different Samuel Adams' beers. The whiskies, currently a rye and a single malt (launched in December 2017) are released under the Putnam New England label. Apart from the distillery with its 150-gallon Vendome copper pot still, the facility consists of a shop, tasting room and an event space.

Liquid Riot Bottling Co.

Portland, Maine, founded in 2013

liquidriot.com

When Liquid Riot opened its doors, it was Maine's first brewery/distillery/resto-bar. At the waterfront in the Old Port, Liquid Riot produces an extensive range of beers and spirits which include bourbon, rye, oat, single malt, rum, vodka and agave spirit. Distillation is made in a German hybrid still with a 5 plate rectification column.

Old Line Spirits

Baltimore, Maryland, founded in 2014

oldlinespirits.com

The owners bought the equipment from Golden Distillery when that was about to close down and brought it to Baltimore. Distilling started in 2016 and a couple of months prior, the first Old Line single malt, two to three years old and obviously from the Golden Distillery production, was released. A peated version was launched in 2017 and a sherry cask finish appeared in autumn 2018.

Cannon Beach Distillery

Cannon Beach, Oregon, founded in 2012

cannonbeachdistillery.com

The owner's philosophy about whisky making is never to make the same spirit twice. All the whiskies are made in small batches with a new release every 2-4 months. Distillation takes place in a 380 litre Vendome still with a 6-plate column.

Witherspoon Distillery

Lewisville, Texas, founded in 2011

witherspoondistillery.com

The main products from this distillery are bourbon, rum and Bonfire (a cinnamon-infused rum), but they also make small runs of Witherspoon Single Malt which is generally aged between 1 and 2 years. The whiskey is distilled in two 1,110 litre stills and the single malt is matured in new American oak and finished in rum casks.

2nd Street Distilling Co

Walla Walla, Washington, founded in 2011

2ndstreetdistillingco.com

Formerly known as River Sands Distillery, the company has been around since 1968 but the distillery only started in 2011. Different types of gin and vodka are produced, as well as a single malt – R J Callaghan. It is aged for 1,5 years in charred American oak and then finished for 6 months in Hungarian oak. In 2016 a 100% malted rye, Reser´s Rye, was also released.

ASW Distillery

Atlanta, Georgia, founded in 2016

aswdistillery.com

The distillery is equipped with two traditional Scottish copper pot stills but with the American twist of fermenting and distilling on the grain. Among the latest releases are Duality, made from 50% malted barley and 50% malted rye with both grains fermented and distilled in the same batch, Ameireaganach Single Malt, the heavily peated Tire Fire, the triple distilled Druid Hill and Badger, released together with Monday Night Brewing.

Dallas Distilleries Inc.

Garland, Texas, founded in 2008

dallasdistilleries.com

The distillery is primarily focused on whiskey. The first products in their Herman Marshall range were launched in 2013. It was a bourbon and a rye and was later followed by a single malt. An unusual feature at the distillery is the open top fermenters which are made from cypress wood.

San Diego Distillery

Spring Valley, California, founded in 2015

sddistillery.com

A distillery focused almost entirely on whiskey. In March 2016 the first six whiskies were released; a bourbon, a rye and an Islay peated single malt. The next whiskey to appear was a single malt made from seven different types of brewing malt. Due to a fire in the distillery in autumn 2017, it was moved, and expanded, to a new location in spring 2018.

Alley 6 Craft Distillery

Healdsburg, California, founded in 2014

alley6.com

A small craft distillery in Sonoma county with rye whiskey as the main product. The first bottles were released in summer 2015 followed by a single malt in May 2016. The owners are experimenting with a range of different barley varieties, mainly from Germany and Belgium.

Arizona Distilling Company

Tempe Arizona, founded in 2012

azdistilling.com

The first release from the distillery was a bourbon sourced from Indiana. The ensuing releases, which started with Desert Durum made from wheat, have all been produced in their distillery. Humphrey's – a single malt – was first released in late 2014. The distillery is one of few using open top fermenters.

Gray Skies Distillery

Grand Rapids, Michigan, founded in 2014

grayskiesdistillery.com

In 2014, the owners bought an industrial building for their grain-to-glass distillery and a year later, the first spirit was distilled. The equipment is made up of a 1,800 litre mash kettle, four fermenters and a 2,500 litre pot still with an attached column. The first bottle of Michigan Single Malt, made from a combination of cherry wood smoked malt, peanut butter toast malt and distillers malt, appeared in November 2016.

Spirit Hound Distillers

Lyons, Colorado, founded in 2012

spirithounds.com

Rum, vodka and sambucca are on the production list, but the distillery´s signature spirits are gin and malt whiskey. The barley for the whiskey is grown, malted and peat-smoked in Alamosa by Colorado Malting and the whiskey released so far is straight, i.e. at least two years old. The first bottles hit the shelves in summer 2015 and the first 4 year old was released in summer 2018.

Dorwood Distillery

Buellton, California, founded in 2014

dorwood-distillery.com

The distillery (which recently changed its name from Brothers Spirits) started producing malt whisky in 2016. The barley is dried using mesquite smoke and the triple distillation takes place in two reflux stills. The vast majority of the releases so far have been unaged but several barrels have been laid down for maturation.

Idlewild Spirits

Winter Park, Colorado, founded in 2015

idlewildspirits.com

Production of the first batch of malt whiskey was in June 2016. For maturation they has moved from 5 gallon barrels, via 10 and 30 gallons to the full-size 50 gallon barrels that they use today. Fermentation and distillation being on the grain add to the over-all character. Their Colorado Single Malt was released in 2018.

Timber Creek Distillery

Crestview, Florida, founded in 2014

timbercreekdistillery.com

Fermentation and distillation is off the grain and, surprisingly, they use a traditional worm tub to cool the spirits – a technique that has become rare even in Scotland. Currently they have whiskies maturing made from corn, wheat, rye, barley and oat. The latest batch of Florida Single Malt was released in October 2017.

Lyon Distilling Co.

Saint Michaels, Maryland, founded in 2013

lyondistilling.com

Focusing on whiskey and rum the distillery is equipped with a 2,000 litre mash tun, stainless steel fermenters and five small pot stills. The first, unaged, malt whiskey was released in late 2015 and the first aged release came one year later.

Dirty Water Distillery

Plymouth, Massachusetts, founded in 2013

dirtywaterdistillery.com

Starting with vodka, gin and rum, the distillery expanded into malt whiskey in 2015. The first release, Bachelor Single Malt, came in 2016 and was followed by Boat For Sale Malt Whiskey which had been made using a beer from Independent Fermentations.

Long Road Distillers

Grand Rapids, Michigan, founded in 2015

longroaddistillers.com

Apart from vodka, gin and aquavit, five styles of whiskey have been released - bourbon, wheat, corn, rye and a 6 month old whiskey made from 51% malted barley and 49% un-malted.

Motor City Gas

Royal Oak, Michigan, founded in 2014

motorcitygas.com

The owners have an experimental approach to whiskey making and use unusual and old grains (Maris Otter and Golden Promise), different yeast strains and unusual woods. The expressions so far have been both unpeated and heavily peated.

Brickway Brewery & Distillery (former Borgata)

Omaha, Nebraska, founded in 2013

www.drinkbrickway.com

Omaha´s first combined brewery and distillery since prohibition. All the wash for the distillation comes from their own brewery and distillation takes place in a 550 gallon Canadian wash still, and a 400 gallon spirit still from Forsyth´s in Scotland. The owners are focused on single malt whiskey but they also produce smaller amounts of bourbon and rye as well as gin and rum. Their first whisky, Borgata American Single Malt White Whisky, was released in 2014 and there is now also an aged version under the name Brickway Single Malt Whisky. A recent expansion of the distillery made it possible to quadruple the whisky production.

KyMar Farm Winery & Distillery

Charlotteville, New York, founded in 2011

kymarfarm.com

Mainly producing wine, liqeurs and apple brandy. Recently though, a whiskey made from 100% malted barley and distilled in a 300 gallon hybrid and an 80 gallon alembic still, was released. After a 6 to 9 months maturation, the spirit is moved to a solera system for blending with older batches to ensure consistency

Coppersea Distilling

New Paltz, New York, founded in 2011

coppersea.com

A "farm-to-glass" distillery with the barley malted on site. Open-top wooden washbacks and direct-fired alembic stills. One of the things that make Coppersea stand out is that they don´t dry the malted barley but instead produce a mash from green, unkilned barley, The one year old Big Angus is made from 100% green barley and in December 2018 a 100% malted straight rye was released.

Old Home Distillers

Lebanon, New York, founded in 2014

oldhomedistillers.com

The distillery produces bourbon, corn whiskey and malt whiskey. The mash is fermented on the grain for 4-5 days, distillation takes place in a 100 gallon hybrid column still and the spirit is matured in charred, new American oak for a minimum of seven months.

StilltheOne Distillery Two

Port Chester, New York, founded in 2010

stilltheonedistillery.com

Different kinds of whiskey are produced in a 250 gallon pot column still from Arnold Holstein. The only single malts released so far are "287" from a pale ale and "9A" made from a stout.

Pioneer Whisky (formerly known as III Spirits)

Talent, Oregon, founded in 2014

pioneerwhisky.com

Focusing mainly on single malts, the distillery changed its name in December 2018. Currently there are two single malts in the range; Oregon Highlander made from a grain bill of brewer´s malt, Munich malt and crystal malt and Islay Style Peated Whisky produced from 100% heavily peated malt from Scotland.

Telluride Distilling

Telluride, Colorado, founded in 2014

telluridedistilling.com

Vodka and malt whiskey is produced in a distillery equipped with open top fermenters and a column still. Maturation is in new charred oak for two years followed by 6 months in port barrels. The first single malt was released in July 2016.

Brickway Distillery

Amalga Distillery

Juneau, Alaska, founded in 2017

amalgadistillery.com

The distillery uses a 250 gallon pot still from Vendome and they are also floor malting their own barley, some of it grown in Alaska. The first single malt will be released in 2020 but both vodka and gin have already been launched.

Fainting Goat Spirits

Greensboro, North Carolina, founded in 2015

faintinggoatspirits.com

First spirits on the shelves for this distillery, as for many others, were gin and vodka. In December 2017, Fisher´s single malt whiskey was launched as a 2 year old with the latest batch being released in July 2019. Fisher´s Straight Rye has also been released.

Andalusia Whiskey

Blanco, Texas, founded in 2016

andalusiawhiskey.com

Focusing entirely on whiskey production, the owners use a 56,000 gallon tank to collect rain water for the production. The spirit is double-distilled in a 250 gallon pot still and the first single malts were released in late 2016 - Stryker, where mesquite and oak have been used to dry the barley and the lightly peated Revenant Oak. This was followed up end of 2017 by Andalusia Triple-Distilled. There is also a special range with cask-finished whiskies.

Golden Moon Distillery

Golden, Colorado, founded in 2008

goldenmoondistillery.com

The distillery is using four antique stills, dating from the early to mid 1900s for the production. At least 15 different kinds of spirits are distilled, one of them being a single malt, aged for a minimum of one year in new oak and then finished in used oak casks. This was made from a mash done at a local brewery. In July 2019, the first batch of their in-house single malt was distilled.

Big Bottom Distilling

Hillsboro, Oregon, founded in 2015

bigbottomdistilling.com

The company started out as a blender and bottler of sourced whiskey, not least bourbon finished in different wine casks. A distillery was built in 2015 and in June 2018, their first own single malt was released.

Black Heron Spirits

West Richland, Washington, founded in 2011

blackheronspirits.com

The owner started out as a winemaker, then decided to sell the company and open a distillery instead. A wide variety of spirits are produced, including bourbon, a corn whiskey and a limited peated single malt which was first released in January 2017.

Bogue Sound Distillery

Bogue, North Carolina, founded in 2018

boguesounddistillery.com

The distillery is equipped with a 500-gallon still and the first spirits released included gin, vodka and rye. Recently the John A.P. Conoley single malt was added to the range.

Dark Island Spirits

Alexandria Bay, New York, founded in 2015

darkislandspirits.com

The owners produce a wide variety of spirits including gin, vodka and brandy. On the whiskey side there´s wheat, corn and bourbon and in June 2018, the first single malt was released - the 3 year old Eleanor Glen.

Jersey Spirits Distilling Co

Fairfield, New Jersey, founded in 2015

jerseyspirits.com

Apart from gin and vodka, the owners have two bourbon varieties

Bently Heritage Distillery

for sale - Crossroads with a mash bill consisting of corn, rye, wheat and barley and Patriot's Trail which is a high rye bourbon. The first distillation of a single malt was in summer 2018 which will be ready to bottle in two years.

Liberty Call Spirits

Spring Valley, California, founded in 2014

libertycalldistilling.com

The distillery, located outside San Diego, uses a variety of barley varieties for their whiskies, including caramel malts and the rare Maris Otter. Their single malt is called Old Ironsides and there is also a four grain whiskey named Blue Ridge.

Mad River Distillers

Warren, Vermont, founded in 2011

madriverdistillers.com

The distillery was built on a 150 year old farm in the Green Mountains. Focus is on rum, brandy and whiskey. The only single malt so far is Hopscotch which was first released in late 2016 with batch four launched in autumn 2019.

Wanderback Whiskey

Hood River, Oregon, founded in 2014

wanderback.com

The idea for the distillery was inspired by a visit to Boulder, Colorado and a sip of Stranahan's whiskey. The first release of Wanderback single malt (made from four different malts) was in September as a 3 year old. Since then there have been another two releases. All of the whiskey has been distilled at Westland Distillery and then brought back to Wanderback for maturation and blending.

Bently Heritage Distillery

Minden, Nevada, founded in 2016

bentlyheritage.com

A true estate distillery, growing their own grains and floor malting it themselves, Bently Heritage is located in an old mill from the early 1900s. The equipment consists of a Briggs mash tun, oak fermenters, one pair of copper pot stills made by Forsyths and hybrid stills with columns for the distillation of vodka, gin and all whiskies except single malt. The first distillation was in 2018 and the owners have so far released gin and vodka but have also laid down a substantial amount of casks for future whiskey releases.

Rock Town Distillery

Little Rock, Arkansas, founded in 2010

rocktowndistillery.com

Founded in 2010, this is a true grain-to-glass distillery where the grains used for distillation are grown within 125 miles of the property. The backbone of the production is made up of several bourbons, rye, a hickory-smoked wheat whiskey and vodka. For their 9th anniversary in June 2019, they released a 2 year old single malt that had been finished in cognac casks for another year.

Distillery 291

Colorado Springs, Colorado, founded in 2011

distillery291.com

The distillery was founded by photographer Michael Myers who moved with his family from New York to Colorado. The first distillation was in 2011 and the core range now consists of seven different ryes, bourbons and American whiskies. In a limited, experimental range Myers has also in 2018 launched two single malts made from 100% barley.

High Peak Distilling

Lake George, New York, founded in 2016

highpeakdistilling.com

John Carr left his job at Adirondack Brewery in 2016 to start High Peak Distilling but he is still very much involved with his old employer. High Peaks obtains all of their fermented wash from the brewery which is then distilled and matured on site. The first release in spring 2018 was the peated Cloudsplitter Single Malt which was followed by Night Spirit Bourbon in August that year and finally Sugar Moon, a maple syrup flavoured bourbon, in spring 2019. All of their bottlings are at least two years old.

Highside Distilling

Bainbridge Island, Washington, founded in 2018

highsidedistilling.com

A family owned distillery which released its first spirit, a gin, in November 2018. Since then Amaro has also been launched while production of single malt whiskey started in January 2019 with an anticipated release in early 2021.

Laws Whiskey House

Denver, Colorado, founded in 2011

lawswhiskeyhouse.com

Alan and Marianne Laws didn't release their first bottling until 2014 when it was three years old and all of their following releases have been at least two years old. The flagship in the range is the Four Grain Straight Bourbon but they also have rye, corn and a single malt called Hordeum Straight Malt Whiskey on the menu. In summer 2018, investment firm First Beverage Group invested in the distillery to help the owners grow their business.

Loch & Union Distilling

American Canyon, California, founded in 2017

lochandunion.com

A fairly large distillery with an impressive set of two copper pot still for whiskey distillation and a third still designated for gin making – all fabricated by Carl in Germany. The first gin was released in spring 2018 but the inaugural single malt whiskey isn't due for at least another three years.

Orcas Island Distillery

Orcas, Washington, founded in 2014

orcasislanddistillery.com

What could best be described as a retirement hobby, former journalist Charles West founded a distillery on Orcas Island, 80 km north of Seattle. Apple brandy is the main produce but in 2019 he also won the Best American Single Malt Whiskey award from the American Distilling Institute for his West Island Single Malt Whiskey which was first released in 2018.

PostModern Distilling

Knowville, Tennessee, founded in 2017

postmodernspirits.com

While focusing on gin, vodka and liqueur, the owners released their first single barrel single malt whiskey in 2018 with the latest being barrel No. 9. The distillery gets its fermented wash from the nearby Crafty Bastard Brewery.

SanTan Spirits

Chandler and Phoenix, Arizona, founded in 2007 (2017)

santanbrewing.com

With two locations in Arizona, this brewery/restaurant added distilling to its concept in 2017. So far two vodkas and one single malt whiskey (Sacred Stave) have been released with the 6 months old single malt having been matured in Arizona red wine barrels.

Seven Caves Spirits

San Diego, California, founded in 2016

the7caves.com

Apart from gin and rum, Geoff Longenecker also makes whiskey from malted barley. The first release single malt was released in spring 2019.

Vapor Distillery (formerly known as Roundhouse Spirits)

Boulder, Colorado, founded in 2007

vapordistillery.com

Ted Palmer founded the distillery with the ambition to make gin. When he met Scotsman Alastair Brogan in 2014, they decided to change the name of the distillery and also include whiskey making. A 3,800 litre copper pot still was installed in 2015 and now the distillery has two single malt whiskies in the range – an American Oak Boulder American Single Malt Whiskey and a peated version of the same. Bourbon, vodka and gin are also part of the range

Canada 🍁

Shelter Point Distillery

Vancouver Island, British Columbia, founded in 2009

www.shelterpointdistillery.com

In 2005, Patrick Evans and his family decided to switch from the dairy side of farming to growing crops and they bought the Shelter Point Farm just north of Comox on Vancouver Island. Eventually the idea to transform the farm into a distillery was raised and with the help of Scottish investors, the construction work began. The distillery is equipped with a one tonne mash tun, five washbacks made of stainless steel and one pair of stills (a 5,000 litre wash still and a 4,000 litre spirit still). Distillation started in spring 2011 and the barley used for the distillation is grown on the farm. Currently, Evans produces 125,000 litres of spirit (whisky, gind and vodka) per year. In May 2016, the first 5 year old single malt was released and this is still the core expression together with a cask strength version. Recent limited reeleases include Montfort DL 141, made completely from unmalted barley, French Oak Double Barrel and Single Cask Rye.

Still Waters Distillery

Concord, Ontario, founded in 2009

www.stillwatersdistillery.com

Located in Concord, on the northern outskirts of Toronto, the distillery is equipped with a 3,000 litre mash tun, two 3,000 litre washbacks and a Christian Carl 450 litre pot still. The still also has rectification columns for brandy and vodka production. The focus is on whisky but they also produce vodka, brandy and gin. Their first single malt, named Stalk & Barrel Single Malt, was released in April 2013 and it was followed in late 2014 by the first rye whisky. The current range consists of Blue Blend, Red Blend, Rye and Single Malt. A collaboration with BarChef has also resulted in whisky cocktails in a bottle

Victoria Caledonian Distillery

Victoria, British Columbia, founded in 2016

www.victoriacaledonian.com

The distillery was founded by the Scotsman Graeme Macaloney and as a helping hand he had Mike Nicolson, who previously

worked at 18 distilleries in Scotland. In addition, they also acquired the services of the late Dr. Jim Swan, one of the foremost whisky consultants in the world. The distillery is equipped with a one ton semilauter mash tun, 7 stainless steel washbacks, a 5,500 litre wash still and a 3,600 litre spirit still from Forsyth. Some of the barley is malted on site and there is also a craft beer brewery. Distilling started in July 2016 and in late 2017, the owners released the Mac Na Braiche, a 12 months malt spirit. In the range there is also sourced Scotch blended malts under the name Twa Cask Collection. The first 3 year old single malt is expected in autumn 2019. A visitor centre offers tours on several levels of the distillery and the brewery, as well as tutored tastings.

Glenora Distillery

Glenville, Nova Scotia, founded in 1990

www.glenoradistillery.com

Situated in Nova Scotia, Glenora was the first malt whisky distillery in Canada. The first launch of in-house produce came in 2000, a 10 year old named Glen Breton but other expressions have occured - 14, 19, 21 and 25 year olds. Glen Breton Ice, the world's first single malt aged in an ice wine barrel, was launched in 2006 and the latest edition was a 19 year old. A recent limited release is the Ghleann Dubh – a 13 year old peated single malt.

Other distilleries in Canada

Pemberton Distillery

Pemberton, British Columbia, founded in 2009

www.pembertondistillery.ca

The distillery was founded in 2009 with vodka produced from potatoes as the first product. Schramm Vodka, was launched later that year. During the ensuing year, the owner started their first trials, distilling a single malt whisky using organic malted barley. The first release was in 2013 when a limited 3 year old unpeated version was launched. Since autumn 2015, the owners have a regular expression called Pemberton Valley Organic Single Malt Whisky which is released in batches.

Yukon Spirits

Whitehorse, Yukon, founded in 2009

www.twobrewerswhisky.com

All of the whisky produced is made from malted grains but not only barley but also wheat and rye. The first 850 bottles of the 7 year old Two Brewer's Yukon Single Malt Whisky were released in February 2016 and the portfolio is now based on four styles; Classic, Peated, Special Finishes and Innovative.

Okanagan Spirits

Vernon and Kelowna, British Columbia, founded in 2004

www.okanaganspirits.com

The first distillery named Okanagan was started in 1970 by Hiram Walker but it closed in 1995. In 2004, Frank Deiter, established Okanagan Spirits. A distillery was opened in Vernon and, later on, a second one was built in Kelowna. A variety of spirits made from fruits and berries as well as gin, vodka, absinthe and whisky are being produced. Since 2013, there is a single malt in the range – The Laird of Fintry – with different expressions including a cask strength and a rum barrel finish.

Lucky Bastard Distillers

Saskatoon, Saskatchewan, founded in 2012

www.lbdistillers.ca

Founded by Michael Goldney, Cary Bowman and Lacey Crocker, in 2012. The first releases were vodka, gin and a variety of liqueurs.

In summer 2016 the first single malt appeared and this is now released in batches.

Central City Brewers & Distillers

Surrey, British Columbia, founded in 2013

www.centralcitybrewing.com

What started as a brewpub has now grown to one of Canada's largest craft breweries. A much needed expansion followed in 2013 when they moved to a larger facility as well as adding a distillery. Apart from whisky they also produce gin and vodka. The only single malt released so far is Lohin McKinnon Single Malt including special bottlings such as peated, chocolate malt and one finished in tequila barrels.

The Dubh Glas Distillery

Oliver, British Columbia, founded in 2015

www.thedubhglasdistillery.com

The distillery is situated at Gallagher Lake and the whisky is double distilled in an Arnold Holstein still. Even though malt whisky is the main focus, gin is also produced. Apart from Noteworthy Gin, Virgin Spirits Barley (a newmake) has also been released. The first release of a single malt was the peated Against All Odds in June 2019.

The Liberty Distillery

Vancouver, British Columbia, founded in 2010

www.thelibertydistillery.com

Equipped with two copper stills with columns (140 and 220 litres) from Carl in Germany, the distillery produces gin, vodka and a wide range of whiskies all made from organic grain. Sold under the brand name Trust Whiskey, there is Southern made from 100% triple distilled corn, a Single Grain based on 100% unmalted barley, Canadian Rye and unmalted barley matured in either burgundy casks or madeira casks.

Eau Claire Distillery

Turner Valley, Alberta, founded in 2014

www.eauclairedistillery.ca

One of the first whisky distilleries in Alberta in modern times, Eau Claire opened in 2014. Their first limited single malt whisky (1,000 bottles) appeared in December 2017 and Ploughman's Rye whisky has since been released.

Arbutus Distillery

Nanaimo, British Columbia, founded in 2014

www.arbutusdistillery.com

Situated on the southern part of Vancouver Island, the distillery has so far focused on vodka, gin and liqueurs but single malt whisky is also on the agenda. The first 3 year old appeared in December 2018 and was followed by batch 2 in June 2019.

Sheringham Distillery

Sooke, British Columbia, founded in 2015

www.sheringhamdistillery.com

Yet another, new distillery on Vancouver Island which recently moved to Sooke after an initial three years in Shirley. Gin, akvavit and vodka were the first to be bottled while the inaugural release of their Red Fife whisky came in 2019. The owners work on different mash bills with malted barley, corn, rye and wheat.

Shelter Point Distillery   Dubh Glas Distillery

Australia & New Zealand

Australia

Lark Distillery

Hobart, Tasmania, founded 1992

www.larkdistillery.com

In 1992, Bill Lark was the first person for 153 years to take out a distillation licence in Tasmania and he is often referred to as the godfather of modern whisky production in Australia. The success of the distillery forced Bill Lark to bring in investors in the company to generate future growth and since April 2018, Australian Whisky Holdings holds a majority of the shares (56%). The whisky is double-distilled in a 1,800 litre wash still and a 600 litre spirit still and then matured in 100 litre "quarter casks". The old distillery site down in Hobart at the waterfront is now a cellar door and a showcase for Lark whisky. The core products in the whisky range are the Classic Cask at 43% and Cask Strength at 58%. Recent limited releases include Heavily Peated Bourbon Cask, The Wolf Release second edition where casks that had held smoked porter were used to mature the whisky and the portmatured Distiller´s Selection LD1016.

Bakery Hill Distillery

North Balwyn, Victoria, founded 1998

www.bakeryhill.com

The first spirit at Bakery Hill Distillery, founded by David Baker, was produced in 2000 and the first single malt was launched in autumn 2003. Three different versions are available – Classic and Peated (both matured in ex-bourbon casks) and Double Wood (ex-bourbon and a finish in French Oak). As Classic and Peated are also available as cask strength bottlings, they can be considered two more varieties. Limited releases also occur with one of the latest being Sovereign Smoke – Defiantly Peated.

Sullivans Cove Distillery

Cambridge, Tasmania, founded 1994

www.sullivanscove.com

Patrick Maguire has been the head distiller since 1999 and in 2018 he was inducted into the global Whisky Hall of Fame. In 2014 the distillery moved to a new building about four times the size of the current facility and in December 2016, the distillery was taken over by a company led by Adam Sable who was general manager of Bladnoch distillery for two years. The core range from the distillery comprises of American Oak, French Oak (where the barrels had contained port) and Double Cask. There is also Special Cask where the barrels that are used may vary from time to time, Old & Rare with whiskies that are 16 years or older and Limited Edition where the latest release (June 2019) was a French oak white wine cask.

Old Hobart Distillery

Blackmans Bay, Tasmania, founded 2005

www.overeemwhisky.com

After several years of experimenting Casey Overeem opened up his distillery in 2007. The mashing was done at Lark distillery where Overeem also had his own washbacks and the wash was made to his specific requirement. The distillation takes place in two stills (1,800 litres and 600 litres). In 2014, Old Hobart distillery was acquired by Lark Distillery and is now owned by Australian Whisky Holdings. The range consists of Overeem Port Cask Matured, Overeem Sherry Cask Matured and Overeem Bourbon Cask Matured - all three bottled at 43% and 60%. A limited release of single malt matured in red wine casks appeared in autumn 2017.

Hellyers Road Distillery

Burnie, Tasmania, founded 1999

www.hellyersroaddistillery.com.au

Hellyer´s Road Distillery is one of the larger single malt whisky

distilleries in Australia with a capacity of doing 100,000 litres of pure alcohol per year. The distillery is equipped with a 6.5 ton mash tun, a 40,000 litre wash still and a 20,000 litre spirit still. The pots on both stills are made of stainless steel while heads, necks and lyne arms are made of copper. The first whisky was released in 2006 and there are now more than ten different expression in the range, including 10 and 12 year olds, peated as well as unpeated and various finishes. A range of limited releases called Master Series include whiskies up to 16 years old. The distillery also has a visitor centre with more than 40,000 people coming every year.

Great Southern Distilling Company

Albany, Western Australia, founded 2004

www.distillery.com.au

The distillery is located at Princess Royal Harbour in Albany. In 2015, the owners opened a second distillery in Margaret River which will is focused on gin production and in autumn 2018 a third distillery, Tiger Snake in Porongurup, started production. In a near future the combined production will be 400,000 litres of pure alcohol per year. The first expression of the whisky, called Limeburners, was released in 2008 and this is still the core bottling. Included in the range are also American Oak, Port Cask and Sherry Cask, all bottled at 43% as well as Peated which is bottled at 48%. Special editions include Darkest Winter and Heavy Peat.

Starward Distillery

Melbourne, Victoria, founded 2008

www.starward.com.au

The distillery, founded by David Vitale, was moved in October 2016 to a new and bigger site in Port Melbourne. The stills (an 1,800 litre wash still and a 600 litre spirit still) were bought from Joadja Creek Distillery. The first whisky was released under the name Starward in 2013 and the current range consists of Nova (matured in Australian red wine barrels) and Solera (matured in casks that had held apera, the Australian version of sherry). Recent special releases include Two-Fold (made from malted barley and wheat), Charred Red Wine Cask and Seafarer (partly matured on the Cunard ocean liner Queen Elizabeth). In 2015, the distillery was given a financial injection when Diageo´s incubator fund project, Distill Ventures, made a substantial investment in the distillery, increasing production to 250,000 litrs of pure alcohol per year.

Other distilleries in Australia

Nant Distillery

Bothwell, Tasmania, founded in 2007

www.nant.com.au

The distillery was founded by Keith Batt but was later taken over by Australian Whisky Holdings. The distillery is equipped with a 1,800 litre wash still, a 600 litre spirit still and wooden washbacks. Two more washbacks were added recently, almost doubling the distillery´s capacity. The first bottlings were released in 2010 and the current core range consists of Sherry, Port and Bourbon.

William McHenry and Sons Distillery

Port Arthur, Tasmania, founded 2011

www.mchenrydistillery.com.au

Equipped with a 500 litre copper pot still with a surrounding water jacket to get a lighter spirit, production started in 2012. To facilitate the cash flow, a range of different gins is also produced. The first whisky was released in May 2016 while the latest edition is a 5 year old, matured in American oak and finished in French oak.

Launceston Distillery

Western Junction (near Launceston), Tasmania, founded in 2013

www.launcestondistillery.com.au

The equipment consists of a 1,100 litre stainless steel mash tun, stainless steel washbacks, a 1,600 litre wash still and a 700 litre spirit still – both with reflux balls. The newmake is filled into barrels which have previously held bourbon, Apera (Australian sherry) and Tawny (Australian port). The first release, matured in Apera casks, appeared in July 2018 and whisky matured in ex-bourbon and ex-tawny have followed since.

Black Gate Distillery

Mendooran, New South Wales, founded in 2012

www.blackgatedistillery.com

Apart from single malt whisky, the distillery produces vodka and rum. The first launch of a single malt was in early 2015 when a sherrymatured expression was released. More bottlings have followed, including hybrid casks (matured in borth red wine and port casks) and peated single malt.

Old Kempton Distillery

Kempton, Tasmania, founded in 2013

www.oldkemptondistillery.com.au

Established as Redlands Estate Distillery in Derwent Valley, the distillery re-located in 2016 to Dysart House in Kempton and later changed the name to Old Kempton Distillery. The first spirit was distilled in 2013 in a 900 litre copper pot still and another three stills have later been installed. The first whisky was launched in 2015 and this has been followed by several more releases.

Archie Rose Distilling Company

Rosebery, New South Wales, founded in 2014

www.archierose.com.au

The first distillation at Archie Rose was conducted in December 2014. Apart from producing rye whisky and peated and unpeated single malt, the distillery also makes gin and vodka. In July 2019 they released their first whisky - a malted rye. They have also launched a new make made from six different malts as well as gin and vodka.

Timboon Railway Shed Distillery

Timboon, Victoria, founded in 2007

www.timboondistillery.com.au

Wash from a local brewery is distilled twice in a 600 litre pot still. For maturation, resized (20 litres) and retoasted ex-port, tokay and bourbon barrels are used. The first release of a whisky, matured in port barrels, was made in 2010 and some of the latest expressions have been Tom´s Cut, bottled at 58% and Christie´s Cut at 60%.

Castle Glen Distillery

The Summit, Queensland, founded in 2009

www.castleglenaustralia.com.au

Established as a vineyard in 1990, Castle Glen moved on to open up also a brewery and a distillery in 2009. Apart from wine and beer, a wide range of spirits are produced. The first whisky, Castle Glen Limited Edition, was released as a 2 year old in 2012 while the latest was an 8 year old single malt.

Joadja Distillery

Joadja, New South Wales, founded in 2014

www.joadjadistillery.com.au

The first distillation was in December 2014 when the distillery was equipped with just the one still (800 litres), used for both the wash and the spirit run. In 2015, a 2,400 litre wash still was installed together with another four washbacks. Whereas many distilleries in Australia use local ex-wine casks for their maturation, Joadja stands out focusing mainly on ex-sherry casks from Jerez in Spain. The first whisky was released in autumn 2017.

Corra Linn Distillery

Relbia, Tasmania, founded in 2015

www.corralinndistillery.com.au

John Wielstra made the first distillation in his hybrid column still in autumn 2016, using a new, local barley strain that had recently been developed. He is also using his own yeast and smokes his barley using dried kelp instead of peat. The first release of single malt was in December 2018.

Shene Distillery

Pontville, Tasmania, founded in 2015

www.shene.com.au

Damian Mackey started distilling whisky in a small shed already in 2007. In 2016 the opportunity came for him to move his production to the Shene Estate at Pontville, 30 minutes north of Hobart. With four stills and a capacity of 300,000 litres this is one of the largest distilleries in Australia. The whisky is triple distilled and the first release of Mackey single malt was in August 2017.

Tin Shed Distilling Co.

Welland (Adelaide), South Australia, founded in 2013

www.iniquity.com.au

The owners opened their first distillery, Southern Coast Distillers,

Corra Linn Distillery

in 2004 with a release in 2010. Eventually it was closed and the current distillery started production in 2013. The first single malt, under the name Iniquity, was launched as a 2 year old in 2015 and batch 16 was released in June 2019.

Mt Uncle Distillery

Walkamin, North Queensland, founded in 2001

www.mtuncle.com

The owners started out by producing gin, rum and vodka - all of which soon became established brands on the market. Their first single malt, The Big Black Cock, was released in April 2014, and was produced using local Queensland barley and matured for five years in a combination of French and American oak.

Loch Distillery

Loch, Victoria, founded in 2014

www.lochbrewery.com.au

This combined brewery and distillery began producing whisky in March 2015. The wash used for the whisky production comes from their own brewery. Their own gin was soon released and the first single malt was launched in July 2018.

Fanny´s Bay Distillery

Weymouth, Tasmania, founded in 2015

www.fannysbaydistillery.com.au

The distillery is equipped with a 400 litre copper pot still, a 600 litre mash tun and a 300 litre washback with a 7-8 day fermentation. The whisky starts in 20 litre port barrels and is then finished in small bourbon casks. The first whisky was released in May 2017 and it is now available in three versions - bourbon, sherry and port.

Applewood Distillery

Gumeracha, South Australia, founded in 2015

www.applewooddistillery.com.au

With a background in wines and perfumes, Laura and Brendan Carter opened a distillery in the Adelaide Hills in 2015. To start with, gin, eau de vie and liqueurs were on the menu, but in summer 2015, whisky production was added. Several young malt spirits have been released but so far no whisky.

Killara Distillery

Hobart, Tasmania, founded in 2016

www.killaradistillery.com

Kristy Booth is the daughter of Bill Lark, often referred to as the godfather of Australian whisky and after 17 years working in her father´s distillery, she opened her own in summer 2016. The first whisky distillation was in August 2016. The succesful Apothecary gin has been on the market for a while and in November 2018 it was time for the first single malt release - a cask strength matured for two years in an ex-tawny port cask.

Devil´s Distillery

Moonah, Tasmania, founded in 2015

www.devilsdistillery.com.au

Using an 1800 litre copper pot still, the distillery started production of malt whisky in 2015. They also have moonshine in their range. The first release of their Hobart single malt was in August 2018 and the latest addition was port finish in June 2019.

Adams Distillery

Perth, Tasmania, founded in 2016

www.adamsdistillery.com.au

This distillery looks like one of the most interesting whisky projects in Australia today. After less than two years, all the

equipment was up for sale to make way for a huge new distillery. The new distillery started production in March 2019 with a capacity of doing 1500 litres of new make daily. Around the same time, the first 2 year old single malt was also released and it was followed more whisky matured in port, sherry and pinot noir casks.

Spring Bay Distillery

Spring Beach, Tasmania, founded in 2015

www.springbaydistillery.com.au

A small, family-owned distillery, located on the east coast of Tasmania and equipped with a 1200 litre pot still. The distillery was expanded in June 2019 with a 2,500 litre wash still. The first spirit released was a gin followed in autumn 2017 by the first single malt. More whiskies, matured in port, sherry and bourbon casks, have appeared since then.

Bellarine Distillery

Drysdale, Victoria, founded in 2017

www.bellarinedistillery.com.au

Located at the unlikely address Scotchman´s Road, the distillery is equipped with four stills, producing both gin and malt whisky. Gin is in the shops but the first whisky has yet to be released.

Backwoods Distilling

Yackandandah, Victoria, founded in 2017

www.backwoodsdistilling.com.au

The distillery is equipped with a 1200 litre copper ot still with an attached column. The first distillation was in January 2018 and the first release is expected in early 2020.

Wild River Mountain Distillery

Wondecla, Queensland, founded in 2017

www.wildrivermountaindistillery.com.au

This is one of Australia´s highest elevated distilleries, located in the Atherton Tablelands in North Queensland at a height of 870 metres. Distillation started in 2017 and in August 2019, the first single malt was released. Batch one of the Elevation Single Malt was lightly smoked and had matured in a combination of ex-Tennessee barrels and Australian red wine casks.

Riverbourne Distillery

Jingera, New South Wales, founded in 2016

www.riverbournedistillery.com

Located at the head of the Molonglo River, close to Canberra, the distillery started producing whisky, rum and vodka in February 2016. The first two single malts, released in June 2018, were named The Riverbourne Identity and The Riverbourne Supremacy - an obvious nod to the movies about Jason Bourne.

The Aisling Distillery

Griffith, New South Wales, founded in 2015

www.theaislingdistillery.com.au

Since the start, around 250 barrels have been filled at this distillery which is 100% dedicated to producing malt whisky. The first release is expected in 2019.

Darby-Norris Distillery

Kelso, Tasmania, founded in 2018

www.darbynorrisdistillery.com.au

A small distillery which started production in spring 2018. Gin and vodka have been released but the first single malt isn´t expected at least until 2020.

Coburns Distillery

Burrawang, New South Wales, founded in 2017

www.coburnsdistillery.com.au

Mark Coburn started production in spring 2017 and has so far released several versions of his gin. The single malt turned two years old in July 2019 but has yet to be released. Coburns is one of very few Australian distilleries with its own peat bog for smoking the barley. Plans for the future include having no less than a set of five 5,000 litre pot stills.

Baker Williams Distillery

Mudgee, New South Wales, founded in 2012

www.bakerwilliams.com.au

For the first six years, the owners were focusing on producing gin, vodka and schnapps – spirits that are still the base in their business. The first whisky was Lachlan, released in spring 2018 and made from barley, wheat and rye (all malted). A second batch appeared in April 2019.

Corowa Distilling Co.

Corowa, New South Wales, founded in 2010

www.corowawhisky.com.au

Situated in a restored flour mill from 1924, the distillery was founded by Dean Druce and with Beau Schlig as Master Distiller. The distillery is focused entirely on single malt whisky using barley from Dean's own estate. Corowa started distilling in March 2016 and to help fund the first years before their own whisky was mature, the owners released a series of single casks from Ben Nevis in 2017. The inaugural release from their own production was First Drop (aged in port barrels) in August 2018 followed by Bosque Verde (also port), Quicks Courage (PX sherry) and Mad Dog Morgan (muscat).

Manly Spirits Co. Distillery

Brookvale, Sydney, New South Wales, founded in 2017

www.manlyspirits.com.au

Equipped with two copper pot stills (1,500 l and 1,000 l), the distillery has already launched gin, vodka and a white dog malt spirit. The first single malt whisky named North Fort is due for release in 2020.

Shipyard Distillery

West Gosford, New South Wales, founded in 2016

www.shipyarddistillery.com

In a Hoga copper still from Portugal two types of whisky are produced – single malt and sour mash – and then matured in American oak or Australian ex-wine casks.

Noosa Heads Distillery

Noosaville, Queensland, founded in 2018

www.noosaheadsdistillery.com

Equipped with a 2,000 litre copper pot reflux still, the distillery launched its first products in spring 2019 – gin, vodka and a white malt. Single malt whisky is maturing but won't be released for another few years.

5 Nines Distilling

Uraidla, South Australia, founded in 2017

www.5ninesdistilling.com.au

The owners, David Pearse and Steven Griguol, built their own equipment including a copper pot still and a mash tun and started distilling malt whisky in 2017. Primarily made with local barley, the whisky is still maturing but several gins have been released.

Fleurieu Distillery

Goolwa, South Australia, founded in 2016

www.fleurieudistillery.com.au

A local craft beer brewery was turned into a distillery proper in 2016 but whisky had been distilled already since 2014. Gareth and Angela Andrews released their first single malt in December 2016 and it has been followed by several more, both peated and unpeated.

The McLaren Vale Distillery

Blewitt Springs, South Australia, founded in 2016

www.themclarenvaledistillery.com.au

The distillery was founded by John Rochfort, the previous CEO at Lark Distillery in Tasmania and is now run by the Rochfort family together with business partner Jock Harvey. Focusing solely on malt whisky, the distillery has a capacity to make 50,000 litres per year. The only release so far, was the Bloodstone Collection in 2017. Twenty different malt spirits (not yet matured for two years) were launched to showcase different types of oak and how different wines from South Australian winemakers affected the spirit.

Settlers Artisan Spirits

McLaren Vale, South Australia, founded in 2015

www.settlersspirits.com.au

Until now the distillery has been concentrating on gin but with a new pot still in 2018, whisky distillation has tripled. The only single malt release so far is the port matured Settlers Single Malt.

Lawrenny Distilling

Ouse, Tasmania, founded in 2017

www.lawrenny.com

With a head distiller previously working for Lark and Archie Rose, Joe Dinsmoor, the distillery has initially been releasing vodka and gin. The first distillation of malt whisky was in November 2017 so the first release is not due at least until end of 2019.

Sawford Distillery

Kingston, Tasmania, founded in 2018

www.sawforddistillery.com

The distillery is run by Jane Sawford and her husband Mark. Before Jane was married, her last name was Overeem and she has been deeply involved in the Australian whisky business for more than a decade. She learned distilling at her father's, Casey, distillery and when Lark Distillery took over the operations she was the Sales and Marketing Manager. Now she has started a distillery of her own together with her husband. No whiskies have yet been released but Lady Jane Whisky has been registered as a trade mark.

White Label Distillery

Huntingfield, Tasmania, founded in 2018

www.whitelabeldistillery.com.au

The distillery is equipped with no less than 16 stainless steel washbacks (4,000 litres each) and one pair of copper pot stills. They work mainly as a contract distiller and brewer. Jane Sawford (nee Overeem) is helping out with marketing and sales and her father Casey overlooks distilling and maturation of the whisky.

Turner Stillhouse

Grindelwald, Tasmania, founded in 2018

www.turnerstillhouse.com

Founded by ex-Californian Justin Turner, the distillery has so far been focusing on gin. A designated whisky copper still was installed in summer 2019 and the first release of a single malt is expected in 2021.

Backwoods Distilling

Yackandandah, Victoria, founded in 2017

www.backwoodsdistilling.com.au

Leigh and Bree Atwood are using locally grown barley and rye for their whiskies and the first distillation was in January 2018. In April 2019, Wild Rye, a rye spirit, was released and the first whisky releases are due in the beginning of 2020.

Chief´s Son Distillery

Somerville, Victoria, founded in 2017

www.chiefsson.com.au

The distillery is owned by the McIntosh family and is equipped with a 4,000 litre copper pot still which makes it a bit larger than most craft distilleries in Australia. The production is all about malt whisky and the first release appeared in March 2019. They currently have a core range made up of three varieties matured in French oak – the lightly peated 900 Standard, the 900 Sweet Peat and 900 Pure malt – and also the 900 American Oak. All of them bottled at 45%.

Kilderkin Distillery

Ballarat, Victoria, founded in 2016

www.kilderkindistillery.com.au

The distillery is owned by Chris Pratt and Scott Wilson-Browne and the name Kilderkin refers to the small type of casks (holding circa 70 litres) they use for maturation. The distillery is equipped with one pair of copper pot stills and have so far released four different gins. The first malt whisky release is expected in late 2019.

Geographe Distillery

Myalup, Western Australia, founded in 2008

www.geographedistillery.com.au

Apart from gin and limoncello, Steve Ryan is also distilling malt whisky. The latest release of Bellwether Single Malt was a 4 year old in November 2018. Medium peated malt from Baird's in Inverness was used and the whisky had matured ex-tawny casks.

New Zealand

Thomson Whisky Distillery

Auckland, North Island, founded in 2014

www.thomsonwhisky.com

The company started out as an independent bottler but in 2014, the owners opened up a small distillery based at Hallertau Brewery in North West Auckland. The wash for the distillation comes from the brewery. First release was in February 2018 and the current range is made up by Two Tone (rye and barley), Manuka Smoke and South Island Peat.

Cardrona Distillery

Cardrona (near Wanaka), South Island, founded in 2015

www.cardronadistillery.com

Building of the distillery started in January 2015 and in October the first distillation was made. The distillery is equipped with 1.4 ton mash tun, six metal washbacks, one 2,000 litre wash still and a 1,300 litre spirit still. The two pot stills were made by Forsyth's in Scotland. The production capacity is one barrel per day and the whisky is matured in sherry casks and bourbon barrels. The owners have released barrel-aged gin and single malt vodka but the first single malt whisky will probably not be released until 2025.

Lammermoor Distillery

Ranfurly, South Island, founded in 2018

www.lammermoorstation.com.nz

Founded by John and Susie Elliot who have been farmers at the huge (5,200 hectares) Lammermoor Station for thirty years. What started out as a hobby has now turned into yet another part of the business at the farm. Lammermoor is a true "grain-to-glass" distillery or as it´s called in New Zealand "paddock-to-bottle". The Elliots control every step from growing and malting the barley through to distilling and maturation. A gin has already been released while a single malt whisky is still a few years away

Cardrona Distillery

Asia

India

Amrut Distilleries Ltd.

Bangalore, Karnataka, founded in 1948

www.amrutdistilleries.com

The family-owned distillery, based in Kumbalgodu outside Bangalore, south India, started to distil malt whisky in the mid-eighties. The equivalent of 50 million bottles of spirits (including rum, gin and vodka) is manufactured a year, of which 1,4 million bottles is whisky. Most of the whisky goes to blended brands (including the newly released MaQintosh Silver), but Amrut single malt was introduced in 2004. It was first launched in Scotland, but can now be found in more than 40 countries. It wasn´t until 2010, however, that the brand was launched in India. The distillery was expanded with two more stills in autumn 2018 and now has the capacity of producing 800,000 litres of pure alcohol. The fermentation time for the single malt is 140 hours and the barley is sourced from the north of India, malted in Jaipur and Delhi and finally distilled in Bangalore before the whisky is bottled without chill-filtering or colouring. On the 9th of May 2019, the owner and chairman of the distillery, Sri. Neelakanta Rao Jagdale, passed away at the early age of 66 and was succeeded by his son Rakshit who had been instrumental in launching the Amrut single malt globally in the early days of the new millenium.

The Amrut core range consists of unpeated and peated versions bottled at 46%, a cask strength and a peated cask strength and Fusion which is based on 25% peated malt from Scotland and 75% unpeated Indian malt. The latter, which is the biggest seller and was awarded Best World Whisky in 2019, was recently re-branded with a new design as were all the other expressions in the core range. Special releases over the years include Two Continents, where maturing casks have been brought from India to Scotland for their final period of maturation, Intermediate Sherry Matured where the new spirit has matured in ex-bourbon or virgin oak, then re-racked to sherry butts and with a third maturation in ex-bourbon casks, Kadhambam which is a peated Amrut matured in ex Oloroso butts, ex Bangalore Blue Brandy casks and ex rum casks and Portonova with a maturation in bourbon casks and port pipes. New editions of

them all are released from time to time. A big surprise for 2013 was the release of Amrut Greedy Angels, an 8 year old and the oldest Amrut so far. That was an astonishing achievement in a country where the hot and humid climate causes major evaporation during maturation. In 2015 it was time for an even older expression, 10 years old, and in 2016, a 12 year old, the oldest whisky from India so far, was released. Two more releases of the 10 year old appeared in 2019. The highly innovative Spectrum has now reached its fourth release, this time matured in casks made of four varieties of oak. Other limited releases include the second version of Double Cask, a 5 year old combination of ex-bourbon and port pipes, the 100% malted Amrut Rye Single Malt – the first rye whisky from the company, Amalgam comprising of Amrut as well as single malts from Scotland and Asia (with a peated version introduced in late 2018) and Con-fusion – a special bottling for members of Amrut Fever. Other recent releases include a Madeira Cask Finish and Port Pipe Peated.

John Distilleries Jdl

Goa, Konkan and Bangalore, Karnataka, founded in 1992

www.pauljohnwhisky.com

Paul P John, who today is the chairman of the company, started in 1992 by making a variety of spirits including Indian whisky made from molasses. Their biggest seller today is Original Choice, a blend of extra neutral alcohol distilled from molasses and malt whisky from their own facilities. The brand, which was introduced in 1995/96 has since made an incredible journey. It is now one of the biggest whiskies in the world with sales of 138 million bottles in 2018. Another brand is Bangalore Malt which was the fastest growing spirit in the world in both 2016 and 2017. This is a simpler version of Original Choice and 61 million bottles were sold in 2018 – an increase of 44% compared to the year before!

John Distilleries owns three distilleries and produces its brands from 18 locations in India with its head office in Bangalore. A visitor centre was recently opened at their distillery in Goa. The basis for their blended whiskies is distilled in column stills with a capacity of 500 million litres of extra neutral alcohol per year. In 2007 they set up their single malt distillery which was equipped with one pair of traditional copper pot stills but in 2017, another pair of stills were added, doubling the capacity to 1.5 million litres per year. The company released their first single malt in autumn 2012 and this was followed by several single casks. In 2013 it was time for two core expressions, both made from Indian malted barley. Brilliance is unpeated and bourbon-matured while Edited, also matured in bourbon casks, has a small portion of peated barley in the recipe. In 2015 the third core expression was released. It was a 100% peated bottling called Bold, bottled at 46% and in spring 2019 yet another "flagship" bottling was added by the way of the unpeated, bourbon matured Nirvana, bottled at 40%. At the beginning of 2014, two cask strength bottlings were released; Select Cask Classic (55,2%) and Select Cask Peated (55,5%) and since then, more Select expressions have been revealed – Oloroso and Pedro Ximenez. Recent limited releases include two 7 year old single malts; Mars Orbiter, a peated whisky matured in American oak and Kanya, unpeated from American oak.

In October 2018, Sazerac, owners of brands such as Buffalo Trace, Pappy van Winkle and Southern Comfort, bought a 23% stake in John Distilleries and in 2019 they acquired another 20%. Paul John, chairman and managing director of John Distilleries still holds a 57% stake in the company.

Other distilleries in India

Rampur Distillery

Rampur, Uttar Pradesh, founded in 1943

www.rampursinglemalt.com

This huge distillery is situated west of Delhi. It was purchased in 1972 by G. N. Khaitan and is today owned by Radico Khaitan, the fourth biggest Indian liquor company. The distillery has a capacity

Neelakanta Jagdale, Chairman of Amrut, passed away in 2019

of producing 75 million litres of whisky based on molasses, 30 million litres of grain whisky and 460,000 litres of malt whisky per year. The first whisky brand from Radico was 8PM, which in 2017 sold 84 million bottles. The first single malt release, the un-chill filtered Select without age statement, appeared in May 2016 and since then Double Cask matured in a combination of ex-bourbon and European oak sherry casks and PX Sherry (American oak with a PX sherry finish) have been released.

McDowell´s Distillery

Ponda, Goa, founded in 1988 (malt whisky)

www.diageoindia.com

Established in the late 1800s, the distillery produces the second best selling Indian whisky with 348 million bottles sold in 2018. Owned by Diageo since 2014, the distillery also produces a very small amount of single malt whisky.

Mohan Meakin

Solan, Himachal Pradesh, founded in 1855

www.mohanmeakin.com

Founded as a brewery in 1820, possibly by Edward Dyer and incorporated as a company in 1855. It was taken over by H G Meakin in 1887 and finally, Narendra Nath Mohan acquired the business in 1949. New breweries were built during the 1970s and 1980s and today, the company is making beer, whisky and rum. Their most famous brands are Old Monk rum and Solan No. 1 whisky. Their first general launch of a single malt whisky under the name Solan Gold Single Malt appeared in 2019 and has matured for at least four years. The location at the foothills of the Himalayas gives an angel´s share of 2-3% per year compared to Amrut in Bangalore facing 10-12%.

Khoday

Bangalore, Karnataka, founded in 1906

www.khodayindia.com

Khoday is a company working in many areas, including brewing and distillation. The IMFL whisky Peter Scot was launched by the company already in 1968 and in spring 2019, the Peter Scot Black Single Malt was launched.

Israel

The Milk & Honey Distillery

Tel-Aviv, founded in 2013

www.mh-distillery.com

Israel´s first whisky distillery, equipped with a 1 ton stainless steel mash tun, four stainless steel washbacks and two copper stills (with a capacity of 9,000 and 3,500 litres each). The current production is 200,000 litres of pure alcohol while the capacity is 800,000. The first distillation was in March 2015 and in February 2016, the first in-house whisky production took place. The first, limited 3 year old single malt, made before the final equipment was installed, was released in August 2017. The first commercial release of a single malt whisky will be at the end of 2019. Meanwhile, Levantine Gin and Roots, a herbal liqueur, hav also been released. The distillery has a visitor centre offering a large variety of tours and workshops.

Golani Distillery

Katzrin, founded in 2014

www.golanispirit.com

Founded by Canadian expat David Zibell, the distillery is equipped with two artisanal copper stills and the whisky is matured in wine casks from the nearby Golan Heights Winery. A young two-grain whisky was released in 2016 while the distillery´s first single malt aged more than three years, appeared in late 2017 as a single cask. A number of different single casks have then been released. The distillery is in a stage of expansion with a new mash tun and more stills coming in. Meanwhile, David has also opened yet another distillery (Yerushalmi) for peated production and even started his own production of copper stills.

Pakistan

Murree Brewery Ltd.

Rawalpindi, founded in 1860

www.murreebrewery.com

Started as a beer brewery, the assortment was later expanded to include whisky, gin, rum, vodka and brandy. The core range of single malt holds two expressions – Murree´s Classic 8 years old and Murree´s Millenium Reserve 12 years old. There is also a Murree´s Islay Reserve, Vintage Gold.

The tasting room at Milk & Honey Distillery

Taiwan

Kavalan Distillery

Yanshan, Yilan County, founded in 2005
www.kavalanwhisky.com

On the 11th of March 2006 at 3.30pm, the first spirit was produced at Kavalan distillery. This was celebrated in a major way a decade later when guests and journalists from all over the world were invited for the 10th anniversary. But it was not just to celebrate 10 years of whisky production but also to witness the recent expansion of the distillery which has made Kavalan one of the ten largest malt whisky distilleries in the world! This rapid development may even have surprised the founder, entrepreneur and business man Tien-Tsai Lee, and his son, the current CEO of the company Yu-Ting Lee. Early on, it was decided that expertise from Scotland was needed to get on the right track from the beginning. Dr. Jim Swan was consulted early on and he, together with the master blender, Ian Chang, developed a strategy including production as well as the future maturation. Jim Swan sadly passed away in early 2017.

The distillery lies in the north-eastern part of the country, in Yilan County, one hour´s drive from Taipei. Following the expansion in 2016, the distillery is equipped with 5 mash tuns, 40 stainless steel washbacks with a 60-72 hour fermentation time and 10 pairs of lantern-shaped copper stills with descending lye pipes. The capacity of the wash stills is 12,000 litres and of the spirit stills 7,000 litres. Kavalan only uses a very narrow cut from the spirit run, leaving more foreshots and feints to accommodate a complex and rich flavour profile. The spirit vapours are cooled using shell and tube condensers, but because of the hot climate, subcoolers are also used.

On site, there are two five-story high warehouses, with a third expected to be completed in 2019, and the casks are tied together due to the earthquake risk. The climate in this part of Taiwan is hot and humid and on the top floors of the warehouses the temperature can reach 42°C. Hence the angel´s share is dramatic – no less than 10-12% is lost every year. At the moment, Kavalan are doing experiments aiming to reduce the angel's share to below 10%, hoping for positive results in 2020. The distillery has its own cooperage where the preparation of the STR (shave-toast-rechar) casks plays an important part for the final character of the whisky.

The brand name, Kavalan, derives from the earliest tribe that inhabited Yilan, the county where the distillery is situated. Since the first bottling was released in 2008, the range has been expanded and now holds more than 19 different expressions. The best seller globally is Classic Kavalan, bottled at either 40% or 43%. In 2011, an "upgraded" version of the Classic was launched in the shape of King Car Conductor – a mix of eight different types of casks, un chill-filtered and bottled at 46%. A port finished version called Concertmaster was released in 2009 and, later that year, two different single cask bottlings were launched under the name Solist – one ex-bourbon and one ex-Oloroso sherry. It was the launch of these two that made the rest of the world aware of Taiwanese whisky.

More expressions in the Solist series have been added and the range now consists of (apart from Bourbon and Sherry) Fino, Vinho Barrique (using Portuguese wine barriques), Manzanilla, Amontillado, PX, Moscatel, Brandy and Port. All of these are bottled at cask strength but in 2012 two versions bottled at 46% were also introduced – Bourbon Oak and Sherry Oak. Other releases include Podium, Distillery Reserve Rum Cask and Distillery Reserve Peaty Cask. The latter, exclusively available at the distillery visitor centre, obtains its smoky flavour from maturation in ex-Islay casks. The distillery has produced whisky from peated barley (10ppm) as well, and plan to launch the first bottlings from that production within 2-4 years. In May 2018 a new core expression was released – Distillery Select – which is intended to work as the entry level to the brand. Two 10th anniversary limited editions bottled at 57,8% were released in autumn 2018, matured in ex-Bordeaux wine casks from Margaux and Paulliac. In autumn 2019 a sherry finished version of the Concertmaster appeared and there are plans to launch another limited bottling end of 2019 to celebrate the 40th anniversary of the foundation of King Car Group. Whisky is, of course, the main product for Kavalan but production of gin is also carried out in the four sets of Holstein stills that were installed already in 2008. The first release in a series of triple distilled gins appeared in early 2019.

In May 2019, Kavalan opened their first designated 'Cask Strength Whisky Bar' in the busy Zhongshan District of Taipei. Recreating the inside of the distillery´s warehouse, it is the only bar in the world to carry the full range of Kavalan whisky. Guests can order whiskies straight from the cask and the bar also uses special effects to illustrate the environmental impact on the flavour of the whisky.

Kavalan is being exported to more than 60 countries and apart from Taiwan, Europe and the US are the most important markets. There is an impressive visitor centre on site with no less than one million people coming to the distillery every year. The owning company, King Car Group, with 3,000 employees, was already founded in 1956 and runs businesses in several fields; biotechnology and aquaculture, among others. It is also famous for its ready-to-drink coffee, Mr. Brown.

Kavalan´s new 'Cask Strength Whisky Bar' in Taipei

Other distilleries in Taiwan

Nantou Distillery

Nantou City, Nantou County, founded in 1978
(whisky since 2008)

en.ttl.com.tw

Nantou distillery is a part of the state-owned manufacturer and distributor of cigarettes and alcohol in Taiwan – Taiwan Tobacco and Liquor Corporation (TTL). Between 1947 and 1968 it exercised a monopoly over all alcohol, tobacco, and camphor products sold in Taiwan. It retained tobacco and alcohol monopolies until Taiwan's entry into the WTO in 2002.

There are seven distilleries and two breweries within the TTL group, but Nantou is the only with malt whisky production. The distillery is equipped with a full lauter Huppmann mash tun with a charge of 2.5 tonnes and eight washbacks made of stainless steel with a fermentation time of 60-72 hours. There are two wash stills (9,000 and 5,000 litres) and two spirit stills (5,000 and 2,000 litres). The owners are currently looking to expand the distillery with another three pairs of stills. Malted barley is imported from Scotland and ex-sherry and ex-bourbon casks are used for maturation. Nantou Distillery also produces a variety of fruit wines and the casks that have stored lychee wine and plum wine are then used to give some whiskies an extra finish. Until recently, the spirit from Nantou has been unpeated, but in 2014, trials with peated malt brought in from Scotland were made. The main product from the distillery is a blended whisky which comprises of malt whisky from Nantou, grain whisky from Taichung distillery and imported blended Scotch. In 2013, two cask strength single malt whiskies were launched – one from bourbon casks and the other from sherry casks. The next expressions were Omar single malt, where several versions have been released, matured in either sherry or bourbon casks. Recent Omar expressions include a 3 year old Peated Cask Strength, a single malt finished in orange brandy casks and a 10 year old matured in PX sherry casks. There is also a blended malt whisky called Yushan.

Omar - the single malt that made Nantou known to the world

Africa

South Africa

James Sedgwick Distillery

Wellington, Western Cape, founded in 1886 (whisky since 1990)

www.threeshipswhisky.co.za

Distell Group Ltd. was formed in 2000 by a merger between Stellenbosch Farmers' Winery and Distillers Corporation, although the James Sedgwick Distillery was already established in 1886. The company produces a huge range of wines and spirits including the popular cream liqueur, Amarula Cream. James Sedgwick Distillery has been the home to South African whisky since 1990. The distillery has undergone a major expansion in the last years and is now equipped with one still with two columns for production of grain whisky, two pot stills for malt whisky and one still with six columns designated for neutral spirit. There are also two mash tuns and 23 washbacks. Grain whisky is distilled for nine months of the year, malt whisky for two months and one month is devoted to maintenance. Three new warehouses have been built and a total of seven warehouses now hold more than 150,000 casks. There is also a highly awarded visitor centre on site.

In Distell's whisky portfolio, it is the Three Ships brand, introduced in 1977, that makes up for most of the sales. The range consists of Select and 5 year old Premium Select, both of which are a blend of South African and Scotch whiskies. Furthermore, there is Bourbon Cask Finish, the first 100% South African blended whisky and the 10 year old single malt. The latter was launched for the first time in 2003 and the latest release (Vintage 2006) was launched in 2018. A new range called Master's Collection was introduced in 2015 with a 10 year old PX finish as the first release. This was followed in 2017 by a 15 year old pinotage cask finish and in September 2018 by an 8 year old, lightly peated oloroso finish. Apart from the Three Ships range, Distell also produces South Africa's first single grain whisky, Bain's Cape Mountain, which has been awarded the world's best grain whisky on several occasions.

In 2013 the Distell Group acquired Burn Stewart Distillers, including Bunnahabhain, Tobermory and Deanston distilleries as well as the blended whisky, Scottish Leader. The man who tirelessly worked to bring the Three Ships single malt to the market, was Andy Watts. After 25 years as the distillery manager, he has now a role in the company where he is responsible for overseeing Distell's whisky portfolio as well as being the Three Ships' Master Distiller.

Andy Watts, Master Distiller at James Sedgwick Distillery

South America

Argentina

La Alazana Distillery

Golondrinas, Patagonia, founded in 2011

www.laalazanawhisky.com

Located in the Patagonian Andes, the first whisky distillery in Argentina concentrating solely on malt whisky production was founded in 2011 and the distillation started in December of that year. The founders were Pablo Tognetti, an old time home brewer, and his son-in-law, Nestor Serenelli but end of 2014, Pablo Tognetti withdrew from the company. Today it´s Nestor and his wife Lila who own and run the distillery. They are both big fans of Scotch whisky and before they built the distillery, they toured Scotland to visit distilleries and to get inspiration. The owners are firm believers in the "terroir" concept where local barley and water and, not least, climate will affect the flavour of the whisky. The distillery is equipped with a lauter mash tun, four stainless steel 1,100 litre washbacks with a fermentation time of 4 to 6 days and two stills and there are now plans to icrease capacity. For the last two years, the owners have been growing their own barley and also do the malting using local peat which means a 100% Patagonia single malt is now maturing in the warehouses. The house style is light and fruity but they have also filled several barrels with peated whisky. The first, limited release was made in November 2014 and in 2017, the first peated bottling, the 4 year old Haidd Merlys was launched.

Other distilleries in Argentina

Distillery Emilio Mignone & Cia

Luján, Buenos Aires province, founded in 2015

www.emiliomignoneycia.com.ar, www.emyc.com.ar

With the first distillation in November 2015, this became the second whisky distillery in Argentina. The distillery is equipped with a 300 litre open mash tun, a 250 litre washback with a 72-96 hour fermentation cycle and two stills, directly fired by natural gas. The first of their Classic Pampa Single Malt Whisky was released in October 2019 and was followed by a peated version. The owners are working on a second and larger distillery (10,000 litres) in Lago Puelo which should be up and running in autumn 2020.

Madoc Distillery

Dina Huapi, Rio Negro, founded in 2015

www.madocwhisky.com

The owner is one of the founders of the first Patagonian distillery, La Alazana. In 2015, he left the company and brought with him some of the equipment, as well as part of the maturing stock to build a new distillery in Dina Huapi. The existing equipment with a lauter mash tun, a washback and a copper pot still was complemented by a wash still and the first distillation took place in September 2016 and a single malt bottled at 40% has been released.

Brazil

Union Distillery

Veranópolis, founded in 1972

www.maltwhisky.com.br

The company was founded in 1948 as Union of Industries Ltd to produce wine. In 1972 they started to produce malt whisky and two years later the name of the company was changed to Union Distillery Maltwhisky do Brasil. In 1986 a co-operation with Morrison Bowmore Distillers was established in order to develop the technology at the Brazilian distillery. Most of the production is sold as bulk whisky to be part of different blends, but the company also has its own single malt called Union Club Whisky with an 8 year old as the oldest expression.

Muraro Bebidas

Flores da Cunha, founded in 1953

www.muraro.com.br

This is a company with a wide range of products including wine, vodka, rum and cachaca and the total capacity is 10 million litres. Until recently, the blend Green Valley was the only whisky in the range. In 2014, however, a new brand was introduced. It has the rather misleading name Blend Seven but it appears to be a malt whisky even though it seems that essence of oak is part of the recipe. The main market for the whisky is The Carribean.

Lila and Nestor Serenelli, owners of La Alazana Distillery

The Year
that was

Including the subsections:
The big players | The big brands | Changes in ownership
New distilleries | Bottling grapevine

The time scale spanning a period is important to know when analyzing trends. This becomes obvious when looking in detail at global alcohol consumption. In 2017, it had increased by 10% per adult compared to 1990. However, within that timespan, if singling out the shorter period from 2012 and adding on 2018, the global alcohol consumption has declined. The figures you may use depends on your agenda.

What is true though is that the tiny increase from the previous year in global alcohol consumption (+0,1%) that was noted in 2017, turned into a decrease of no less than 1.6% in 2018 when a total of 27.6 bn nine-litre cases of alcohol were consumed. The losers were beer and wine. Beer was down by 2.2%, mainly due to decreases in China, USA and Brazil, and wine lost 1.6% in volume but gained in value thanks to consumers buying more expensive brands.

Consumption of spirits, on the other hand, increased by 1.1% in terms of volume. Most spirit categories gained volumes in 2018 with whisky leading the way by adding another 10 million cases. The gin bubble hasn't burst yet and especially pink gin is growing. The gin category grew by 8.3% compared to 2017. Vodka, however, which was down by 2.6%, is in its sixth year in a row of declining sales. The only case for joy in the vodka business is an increase in higher-priced brands. The combined category of cognac and brandy show positive growth which implies that, especially for cognac, they have made a comeback from the sluggish years between 2012 and 2104 when the Chinese market for expensive, foreign spirits crashed. Rum is slowly but steadily increasing but has still to reach its full potential while other cane spirits, especially Brazilian cachaca, is slipping. A significant increase, although from a small baseline, was tequila and other agave spirits which grew by 5.5% in 2018.

The largest spirits category is still 'Other spirits' which in terms of volumes mainly consists of baijiu, a Chinese spirit usually distilled from fermented sorghum, soju, the Korean national spirit made from rice, wheat, barley or potatoes and shochu from Japan, made from a variety of plants including sweet potato, barley, rice and buckwheat. Baijiu alone stand for almost 40% of global spirits' consumption beating the combined volumes of whisky, rum, gin, tequila and vodka!

The global spirits sales in 2018 in 9-litre cases and broken down into category looks as follows:

CATEGORY	CASES	CHANGE
Other spirits	1.2bn	+1.1%
White spirit	399m	+0.4%
Whiskies	375m	+2.8%
Brandy and Cognac	169m	+1.4%
Rum	151m	+2.0%
Liqueurs	112m	+1.0%
Tequila and Mezcal	37m	+6.6%
Total	2.4bn	+1.1%

If we focus just on spirits sold globally (baijiu, soju and schochu are almost entirely sold in their respective home markets) and take into account that the category 'White spirits' is made up of both vodka, gin and other varieties, whisky is the biggest single category in the world both in terms of volumes and values.

In 2018, for the third year in a row, volumes and values of Scotch whisky export went up. In fact both figures reached a record high. Blended Scotch is still by far the biggest sub-category but showed a decline in 2018 following a strong 2017. Single malt Scotch on the other hand made up for the loss. The recent development for that subcategory is nothing short of astonishing. If we look at the last six years (since 2012) single malt export values have gone up by no less than 67% while the total value of Scotch export has only increased by 10% during the same period.

SINGLE MALT SCOTCH - EXPORT

Value:	+11.3% to £1.3bn
Volume:	+1% to 123.6m bottles

BLENDED SCOTCH* - EXPORT

Value:	-4.7% to £3.0bn
Volume:	+1.7% to 981m bottles

TOTAL SCOTCH - EXPORT

Value:	+7,8% to £4.7bn
Volume:	+3.6% to 1.28bn bottles

* This is bottled blended Scotch, excluding bulk sales.

The European Union

The European Union is still the biggest export market for Scotch whisky, both in terms of volumes and value but this could change already next year. Scotch is currently selling well in North America, and with its momentum it may have reached the top spot in terms of values when we sum up next year. 36% of the export volumes (-3%) and 30% of the values (+2%) went to the EU in 2018. Single and blended malts showed growth while blends decreased.

EU — Top 3

France	volumes	-6%	values	+2%
Spain	volumes	-9%	values	-2%
Germany	volumes	-16%	values	-5%

France has for many years been the dominant market for Scotch in the EU and still is. In terms of volume it is in fact the biggest market in the world with nearly 15% of the totals. The growth has also been steady the last couple of years. Which cannot be said of number two and three. Spain managed, despite a decline in 2018, climb to second place, due to even bigger decreases in Germany. This should, however, be seen in the light of the two previous years' strong surges in Germany. And, if we focus on values, Germany is still the second biggest market in the region.

The situations after the leading three markets is interesting. In place four (and number eleven globally) is Latvia with an increase of more than 1200% in the last decade. With two million people it becomes quite obvious that the majority of the whisky is not consumed locally. Instead, Latvia serves as a hub for exports to Russia. In the following spots we find Poland with a steady increase in both volumes and values, the Netherlands where both figures decreased and finally Italy with a 15% decrease in volumes but an 11% increase in values.

North America

The second largest region in terms of volumes but since a few years back down to place three having been overtaken by Asia in terms of values. Be that as it may, North America showed strong growth during 2018 with volumes up by 8% and values by 13%. USA is if course the dominant market with Mexico and Canada making up the rest.

North America — Top 3

USA	volumes	+7%	values	+13%
Mexico	volumes	+12%	values	+18%
Canada	volumes	+3%	values	+7%

In 2018, USA became the first billion pound export market for Scotch whisky as it grew to £1.04bn. It was a good year for Scotch in the USA with volumes up by 7% and values by 13%. That market makes up the lion's share of the regional import of Scotch whisky (82% of the values and 65% of the volumes) but let's not forget Mexico, a country which in recent years has climbed to fourth place on the global Scotch export list, at least in terms of volumes. The red figures from last year turned black during 2018. When one compares the two countries in terms of the split between malts and blends, it becomes interesting. No less than 35% of the total exports to the USA are made up of malts while the corresponding figure for Mexico is 8%. This clearly shows how consumers in the USA have climbed the aspirational ladder to where malts are held in higher esteem by the experienced drinkers. Canada, in third place, also showed a healthy increase during 2018.

Asia

A few years ago, Asia was the region where Scotch whisky grew year after year and the producers had high hopes for the future. Then, starting in 2012, followed a couple of years where both volumes and values dropped. Since Asia was considered a market of continuous growth, the disappointing figures shelved a number of expansions planned in Scottish distilleries and quite a few of them cut down on production. The gloom persisted for four years until 2016

when exports were on the rise again. This positive trend continued in 2017 and in 2018 it was obvious that Asia was back in the game. Volumes were up by 14% while values climbed by 16%. Asia is now the second most important region in terms of volumes. As opposed to EU and North America which could be defined as mature markets, Asia is still very much an emerging one including the world's two largest countries in the continent.

Asia — Top 3

India	volumes	+26%	values	+34%
Japan	volumes	+49%	values	+29%
Singapore	volumes	-4%	values	+10%

India is definitely the market with the greatest potential in the region. With an increase in volumes in the last decade by 260% it has become the third biggest market in terms of volumes and the eighth in values. With Scotch having just a 1% share of the Indian spirits market, there's room for huge increases. Especially if the current 150% import tariff on Scotch could be lowered. Another reason for the positive outlook is that India traditionally is a "brown spirit" country while in the Chinese market, consumers are used to drinking "white spirit", not least baijiu, the world's most consumed spirit.

In second place this year, we find Japan, a country with

a significant and highly regarded whisky production of its own. However, in the last couple of years, the producers, taken by surprise by the growing demand for Japanese whisky, have had to put their decreasing stock on allocation while removing aged whiskies from their range. The positive 2018 Scotch whisky export figures to the country could very well be a reclection of that situation. Volumes were up by 49% and values by 29%. Singapore is still in the top 3 in Asia and, in fact, the third biggest market for Scotch in the world in terms of values. The caveat, as always when interpreting the numbers for Singapore, is that the vast majority of the whisky is re-exported to other markets in the region, not least to ASEAN countries and China.

Behind the Top 3 we find Taiwan in fourth place. This small country has for many years had a reputation as the single malt market par preference. With the value of malt whisky exports representing 30% of total Scotch whisky exports to the world, the value of Scotch malt exported to Taiwan is 63% of the country's total! But things are not as they used to be. Even though values of Scotch went up by 5% in 2018, the country has been in a long period of economic difficulties which can be detected in the figures.

One cannot leave Asia without talking about China, the most populated country in the world. It is currently number 17 on the Scotch whisky export list in terms of values (+24% in 2018) and number 19 in volumes (+8%). For the big spirits producers, everything was set on Chinese growth until 2012 when the government cracked down on corruption and extravagant gifting. Producers of expensive cognac and whisky saw the sales figures imploding. Six years later the situation has normalized and the new whisky consumers in China are found in the wealthy middle class who want to know about the whisky rather than show off an expensive bottling. The lowering of the import duty on Scotch from 10% to 5% has also helped grow the market.

Central and South America

Planning your Scotch whisky business for this region has proved to be a nightmare in the past decade. The economic volatility makes up for a roller coaster of sales figures with not much to rely on. Figures for 2018 were, however, positive with volumes growing by 11% and values by 10%. The region is very much driven by the sales of blended Scotch. On a global basis, malts (single and blended), make up for 31% of the total value of Scotch. In Central and South America that figure is 4.5%. While increases in malt whiskies, albeit from a small base, were leading the way in 2017 in this region, in 2018 it was time for blends to make a comeback. Volumes for that category were up by 15% and values increased by 11%.

Central & South America — Top 3

Brazil	volumes	+8%	values	+13%
Chile	volumes	+22%	values	+25%
Colombia	volumes	+28%	values	+36%

For the second year in a row, the biggest market Brazil showed positive figures which rendered the country a 10th place on the global sales list in terms of volumes. The increase is impressive given the fact that Brazil experienced

a recession in 2015 and 2016 and the economy only grew by 1% per year in 2017 and 2018. Beer accounts for 92% of total alcohol sales and the domestic Cachaça makes up for most of the remaining 8%. With more than 200 million people and a youthful demographic, Brazil is still uncharted terrain for Scotch whisky producers.

If we look at number two and three in the region, volumes of Scotch exports to Chile has grown by 67% in the last five years and for Colombia the corresponding figure is 93%. Chile is considered South America's most stable and prosperous nation and the Colombian economy is also improving. A striking example of the importance for Scotch whisky exports having a solid economy with growth is Venezuela. In 2007, it was the 6th biggest market in the world for Scotch in terms of volumes. Eleven years later, in 2018, just over 2% of it was left, the total economic collapse since Maduro took over in 2013 being responsible for that.

Africa

Following an excellent 2017 with double digit growth in volumes as well as values, the region was one of two showing red figures in 2018. Volumes went down by 13% and values decreased by 8%. The continent is often mentioned as the next future market for Scotch, but to be honest, the positive figures for 2017 was a one-off. The track record for the region hasn't been that positive in recent years.

Africa — Top 3

South Africa	volumes	-20%	values	-18%
Kenya	volumes	+28%	values	+18%
Nigeria	volumes	-14%	values	+48%

One thing to remember when looking at the figures for Africa is that exports to South Africa make up 65% of the totals in terms of volume and 54% in terms of value. If the economy deteriorates in that country it will affect the whole region on a grander scale and in 2018, following a very positive 2017, exports of Scotch to South Africa again went down quite severely.

Looking at the rest of the region, there are currently few countries where producers of Scotch actually have a foothold. Kenya is number one with a total volume that equals that of Denmark. Other countries with a reasonable import of Scotch are Nigeria, Morocco and Angola but contrary to the case of Kenya, their imports decreased with double digit figures in 2018.

Middle East

Values were up by 6% and volumes by 2% in this region which has increased in importance in recent years. In 2018, the growth was in blended Scotch and blended malt while at the same time single malts went down.

Middle East — Top 3

UAE*	volumes	+16%	values	+13%
Lebanon	volumes	+1%	values	+3%
Israel	volumes	-8%	values	-2%

* United Arab Emirates

The biggest market in the region is the United Arab Emirates with 55% of the volumes and 57% of the values. Lebanon in second place continues to do well although with small increases compared to last year while Israel, following a year with extreme growth, decreased in 2018. A major share of the sales to UAE are actually destined for duty free sales or for re-exporting to other countries.

Australasia

Until 2016, Australasia was the second smallest Scotch export region but some good figures in recent years have made it possible to surpass non EU-members. In 2018, values were up 8% and volumes by 1%.

Australasia — Top 3

Australia	volumes	+2%	values	+11%
New Zealand	volumes	-4%	values	-10%
New Caledonia	volumes	-30%	values	-39%

Unsurprisingly, Australia is the dominating market with 93% of the value and during the past 15 years that market has been stable, which can also be said about New Zealand in second place.

Eastern Europe

The second smallest of all regions (1.8% of the volumes and 0.6 % of the values) is labelled Eastern Europe but with many countries in this geographical region belonging to the EU, their import are hidden in that larger group's. Here, we are talking mainly about Russia and some of the surrounding countries. In the last few years, the figures have fluctuated significantly between years due to export restrictions against Russia because of the conflict in the Ukraine. In 2018, volumes were up 43% while values increased by 30%. The biggest increase was in single and blended malt.

Eastern Europe — Top 3

Russia	volumes	+59%	values	+68%
Georgia	volumes	+9%	values	+14%
Ukraine	volumes	+104%	values	+140%

The biggest market in the region is Russia but the exact volumes of whisky landing in that market are sometimes difficult to interpret. Large volumes of Scotch whisky destined for Poland and, especially, Latvia will eventually be re-exported to Russia.

European non EU-members

The smallest of the nine regions, responsible for only 1.6% of the volumes. On the other hand, a large portion of that is made up of malts which makes the share of the total values almost 2%. Volumes were down by 7% in 2018 and values down by 10%.

Europe (non-EU) — Top 3

Turkey	volumes	+-0%	values	-13%
Switzerland	volumes	-23%	values	-14%
Andorra	volumes	-5%	values	+1%

Turkey is by far the biggest market with approximately 60% of the totals. The country has for a long time favoured a preferential, three-tiered tax system where the domestic Raki is taxed substantially lower than many imported spirits including Scotch. This system, however, was abolished in 2018 but that didn't seem to affect the figures for imported Scotch.

The big players

Diageo

Being the undisputed leader of the spirits industry, the results from Diageo are obviously scrutinized by reporters of both economy and the whisky world. The figures for the year ending 30th June 2019 showed a rise in sales by 5.8% to £12.87bn while the operating profits went up by 9.5% to £4bn. The CEO, Ivan Menezes, commented by saying "Diageo has delivered another year of strong performance. Organic volume and net sales growth was broad based across regions and categories, with new product innovation being a strong contributor."

Diageo's business can be divided into five different geographical regions where North America represents 35% of the company's sales followed by Europe and Turkey (23%), Asia Pacific (21%), Africa (12%) and latin America and Caribbean (9%). All regions showed positive growth in the past year.

Scotch is the biggest part of Diageo's business with 25% of the net sales followed by beer (16%) and vodka 11%. The majority of the brands are then divided into Global Giants representing 41% of the net sales (Johnnie Walker, Smirnoff, Baileys, Captain Morgan, Tanqueray and Guinness), Local Stars representing 20% of sales (for example Buchanan's, J&B, Windsor, Black & White and Old Parr) and Reserve representing 19% of sales (for example all Scotch malts and Bulleit).

If we look at single labels, Johnnie Walker, the best selling Scotch in the world, increased net sales by 6% selling 227 million bottles last year which ranks the brand as the 5th best selling whisky in the world with four Indian whiskies in the lead. Part of this year's success for the brand is due to the White Walker launched in autumn 2018 to coincide with the upcoming final season of the Game of Thrones television series. In spring 2019, the theme was followed up by the release of nine single malts attributed to the various houses in Game of Thrones and finally, in autumn 2019, two more Johnnie Walker expressions were released – A Song of Ice and A Song of Fire.

Behind Johnnie Walker is J&B (37 million bottles in the last year) and this is a brand that has lost 16% of its volumes in the past five years. Black & White in third place, however, has gained 93% in the same period selling 32 million bottles and if this trend continues, it may well have

Diageo´s Game of Thrones theme was a boost not just for Johnnie Walker but also for the nine single malts that were a part

surpassed J&B in a couple of years or so.

Sales of Diageo's Scotch malts grew by 12% in the last year and the biggest brand by far is The Singleton selling around 6.5 million bottles and this is followed by Talisker, Cardhu, Lagavulin and Oban.

Through their incubator company Distill Ventures, Diageo has invested in several smaller brands in the past five years yet at the same time they have been keen on getting rid of brands that no longer fit in with their strategy. In autumn 2018, no less than 19 value spirits brands were sold to Sazerac in a deal worth $550m.

In April 2019, Diageo formed a joint venture with baijiu producer Jiangsu Yanghe Distillery Co to launch a new whisky, Zhong Shi Ji, in the Chinese market.

In April 2018, Diageo announced a £150m investment in whisky tourism in Scotland. It will be centred around a Johnnie Walker experience in Edinburgh but all their distillery visitor centres will be upgraded, in particular the ones at Glenkinchie, Clynelish, Caol Ila and Cardhu.

Pernod Ricard

In the end of August Pernod Ricard presented a strong result for the full year 2018/2019. Organic sales grew by 5% to €9.18bn while the profit from recurring operations increased by 9% to €2.58bn. The Chairman and CEO of the company, Alexandre Ricard, commented the results by stating "FY19 was an excellent year demonstrating clear business acceleration while investing for longterm value creation."

But despite the good results there are reasons for concern. Pernod Ricard has been targeted by activist investor Elliott Management controlled by billionaire Paul Singer. In December 2018, the hedge fund took a stake in the company "in excess of 2.5%" stating that there was room for improvement in the French company and also noted that Pernod Ricard's track record in mergers and acquisitions hade been "disappointing".

The company has divided their market into three regions where Asia/Rest of the world is the biggest in terms of sales (+12% last year), followed by Europe (+1%) and Americas (+2%). In Asia, China and India performed exceptionally well (+21% and +20% respectively) with the growth in China driven mainly by Martell and in India by Seagram's Indian whiskies.

Looking at their strategic international brands, ten out of thirteen showed growth (Absolut, Ricard and Malibu being the exceptions) and especially Martell (+18%) and Royal Salute (+16%) excelled. The Scotch blends did well, Ballantine's +7% and Chivas Regal +6% as did The Glenlivet single malt which increased its volumes by 9% to 14.4 million bottles. Jameson, the Irish whiskey category leader continues to show strong growth, up 6% with 92 million bottles sold.

In connection with the financial presentation, the company also unveiled plans to build a malt whisky distillery in China - the first distillery in the country from an international spirits company. The distillery will be situated at Emeishan, south of Chengdu and the plan is to have it operational in 2021.

Edrington

For the fiscal year ending 31 March 2019, the company showed a sales increase of 9% to £679.8m and an increase in earnings before interest and tax (EBIT) of 8% to £223.2m. All in all a successful year and slightly better than 2017/2018.

The increasingly dominant strategy for the company is now about premiumisation or, as the new CEO Scott McCroskie states in his chief executive report, "Central to the Edrington 2025 strategy is our new vision – we will give more by building the world's leading portfolio of exceptional super premium spirits."

The blend Cutty Sark was sold to La Martiniquaise-Bardinet, owners of Glen Moray and Label 5 in November 2018, in line with the new strategy. Just a few months later Glenturret distillery was off-loaded to Lalique Group from Switzerland. That deal meant that Edrington no longer has a Famous Grouse Experience to support their only remaining blended Scotch. Famous Grouse is still a major brand – category leader in Scotland an six other markets – and while many of their competitors struggle on the same difficult market, it remains to be seen if a blend will fit in with Edrington's product line in five years from now. The outcome for Famous Grouse in the past year was a decline of 8%.

Macallan on the other hand, with a brand new distillery to back them up, grew by 6%. Highland Park also showed good figures and the owners invested in a designated distillery shop in Kirkwall in summer 2019. Glenrothes, finally, has two complete new ranges since last year and Edrington are looking to make that their next prestige brand.

The best performer in the company's portfolio during last year wasn't a whisky, however. Instead Brugal rum was the winner. Sales of the brand grew 26% and a new distillery became operational in June 2019. Edrington took over the brand in 2008 and it was only five years ago that the owners were forced to write down the value of the brand due to "extreme difficulties in its key markets".

Ian Curle, who took the helm at Edrington 15 years ago, stepped down as CEO in 2019 and was succeeded by Scott McCroskie.

Gruppo Campari

Few, if any, of the large spirits companies are so dependent on one brand as the Campari group. Aperol, once a niche product in northern Italy, has become the companies biggest asset. That doesn't mean there are no other famous brands in the portfolio. The brands in the company are divided into three main groups – Global Priorities (56% of the sales including Aperol, Campari, Grand Marnier, SKYY and Wild Turkey), Regional Priorities (17% with, among others, Glen Grant, Forty Creek and Cinzano) and Local Priorities (12%).

The company results for 2018 showed a decrease in reported sales by 2.4% to €1.71bn while gross profits were up by 1.6% to €1.03bn. Strong sales of Aperol (+28%) not least in the brand's core markets (Italy, Germany, Austria and Switzerland) was the main driving force helped by strong sales of Campari and Wild Turkey. SKYY vodka on the other hand, was down by 8% due to a weakness in the

US market. The only Scotch whisky in the portfolio, Glen Grant, was down by almost 6% and the company has now decided to focus on aged expressions of the brand (12 years and up) in order to increase attention.

The Americas is by far the biggest market for Campari Group (44%) followed by Southern Europe, Africa and the Middle East (28%), North, Central and Eastern Europe (21%) and Asia Pacific (7.5%). The company has been known to acquire other drinks companies and in 2018 they bought the cognac producer Bisquit while at the same time offloading the Italian Lemonsoda business.

Beam Suntory

Beam Suntory Inc, a part of Suntory Holdings' alcoholic beverage operations which also includes beer and wine, is the world's third largest drinks group in the world after Diageo and Pernod Ricard. For 2018, revenues for the entire division increased by 3.7% to JPY749 billion while operating income grew 4.8% to JPY133 billion.

Suntory's portfolio, before taking over Beam, included Japanese brands such as Yamazaki, Hakushu, Hibiki and Kakubin but also Bowmore, Auchentoshan, Glen Garioch and the Irish blend Kilbeggan. With the deal in 2014, a range of other spirit brands were added to the list; Jim Beam bourbon, Teacher's blended Scotch, the two single malts Laphroaig and Ardmore, as well as Canadian Club and Courvoisier cognac.

If we look at the different brands, Jim Beam, the world's most sold bourbon, was up 10% and sold 116 million bottles. The development in Japan is particularly interesting. In 2012, before the take-over, the brand sold 360,000 bottles. Six years later sales were 9.6 millions! The company's other major bourbon, Maker's Mark increased during 2018 by 12% to 26 million bottles. Courvoisier and Canadian Club also fared well and were up by high-single digits. Teacher's blended Scotch on the other hand fell by 11% to 16 million bottles but is still doing well in India.

In April 2019, Matt Shattock resigned as CEO of Beam Suntory and was replaced by Albert Baladi.

Brown Forman

What could have been a great year for Brown Forman ended up as a good year. One of the reasons for that was President Trump's trade war against Europe with increased taxes on steel and aluminium which, in its turn, resulted in duties on American products, including bourbon, imported to Europe. Even if Brown Forman has 47% of its business in USA, still as much as 18% of sales goes to four large EU markets: Great Britain, Poland, Germany and France.

The figures for the fiscal year ending April 2019, showed net sales increasing by 2% to $3.32bn while net income increased by 17% to $835m. The backbone of Brown Forman's business is Jack Daniel's – the most sold American whiskey in the world and in sixth place of all whiskies that are produced. The Jack Daniel's family of brands grew by 4% and sold 160 million bottles. It is clear though that the growth is now in premium bourbons, a category where Brown Forman's Woodford Reserve, introduced in 1997, did very well. The sales growth for that brand was 22% and it is now close to selling one million cases in a year.

Other brands in the portfolio include Finlandia vodka (-1%), El Jimador tequila (+9%) and, since spring 2016, BenRiach Distillery Company with BenRiach, Glen-Dronach and Glenglassaugh. With no figures revealed, Lawson Whiting, CEO of Brown Forman says the BenRiach Distillery Company is seeing "dramatic growth". The company also has an interest in Irish whiskey through the ownership of Slane Irish Whiskey.

Rémy Cointreau

Another successful year was closed on 31st March 2019 for Rémy Cointreau. Sales grew by almost 8% to €1.22bn while net profits were up by 6% to €157m. The company is divided into two divisions where cognac, 'House of Remy Martin', represents around 75% of the total sales. The remaining 25% consists of 'Liqueurs & Spirits' with brands such as Cointreau, Metaxa and Bruichladdich.

The main contributor to the fine result was a strong demand for cognac in China, especially during the New Year in February, but also a good performance in another key market, the USA. The long term outlook for the Chinese market is, however, uncertain due to fears of an economic slowdown in the country and the trade tensions between China and USA.

Without revealing any sales figures, Bruichladdich was reported to have had an "excellent" year. Bruichladdich is not the only whisky in the company. A few years ago the French distillery Domaine des Hautes Glaces and Westland Distillery, based in Seattle, also joined the stable.

The big brands
Blended Scotch

Until 2008, Johnnie Walker was the most sold whisky in the world. The year after, the Indian brand Bagpiper pushed it to second place. In 2010 and 2011, it was again the world leader but since 2012, various Indian whiskies have skyrocketed in the charts and today Johnnie Walker is in fifth place. If we focus on the definition of whisky in the EU and North America where whisky can't be made from molasses, like all the big brands in India are, Johnnie Walker is, and has been for many years, the indisputable leader. An increase of 3% in 2018 resulted in 227 million sold bottles. Sales during 2018 were boosted not least by the new expressions launched in connection with the popular TV series Game of Thrones. In the North American market alone, net sales were up by 6% thanks to the campaign and the launch of White Walker. In India, Johnnie Walker Red showed a double digit increase supported by the company's introduction of a new cocktail – Johnnie Ginger.

Number two in blended Scotch – no surprise here either

A new global campaign, Success is a Blend, has been launched to promote sales of Chivas Regal

– is Ballantine's which in the last five years has increased by 25% (8% in 2018 alone) and sold 89 million bottles. Especially strong growth was seen in Russia, Latin America and Asia. One of their moves to increase brand awareness has been to release bottlings of the signature single malts that make up the blend – Miltonduff, Glenburgie and Glentauchers.

In place number three, we find Grant's where the growth in the last five years is just 4.5%. The brand sold 55 million bottles in 2018. In order to move forward, a new campaign was launched last summer including packaging redesign and new expressions. Hard on the heels of Grant's, with 54 million bottles sold in 2018, is Chivas Regal. Even if the brand showed a healthy increase in 2018 (+7%), Chivas Regal has had a similar history of stagnation in the last five years as Grant's. The growth last year came mainly from a relaunch in China with Turkey, India and Japan also showing positive signs. A new and contemporary campaign for Chivas Regal called *Success is a blend* was launched globally in autumn 2018. The message for the campaign is that blend is better, in life and in Scotch.

In place number five we find a brand with an excellent sales development in recent years. Fifteen years ago, Lawson's sold 14 million bottles. In 2018, that figure had gone up to 40 million. In Mexico alone, the brand is selling 8 million bottles! In recent years, the brand has even surpassed what used to be the owners' (Bacardi) number one blend, Dewar's.

Number six is J&B, a brand which was the second biggest blend in the world up until 2006, but has since then been in a downward spiral. In its heyday it sold 75 million bottles but in 2018 that figure had gone down to 37 million bottles.

In places seven and eight we have two brands that sold around 36 million bottles in 2018. One of them is William Peel which is owned by Marie Brizard Wine & Spirits, a company which has been facing some difficult times lately. Tough competition in three of their key markets (France, USA and Poland) called for action but a proposal to sell off a selection of brands was cancelled. The other brand is Famous Grouse, which in 2018 showed sales figures similar to those ten years ago. The owners, Edrington, recently sold Glenturret distillery including The Famous Grouse Experience and with the company's new direction of "building the world's leading portfolio of exceptional super premium spirits" it remains to be seen if there is a future place in their portfolio for Famous Grouse.

Bacardi, owners of Dewar's, have seen sales volumes for their flagship blend go down more or less continuously in the past decade. Perhaps a change is imminent. Sales figures increased in 2018 by 6% to reach 34 million bottles and a ninth place on the Top 10 list. A global campaign called *Live True*, which will run for fine years, was launched in October 2018. In it the company puts a "renewed focus on innovation".

Finally, in place ten, we have a classic brand which has been away from the limelight for quite some time. Black & White was established in 1884 and had its heydays from 1920 up until the early 1960s when the decline began. In the last decade though, sales have increased by 230% (13% during 2018) and have now reached 32 million bottles. The impressive comeback is due to an increased demand for the whisky in Latin America and India.

Single Malt Scotch

Blended whisky is still by far the biggest category in Scotch and will likely remain so for many years to come but single malts have become hugely important to most producers in the last decade. Some producers have even given up blends in order to focus on malts. Let me show the development by comparing figures from 2006, the first year I published the Malt Whisky Yearbook, with where we are today. In 2006, malts represented 7% of the total volume of Scotch whisky being exported. The share of the value was 16%. In 2018, volume had gone up to 9,7% of the totals but the impressive figure is the share of values where the malt contribution now lies at 27.7%!

In 1963 Glenfiddich Straight Malt was launched as the first single malt that was promoted on a global scale and by means of carefully planned marketing. Every year since then, with one exception, Glenfiddich has been the best-selling single malt. In 2018, Glenfiddich managed to sell 16.8 million bottles, up from the year before by 11%. Their biggest competitor, Glenlivet, followed in second place with 14.4 million, an increase by 9%. The one year when Glenfiddich was second was in 2014 when Glenlivet managed to top the list. Both distilleries seem focused on future growth judging from grand expansion plans where Glenlivet recently reached a capacity of 21 million litres while the people at Glenfiddich are working on a new distillery that in a year or so will bring their total capacity to 20 million litres.

The two brands are also the only two single malts that have managed to sell more than 1 million cases (12 million bottles) in one year. At least officially. There are rumours that number three on the sales list, Macallan, has recently passed the 1 million mark as well but this hasn't been confirmed by the owners.

For Macallan and the rest of the malts on the Top 10 list, no sales figures for 2018 were available at the time of printing so what follows here is the ranking and figures from one year ago. There may be changes between them in the ranking but most likely no new brand has managed to enter the list: Macallan (10.9 million bottles), The Singleton (6.2), Glenmorangie (6), Balvenie (4), Laphroaig (3.7), Aberlour (3.6), Glen Grant (3.6) and Talisker (3).

As usual, let's end up with a look at the top whiskies in North America, India and Ireland.

In North America, Jack Daniel's is the undisputed leader and the sixth most sold whisk(e)y in the world with 160 million bottles. In second place is the most sold bourbon in the world, Jim Beam, which enjoyed a 10% increase during 2018 selling 116 million bottles. It was followed by the Canadian whisky Crown Royal (88 million), the American blended whisky Seagrams 7 Crown (32 million) and in fifth place the bourbon Evan Williams (31 million).

In India, we find eight of the ten most sold whiskies in the world even though they cannot be sold in the EU as whisky since they are made from molasses rather than grain. The top 5 are the same as last year; Officer's Choice (408 million bottles), McDowell's No. 1 (348 million), Imperial Blue (272 million), Royal Stag (259 million) and Original Choice (138 million). Just to give an idea of the huge volumes that Indian whisky represents, these five brands

The sale of Glenturret distillery also meant closing down The Famous Grouse Experience

increased their joint volume since 2017 by 12% which means no less than 153 million bottles in one year!

The Irish whiskey industry is completely dominated by three big brands with Jameson in the top spot with 90 million bottles and a growth since last year of 9%. This is followed by Tullamore Dew, 15.6 million bottles which means an increase of 15% and Bushmill's in third place with almost 10 million bottles.

Changes in ownership

Already in summer 2018, Edrington announced that Glenturret was for sale but, apparently, there was no signed deal to present. It took until December when the Swiss luxury goods company Lalique Group took over the distillery in a 50/50 joint venture together with billionaire Hansjörg Wyss. For their part of the distillery, Lalique paid £15.5m. Details of Wyss´s payment were not disclosed. Offloading Glenturret was a natural decision for Edrington which is now focusing on "building the world's leading portfolio of exceptional super premium spirits." There is little place for a very small distillery like Glenturret with that focus. The new owners, on the other hand, intend to increase production and revamp the entire product range as well as renovate the existing Famous Grouse Experience and turn it into a new visitor attraction which will include a Lalique shop. To assist them with the blending and branding of the whisky, Ken Grier, former Creative Director at Edrington and Bob Dalgarno, Whisky Maker at The Macallan for 17 years, have joined the team.

If the sale of Glenturret was more or less expected, the second deal of the year came as a total surprise. In 2014, the Bulloch family sold Loch Lomond Distillery Company (including Loch Lomond and Glen Scotia distilleries, Glen Catrine Bonded Warehouse and a number of brands) to Exponent Private Equity. The new owners did an outstanding job at relaunching the malt brands and establishing themselves in a number of new markets. They also made substantial investments in their two distilleries. In the beginning of June 2019, the company announced that investment company Hillhouse Capital Management, based in Hong Kong, had taken over Loch Lomond in a deal reputedly being worth around £400m. According to a spokesperson for Exponent, the Asian market has been vital to the company in the past five years of expansion and will continue to be so and Hillhouse, as new owners, will play an integral part in that strategy.

New distilleries
Scotland

My primary reason for publishing the first Malt Whisky Yearbook back in 2005 was to bring order in the vast amount of new bottlings appearing every year. A Yearbook was obviously needed to cover all these new releases. Even with the odd new distillery opening, it didn't cross my mind that the distillery part of the business was in need of any special covering. Well, little did I know! Since the new millennium started, no less than 33 new malt distil-

Douglas Laing´s Clutha Distillery will be the second distillery to open by the river Clyde following Clydeside Distillery which started in 2

leries have been founded in Scotland alone, not to mention the hundreds of distilleries that have been established all around the world.

This part of the book deals with the embryonic distillery projects – they haven't started producing yet and in some cases their story is more of a plan where neither funds have been secured nor planning permission has been granted.

Three distilleries have started production since last year – Ardross, Holyrood and Lagg (read more about them in the New Distilleries chapter, pages 188-189). Another three "new" distilleries that haven't started up yet, are not covered here. They are the closed Brora, Port Ellen and Rosebank – soon to be resurrected. Read more about them on pages 193-195.

Let's start in Edinburgh where Holyrood Distillery opened up this summer. There are, however, another two presumptive distilleries lining up – both of them being constructed in Leith.

One of them, in Graham Street, is planned by Halewood Wines & Spirits who currently exports wines and spirits to 90 countries around the world and also have a shareholding in West Cork Distillers in Ireland. The new distillery will be named after John Crabbie, co-founder of North British grain distillery in 1885 and a notable whisky blender. The company submitted a planning application in May 2018 for the £7m project and the application was granted in October 2018. Meanwhile, the owners have been producing gin, malt whisky and grain whisky in a pilot plant equipped with Holstein stills in Granton.

Patrick Fletcher and Ian Stirling have plans to build a distillery beside Ocean Terminal Shopping Centre and the Royal Yacht Britannia. Once operational, the Port of Leith distillery will be producing 400,000 litres of pure alcohol per year and there will also be a visitor centre with a shop, restaurant and bar. The distillery, which will cost £10m to build, is set to become Scotland's first vertical distillery and plans to start production in autumn 2020.

Let's head over to Islay where Ardnahoe opened in 2018 as the island's ninth distillery. The tenth distillery will probably be Port Ellen if all goes according to plans, yet another distillery just outside Port Ellen will open soon after. Sukhinder Singh, owner of Elixir Distillers and The Whisky Exchange in London, intends to build a distillery just outside the village on the road to Laphroaig. A planning application was filed with the local council in December 2018 and the working name of the distillery is Farkin. The distillery will be equipped with 16 washbacks, two pairs of stills and will have their own floor maltings. The planned capacity for the distillery is 1.2 million litres of alcohol including malt and grain whisky, gin and rum. There will also be a visitor centre on site. It seems plausible production will start in 2021.

On the west coast of the Cowal Peninsula, in the west of Scotland (just north of the isle of Bute), the village Polphail was built in the 1970s to house workers on a planned oil rig construction plant nearby. The plans for the oil rig yard were never realized though and the houses that had already been built turned into a ghost town and were finally demolished in 2016. This is now the unlikely spot for a possible whisky distillery. The person behind it is none other than the previous owner of Loch Lomond distillery, Sandy Bulloch (who in fact provided the funds to demolish the 'ghost town') and the distillery will be named Portavadie after the nearby hamlet. The planning application was approved by Argyll & Bute Council in August 2018 and the plan is to produce both whisky and gin.

Fourteen years ago, there were two working distilleries in the Lowlands – three if you count the intermittent production at Bladnoch. Today there are fifteen with Holyrood in Edinburgh being the latest to open. Following the current trend, there is more to come.

No name has yet been revealed for Gordon & MacPhail´s new distillery in the Cairngorms

R&B Distillers, which opened their first distillery on Raasay in 2017, have postponed the building of a second one in The Borders. In 2015 the company held a public vote to decide the location and Peebles was selected. One of the owners of R&B Distillers is Alasdair Day who launched a blended whisky named The Tweeddale already in 2010.

Plans for another distillery in The Borders have been presented by Mossburn Distillers, owners of the second distillery on Skye, Torabhaig, which opened in 2017. Planning application was submitted in 2016 for the building of, not one, but two distilleries on the site of Jedforest Hotel near Jedburgh. The first of the two, named Jedhart Distillery, to be built will be equipped with three stills and the intention according to the owners is to "focus on small production and educating visitors on the craft of making spirit." The next stage involves building Mossburn Distillery and visitor centre with the capacity of producing 2.5 million litres per year including also grain whisky.

Moving on to the west, The Ardgowan Distillery Company received planning permission in March 2017 to build a distillery on the Ardgowan Estate, 30 miles west of Glasgow. The initial goal was to have the 800,000 litre distillery in production by 2019 but there is still the matter of funding to be resolved. New, revised plans for the distillery were approved by the local council in October 2018. The distillery, with a capacity of 1 million litres per year, will now be fitted with two copper pot stills and six wooden washbacks with a possibility of doubling the numbers in the future.The plan is to start the construction in 2019 with production beginning sometime in 2020. In May 2019, it was announced that the company had appointed Max McFarlane as its whisky maker. McFarlane has a long history in the Scotch whisky industry, among other positions as Master Blender of Highland Park.

Two malt distilleries have opened in Glasgow in the last two years and a third one is underway. Independent bottler

Douglas Laing announced in July 2017 that they had plans to build a distillery on the banks of the river Clyde at Pacific Quay, just opposite the new Clydeside Distillery which opened in 2017. The total cost for the distillery is £10.7m but this also includes a bottling complex, a new corporate head office, a visitor centre, whisky laboratory and archive. Initial capacity will be 100,000 litres of pure alcohol. Planning applications, still pending decision, were submitted to Glasgow City Council in spring/summer 2018 and at that time it was also decided the name of the distillery would be Clutha.

After several years of planning (the final approval was granted in 2010), the Falkirk Distillery Company has managed to build a distillery at Salmon Inn Road, Polmont but distillation hasn't started yet. According to one of the directors, Fiona Stewart, production will commence in autumn 2019.

Campbell Meyer & Co, blenders, bottlers and exporters of whisky, own a 150,000 square ft bonded warehouse in East Kilbride, just south of Glasgow. In spring 2016, it was announced that the company had plans to add a distillery as well. Whether or not whisky production has actually commenced is still uncertain.

Up in Speyside, The Cabrach Trust plans to build a distillery in the village of Cabrach south of Dufftown. The idea is to convert the old Inverharroch Farm to a distillery and heritage centre including a museum of illicit whisky and smuggling. The total cost is estimated £6.5m and it will be operated as a social enterprise. Planning permission was received in September 2017 and the owners have hopes to start the construction during 2019 and to have the centre and distillery operating within a couple of years. Still in Speyside, a new distillery is planned at Craggan, near Grantown-on-Spey in the Cairngorms National Park. Behind the project are none other than Gordon & MacPhail, legendary independent bottler and owner of

Benromach Distillery in Forres. The distillery, which the owners hope to have operational sometime in 2020, will have a capacity of 2 million litres. The architects involved is NORR who also worked on Dalmunach and the recent expansions of Glenlivet and Glenfiddich.

Since Dallas Dhu closed in 1983, the distillery has been preserved as a museum by Historic Environmental Scotland. In 2018 the HES sent out an appeal for interested parties to help redevelop the site and ultimately turn it into a working distillery again. They received more than 70 sub-missions to that plea and have now made a shortlist of six proposals to see which one they will go forward with.

With Glen Wyvis having started production, there still re-mains a few projects north of Inverness to keep an eye on.

Quite possibly, Heather Nelson will be the first woman to found a Scotch whisky distillery. Co-owner of a film and TV production company, Nelson has studied at the Institute of Brewing and Distilling to gain the necessary qualifica-tion. Her planning application to build a distillery on the old World War II airbase at Fearn near Tain was submitted in March 2017 and was approved in just four weeks. The distillery will be equipped with two stills (1,000 and 600 litres respectively) and three washbacks with a capacity of producing 30,000 litres. The start of production has been delayed but the project is still ongoing.

A bit further north, just south of Brora, lies Dunrobin Castle which attracts 85,000 visitors each year. Here, Elizabeth Sunderland, a granddaughter of the former head of Clan Sutherland, and her husband Boban Costin will build a single estate distillery housed in an old powerhouse. Planning permission was granted in late 2016 and the owners hope to have the distillery, which will produce both gin and whisky, up and running in 2019.

Plans for a distillery on Barra has been an ongoing theme in this book for several years now. The latest development was a crowdfunding initiative to raise £2.5m that was launched in early 2019. It was also announced that former master distiller at Burn Stewart and Bladnoch, Ian McMil-lan, would be joining the team as a consultant.

Finally, there are plans for whisky distilleries on both North and South Uist in the outer Hebrides. Jonny Ingle-dew and Kate MacDonald opened a gin distillery on North Uist in spring 2019 and have plans to also open a designa-ted whisky distillery. On South Uist, plans for a commu-nity-run whisky distillery were revealed in 2018. The cost for the distillery is estimated at £10m and with a 300,000 litre capacity, they will also have their own malting floor using local peat to dry the barley.

Ireland & Northern Ireland

In 2013, there were four distilleries in Ireland and Northern Ireland producing whiskey. Today, five years la-ter, there are twenty-five – an amazing development. And, as you will see, there are more distilleries planned. Global Irish whiskey sales in the last five years have increased by more than 40%, albeit from low numbers, but there is no doubt that this category has a momentum right now.

Since last year's book, five more distilleries in Ireland have started producing whiskey: The Dublin Liberties and Roe & Co in Dublin, Killowen in County Down, Clona-kilty in County Cork and Lough Gill in County Sligo. All of them are covered in the section Distilleries around the globe.

But obviously there is more to report. A number of projects are just about to start production while others are still in the construction phase or seeking funds. What is interesting though is that in Malt Whisky Yearbook 2017, I could report on at least twenty new distillery projects while in this edition, the number has gone down to ten. Could it be that we have reached a time where the expansion of Irish whiskey distilleries that we have seen in later years has reached its peak? Time will tell but let's move on to the situation right now.

In County Mayo, Jude and Paul Davis together with Mark Quick have been working on the construction of their Nephin Distillery for quite some time now. A cooperage has already been opened and there is a store in the village selling Nephin merchandise. Unlike many other new distil-lers, Nephin will not source whiskey from other producers to sell under its own name, but prefers to wait until its own whiskey is ready to be bottled. The proposed capacity of the distillery is 500,000 litres and the use of local peat will add to the character of the whiskey.

While still on the west coast, we can report on another project. Sliabh Liag Distillery, in southwest Donegal, started as a gin distillery and the plan was to expand into whiskey production as well. That is exactly what James and Moira Doherty still intend to do but instead of distilling whiskey at the present distillery they will build a new one in the historic town of Ardara, 25 km north east of Sliabh Liag and also move the gin still there. At a cost of €6m the distillery will have an impressive capacity of 400,000 litres of pure alcohol and production could start sometime in 2020. The idea is to make both single malt and single pot still, some of them heavily peated as they would have been in the 19th century. Meanwhile, a sourced blended whiskey named Silkie was released by the owners a few years ago.

The island co-op at Cape Clear, six kilometers off the Cork coast, received planning permission in August 2016 to build a €7m distillery on the island. Unfortunately, one of their major investors pulled out along the way and the owners started a Kickstarter campaign in spring 2019 in order to fund parts of the project.

Gortinore Distillers, based in Waterford, launched their triple-distilled Natterjack Irish Whiskey in 2019. The whiskey had not been distilled by them but now the owner, Aidan Mehigan, along with two friends and his father, have acquired The Old Mill in Waterford with plans to turn that into a distillery.

Local farmer, Liam Ahearn, and his wife, Jennifer Nickerson, chose the Ahearn family farm between Clon-mel and Tipperary as the designated spot for Tipperary Boutique Distillery. They also included Jennifer's father in the business. Stuart Nickerson is well-known to lovers of Scotch after having held the position as Distillery Manager at Glenmorangie and is the mastermind behind the resurrection of Glenglassaugh. In December 2017, the owners managed to secure €5m in a partnership deal with Steelworks Investments. While sourced whiskey has already been released, the owners' planning application for a distillery at Dundrum House Hotel was approved by the local council in summer 2019. Construction of the distillery started in autumn that year and the project is expected to

Liam Ahearn, Jennifer Nickerson and William Ahearn from Tipperary Boutique Distillery

take about nine months to complete. In the Whisky Icons Awards 2019, Jennifer Nickerson became Irish Whiskey Brand Ambassador of the year.

In County Longford, west of Dublin, Peter Clancy in partnership with his brother and sister is planning for a distillery on the grounds of the old post office in Lanesborough. A gin still has already been installed while whiskey stills have been ordered from Italy. Apart from single malt and single pot still, Lough Ree Distillery will also be producing gin and vodka and the first product, Sling Shot Gin, has already been launched.

Neil Stewart is planning to convert a 200 year old mill in Boyle, Co. Roscommon into a whiskey distillery. The local council approved the planning application in autumn 2017 and around €5m will now be invested in the project. Apparently the celebrity actor Chris O´Dowd, a Boyle native and known from the series The IT Crowd and the movie Bridesmaids, is one of the investors.

Finally, in Northern Ireland, there are currently three ongoing projects.

Joe McGirr is the mastermind behind Boatyard Distillery in Enniskillen. The company received its planning permission in 2015 and in spring of the ensuing year, the first still was installed. The beginning of May 2016 marked the first distillation and since then the company has enjoyed some remarkable success with their gin and vodka. Since the

start, whiskey has also been part of the plan but distillation has not yet started.

Michael McKeown, founder of Matt D´Arcy & Company, has been granted a planning permission in summer 2018 for a whiskey distillery in Newry in county Down. Around £7m will be invested in the 100,000 litre distillery and a visitor centre and the plan is to be up and running in mid 2020. Before that sourced whiskey from other distilleries will be released.

The owner of Chateau de La Ligne vineyard in Bordeaux, Terry Cross, has received planning approval to build a distillery within the grounds of Killaney Lodge near Carryduff just south of Belfast. Hinch Distillery including a visitor centre will cost £6m to build and the owners hope to start production in late 2019. The distillery will be equipped with three pot stills (10,000 l, 5,500 l and 2,500 l). Sourced whiskey (no age, 5 and 10 year old) has already been launched to build the brand.

One distillery project in Northern Ireland which will not come to fruition is the one in Derry, where Niche Drinks were granted building approval in 2016. Plans for the large distillery (500,000 litres) were scrapped in November 2018. In April 2018, the American company Luxco acquired Niche Drinks and the idea was to produce a whiskey named The Quiet Man (which is already on the market but made from sourced whiskey) at the new distillery.

Bottling grapevine

In terms of new releases, this year's winner is without doubt Chivas Brothers. I have, throughout the years, pointed out that the company focuses on two brands – The Glenlivet and Aberlour – and have not done so much for the other brands.. A handful of distilleries have been blessed with one or two official bottlings while the rest have been left without any notice. A change was seen a few years ago when cask strength versions from several distilleries were made available for purchase at their distillery visitors' centres. Later, three distilleries were highlighted in the Ballantine's single malt collection – Glenburgie, Glentauchers and Miltonduff but that pales in comparison with what happened this year. In June 2019, Chivas Brothers released their largest ever single malt collection under the name The Secret Speyside Collection. A total of fifteen single malts from four different distilleries were launched. There were three each from Longmorn, Glen Keith and Braes of Glenlivet. These were accompanied by an even bigger surprise – no less than six new bottlings from the closed Caperdonich, three of them peated. All 15 whiskies were aged between 18 and 30 years. An impressive range of whiskies but it didn't stop at that. No less than 33 single casks, bottled at cask strength were also released in the Distillery Reserve Collection, available only at the visitor centres. The release covered no less than seven expressions of Scapa and, the biggest surprise of them all, a 4 year old Dalmunach – the first ever release from that distillery.

Even though Chivas Brothers seem to have stolen the thunder with 48 new single malts in one year, there were plenty of other exciting releases. Balblair completely revamped their range abandoning vintages by releasing four new bottlings carrying age statement. But there were more distilleries changing both design and content of their ranges. Tomatin presented new bottlings for their peated range, Cu Bocan, and three new expressions – Signature, Creation #1 and Creation #2. Arran also launched a new design for their core range as well as releasing the second expression in their Explorer's Series, Lochranza Castle.

In an attempt to show different sides of Balvenie single malt, a range of three bottlings called The Balvenie Stories was launched in May – The Sweet Toast of American Oak, The Week of Peat and A Day of Dark Barley. From their neighbour Glenfiddich came the Grand Cru 23 year old as well as two exclusives for Taiwan, Black Queen and Ice Breaker, both finished in Taiwanese wine casks. In their Private Edition range, Glenmorangie launched the exciting Allta where whisky maker Bill Lumsden had been using wild yeast from the fields surrounding the distillery. They also released Grand Vintage Malt 1991 and Cask 1784, the first of four new distillery exclusives.

Glenrothes were busy last year presenting two completely new ranges and tried to top that in 2019 by launching both a 40 and a 50 year old. From Highland Park came the final instalment in the Viking Legend series, Valfather, as well as Twisted Tattoo, a 16 year old partly matured in ex-Rioja casks, and Triskelion which was a collaboration between the distillery's three master whisky makers – Gordon Motion, John Ramsay and Max McFarlane.

The creators of Jura single malt have been especially energetic lately. Last year the core range was completely overhauled and in 2019, they released four new expres-

sions for duty-free as well as Jura Tide, Jura Time and Jura Two-One-Two. Kilchoman released an STR Cask Matured single malt matured in shaved, toasted and re-charred wine casks while Oban came out with Old Teddy, a distillery exclusive matured in ex-bodega sherry casks.

Diageo did not let their fans down when they launched yet another range of their Special Releases. This year it comprised a 29 year old Pittyvaich, a 30 year old Dalwhinnie, a 14 year old Cardhu, a 12 year old Cragganmore, a 15 year old Talisker, an 18 year old Singleton of Glen Ord, a 26 year old Mortlach and a 12 year old Lagavulin. Just like last year, there were no Port Ellen or Brora in the range but bottlings of them turned up during the year anyway. The oldest bottling ever from Brora, a 40 year old, appeared in August while a 39 year old Port Ellen became the inaugural release in a new range named Untold Stories.

In spring, Diageo teamed up with Game of Thrones just before the final season of the popular TV series, launching a special version of Johnnie Walker (White Walker) and single malts from no less than nine of their distilleries. But there were more bottlings released by Diageo during the year; a 41 year old Singleton of Glendullan, a 10 year old Lagavulin for duty-free, a 41 year old Talisker finished in Manzanilla casks and, finally, the oldest official bottling ever from Mortlach – a 47 year old, the first in a new series called The Singing Stills.

After two years closure for refurbishing, Tobermory opened up and at the same time released a new core expression, a 12 year old. Its sister distillery on Islay, Bunnahabhain, excelled with three finishes – 2007 Port Pipe, 2007 French Brandy and 1988 Marsala and Deanston, the third distillery in the group, joined the theme with no less than five whiskies with an extra maturation – 1997 palo cortado, 2002 organic oloroso, 2006 cream sherry, 2006 fino sherry and, for the US market, a 2012 beer finish.

Glen Scotia launched 2003 rum finish and at the same time announced the release of a 45 year old later in the year while stablemate Loch Lomond came out with a 50 year old. As usual Dalmore had an old expression up their sleeve, this time a 60 year old and, while still on whiskies of substantial age, Tullibardine launched a Vintage 1964 and Craigellachie surprised the market with a 51 year old. The surprise factor for the Craigellachie release increased when it was announced that the whisky was not for sale but would be offered to whisky enthusiasts for free! Those who registered took part in a raffle and the whisky was served in pop-up bars around the world.

Four new versions of Octomore were released by Bruichladdich as well as new editions of Islay Barley, Bere Barley and Organic. Eyebrows were raised at Scotch Whisky Association last year when Glen Moray released a whisky with a cider cask finish. "Not allowed according to the regulations" was their verdict (not even with the new rules implemented from 2019). This year, Glen Moray played by the book and launched a rhum agricole finish.

The new owners of GlenAllachie acted quickly when they took over the distillery a couple of years ago and presented both single casks and a core range within months. In 2019 it was time for a new 15 year old as well as an 8 year old Koval finish, a 10 year old port finish and a 12 year old PX sherry cask finish. GlenDronach launched PortWood with an extra maturation in ex-port pipes, the 16 year old Boynsmill exclusive for duty-free and batch 17 of their single casks while BenRiach presented batch 16 of their single casks, this time with no less than 24 different expressions.

Benromach released two new vintagew (1972 and 1977) in their Heritage Collection, a special version of their Peat Smoke and the new Cask Strength Batch 1 Vintage 2008 which replaced the 100º Proof. In 2016, Bowmore started a new series, highlighting the influence from their famous Vault No. 1 and later, in 2018, the second edition appeared – Peat Smoke. They also launched a 21 year old for duty free and the 36 year old Dragon Edition which was the first of four bottlings exclusive to the Chinese market.

Alan Winchester, The Glenlivet's Master Distiller, has busied himself with puzzling whisky drinkers in recent years by different "mystery bottlings". With minimal information on the label, customers are invited to reflect for themselves about the content and the maturation. This year's release was Enigma, exclusive to the US market, and the consumers were invited to solve a digital crossword puzzle. Details of the whisky will be revealed in late 2019.

Macallan introduced a new duty-free range in late 2018 when they launched Concept No. 1 which had matured first in ex-sherry casks and then for an equal amount of time in ex-bourbon. We also got a new core bottling – The Macallan Estate from the same distillery. The sherry-matured whisky was bottled at 43% and it had been made using barley from the distillery's own estate, Easter Elchies. Tamdhu made their debut in the travel retail exclusives category with Ámbar 14 year old and Gran Reserva First Edition and Tamnavulin released a Sherry Cask Edition of their core bottling as well as a Tempranillo finish for duty-free.

This year's edition of Longrow Red from Springbank had been finished in Pinot Noir casks and the owners also released Hazelburn 14 year old matured in Oloroso casks. Highlighting the history and tradition of Glengoyne

whisky-making, Ian Macleod introduced a new, limited range in 2019 called The Legacy. Chapter One. This inaugural bottling honours Cochrane Cartwright, distillery manager in the 1860s, who introduced the slow distillation and the use of sherry casks that Glengoyne has since been famous for.

Glengyle introduced a heavily peated version of their Kilkerran single malt while Speyside distillery released cask strength versions of their three core bottlings – Tenné, Trutina and Fumare. Three new Aultmore single malts, exclusive to Heathrow Airport, were launched in spring. All three were 22 years old having spent the final eleven years in casks that had held Super Tuscan wine, Châteauneuf-du-Pape and Moscatel wine respectively.

Ardbeg released their first permanent age-stated whisky since 2000 when Ardbeg Ten was introduced. The recipe will change slightly with every yearly batch but the first 19 year old Traigh Bhan, bottled at 46.2%, had been matured in a combination of American Oak and ex-oloroso sherry casks. From the people of Ardbeg there was also the release of this year's Ardbeg Day bottling, Drum, matured in ex-bourbon barrels and then finished in ex-rum casks. As usual, a Committee version was launched in March, bottled at 52% while the general release followed in early June, bottled at 46%.

Let's conclude with bottlings from some of the newer distilleries. Kingsbarns made a sneak debut in 2018 but in early 2019, their first generally available flagship malt, Dream To Dram, was launched. Daftmill continued with more releases including six single casks, all distilled in 2006. Abhainn Dearg released their first 10 year old in late 2018 and Wolfburn released the 4 year old Batch 375 in spring 2019. Glasgow Distillery launched their 1770 peated version and Arbikie released the first rye whisky made in Scotland for more than 100 years.

Laphroaig Ian Hunter 30 years old, Benromach Cask Strength Vintage 2008, Peated Caperdonich 25 years old, The Glenlivet Enigma, The Balvenie A Day of Dark Barley 26 years old, Braes of Glenlivet 30 years old

Independent
bottlers

The independent bottlers play an important role
in the whisky business. With their innovative bottlings, they increase
diversity. Single malts from distilleries where the owners' themselves
decide not to bottle also get a chance through the independents.
The following are a selection of the major companies.
Tasting notes have been prepared by Ingvar Ronde.

Gordon & MacPhail

www.gordonandmacphail.com

Established in 1895 the company, which is owned by the Urquhart
family, still occupies the same premises in Elgin. Apart from being
an independent bottler, there is also a legendary store in Elgin and,
since 1993, an own distillery, Benromach. In 2018, the company
announced that not only would they establish a designated gin distillery on the Benromach site but they were also to build a new malt
whisky distillery at Craggan, close to Grantown-on-Spey. Gordon &
MacPhail has an incredible variety of casks in their warehouses in
Elgin and in 2018, they revamped their portfolio of bottlings. Going
forward, there will be five distinctive ranges; Connoisseurs Choice,
a series well-known to most whisky aficionados, has received a
new look and consists of single malts bottled either at 43% or 46%.
Discovery, a new range unveiled in May 2018, is grouped under
three flavour profiles - smoky, sherry and bourbon. Distillery Labels
is a relic from a time when Gordon & MacPhail released more or
less official bottlings for several producers. Currently 10 distilleries
are represented in the range and the whisky is bottled at either 40 or
43%. Private Collection, a new range, will feature old single malts
including bottlings from closed distilleries. Generations, finally,
was first introduced in 2010. This range comprises the oldest and
rarest whiskies in stock, including previous releases such as Mortlach and Glenlivet 70 year old and, in 2015, the oldest single malt
ever bottled - a 75 year old Mortlach.

Recent releases from Gordon & MacPhail include Longmorn
Twins, a pair of 57 year old single casks matured in European oak
and American oak respectively and a 70 year old Glen Grant distilled in 1948, the oldest from the distillery ever bottled. Some of the
releases in the Private Collection range in summer 2019 were also
spectacular – a 50 year old Dallas Dhu, a 38 year old St. Magdalene
and a 53 year old Longmorn matured in a first fill sherry butt!

Highland Park 2001 17 year old, 57,7%
Nose: Eucalyptus, canned tomatoes,
peaches, apricots, ginger and
candy floss.
Palate: Distinctly herbal with a subtle
smokiness, coriander, sage,
menthol and a pleasant fruitiness.

Caol Ila 2005 14 year old, 45%
Nose: Burnt grass, liquorice, lemon,
muscovado sugar, menthol,
vanilla and chocolate.
Palate: Rich with a peppery start followed by notes of eucalyptus,
fennel, oranges and cocoa.

Berry Bros. & Rudd

www.bbr.com

Britain's oldest wine and spirit merchant, founded in 1698 opened
a new flagship shop in London in 2017. The famous address 3 St
James's Street, where the company has been since the start, was
returned to its appearance of 30 years ago and is now a space for
consultations, meetings and events. The new, and much larger
store, is just around the corner in 63 Pall Mall. A driving force
behind the new shop was CEO Dan Jago who joined the company
in 2015. He left BBR in 2019 and was succeeded by Lizzy Rudd,
a third-generation member of the Rudd family. Berry Brothers had
been offering their customers private bottlings of malt whisky for
years, but it was not until 2002 that they launched Berry's Own
Selection of single malt whiskies. Under the supervision of Spirits

Manager, Doug McIvor, some 30 expressions are on offer every year. Bottling is usually at 46% but expressions bottled at cask strength are also available. The super premium blended malt, Blue Hanger, is also included in the range. In autumn 2014 the Exceptional Casks Collection was launched, comprising of old and rare whiskies and rum. A new series called The Classic Range was released in spring 2018. It´s made up of four bottlings of blended malt; Speyside, Islay, Sherry Cask Matured and Peated Cask Matured. In spring 2019, The Perspective Series was launched including four blended Scotch (21, 25, 35 and 40 years old). BBR sold Cutty Sark blended Scotch to Edrington and obtained The Glenrothes single malt in exchange but in 2017, BBR sold back The Glenrothes to Edrington.

Islay Blended Malt, 44.2%
Nose: Punchy smoke, charcoal, sea-shells and seaweed.
Palate: Smoky vanilla followed by honey, cereals, yellow plums and a hint of pepper.

The Perspective Series No. 1 35 year old, 43%
Nose: Seductive with notes of dried figs, chocolate covered orange peel, menthol, pine apple and black treacle.
Palate: Rich and velvety, dark chocolate, honey, tobacco, dark fruits and berries and cinnamon.

Signatory

Founded in 1988 by Andrew and Brian Symington, Signatory Vintage Scotch Whisky lists at least 50 single malts at any one occasion. The most widely distributed range is Cask Strength Collection which sometimes contains spectacular bottlings from distilleries which have long since disappeared. One good example is a very rare Glencraig 1976, 38 years old. Another range is The Unchill Filtered Collection bottled at 46%. Some of the latest bottlings released include a Tormore 30 year old, a Bruichladdich 28 year old and a Clynelish 23 year old. Finally there is also the Single Grain Collection with a Port Dundas 22 year old and a Cambus 26 year old as some of the latest releases. Andrew Symington bought Edradour Distillery from Pernod Ricard in 2002 and the entire operations, including Signatory, are now concentrated to the distillery in Perthshire.

Ian Macleod Distillers

www.ianmacleod.com

The company was founded in 1933 and is one of the largest independent family-owned companies within the spirits industry. Gin, rum, vodka and liqueurs, apart from whisky, are found within the range and they also own Glengoyne and Tamdhu distilleries. In autumn 2017 they also revealed their plans to resurrect Rosebank Distillery in Falkirk which has been closed since 1993. Their single malt range includes The Chieftain´s, which cover a range of whiskies from 10 to 50 years old while Dun Bheagan is divided into two series – Regional Malts, 8 year old single malts expressing the character from 4 whisky regions in Scotland and Rare Vintage Single Malts, a selection of single cask bottlings from various distilleries. There are two As We Get It single malt expressions – Highland and Islay.

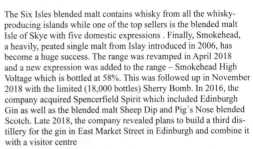

The Six Isles blended malt contains whisky from all the whisky-producing islands while one of the top sellers is the blended malt Isle of Skye with five domestic expressions . Finally, Smokehead, a heavily peated single malt from Islay introduced in 2006, has become a huge success. The range was revamped in April 2018 and a new expression was added to the range – Smokehead High Voltage which is bottled at 58%. This was followed up in November 2018 with the limited (18,000 bottles) Sherry Bomb. In 2016, the company acquired Spencerfield Spirit which included Edinburgh Gin as well as the blended malt Sheep Dip and Pig´s Nose blended Scotch. Late 2018, the company revealed plans to build a third distillery for the gin in East Market Street in Edinburgh and combine it with a visitor centre

Smokehead Sherry Bomb, 48%
Nose: Lovely smoky notes of bonfire and barbecue, raisins, baked apples and dark chocolate.
Palate: A combination of sweet sherry and dry, tannic smoke, smoked salmon, cardamom, dark cherries and treacle.

Pig´s Nose, 40%
Nose: Slightly vegetal fruit notes hit first followed by dusty fruit spice and dried cigar leaves with a flicker of iodine.
Palate: Dried apricots with a warming spice roll onto the palate with chunky fruit cake notes.

Blackadder International

www.blackadder.se

Blackadder is owned by Robin Tucek, one of the authors of the classic whisky book, The Malt Whisky File. Apart from the Blackadder and Blackadder Raw Cask (bottled straight from the cask without any filtration at all), there are also a number of other ranges – Smoking Islay, Peat Reek, Aberdeen Distillers, Clydesdale Original and Caledonian Connections. The company has also been known for bottling unusual expressions of Amrut single malt. All bottlings are single cask, uncoloured and un chill-filtered. Most of the bottlings are diluted to 43-46% but Raw Cask is always bottled at cask strength.

Duncan Taylor

www.duncantaylor.com

Duncan Taylor was founded in Glasgow in 1938 as a cask broker and trading company. In 2001, the company was acquired by Euan Shand in 2001 and operations were moved to Huntly. Duncan Taylor´s flagship brand is the blended Scotch Black Bull, a brand with a history going back to 1864. Black Bull was rebranded in 2009 by Duncan Taylor and the range consists of three core releases – Kyloe, a 12 year old and a 21 year old. There are also limited versions such as 30 year old, 40 year old, Special Reserve, 10 year old rum finish and 10 year old Retro Sherry Cask. The Black Bull brand is complimented by Smokin' which is a blend of peated Speyside, Islay and grain whisky from the Lowlands.

The portfolio also includes The Rarest (single cask, cask strength whiskies of great age from demolished distilleries), Dimensions (a collection of single malts and single grains aged up to 39 years), The Octave (single malt whiskies matured for a further period in small, 50 litre ex-sherry octave casks), The Tantalus (a selection of whiskies all aged in their 40s), The Duncan Taylor Single Range (whiskies aged 30 years or more from closed distilleries), Battlehill (a range of single malts and single grains) and Rare Auld Grain (a selection of rare grain whiskies bottled at cask strength). The blended malt category is represented by Big Smoke, a young peated whisky available in three strengths, 40%, 50% and 60%. Finally, there is also the Duncan Taylor Blends in three versions – Five Stars, 12 year old and 18 year old.

Black Bull 21 year old, 50%

Nose: Pine needles, vanilla, oatmeal crackers, fennel, chocolate, citrus and nougat.

Palate: Delicious notes of dried fruits, chocolate cake, dry smoke, cherry trifle, tiramisu, clove and cinnamon.

Bunnahabhain Octaves 4 year old, 53,8%

Nose: Smoky with notes of ashes, green grass, orange peel and red apples.

Palate: A sweet vanilla note goes well with the dry peat, ginger, cherry preserve, baked carrots and jasmine tea.

Scotch Malt Whisky Society

www.smws.com

The Scotch Malt Whisky Society, founded in 1983 and owned by Glenmorangie Co since 2003, has more than 30,000 members worldwide and apart from UK, there is a network of international branches and partner bars in 19 countries around the world. In 2015, Glenmorangie sold the SMWS to the HotHouse Club and a group of the managers. The idea from the very beginning was to buy casks of single malts from the producers and bottle them at cask strength without colouring or chill filtration. The Society has played a significant role for the interest in single cask Scotch that has exploded in recent decades. The labels do not reveal the name of the distillery. Instead there is a number but also a short description which will give you a clue to which distillery it is. Around 500 casks are bottled every year. The SMWS also arranges tastings at their different venues but also at other locations. In recent years, the range has been expanded to also include single grain, whiskies from other countries as well as rum, gin, cognac and other spirits. Celebrating the society´s 35th anniversary, a pop-up venue was opened in Battersea, London in autumn 2018 (closing July 2019).

Murray McDavid

www.murray-mcdavid.com

The company was founded in 1996 by Mark Reynier, Simon Coughlin and Gordon Wright and in 2000, they also acquired Bruichladdich distillery. In 2013 Murray McDavid was taken over by Aceo Ltd. and a year later they signed a lease for the warehouses at the closed Coleburn distillery for storing their own whiskies as well as stock belonging to clients. The bottlings are divided into six different ranges; Mission Gold (exceptionally rare whiskies bottled at cask strength), Benchmark (mature single malts bottled at 46%), Mystery Malt (single malts where the distillery is not revealed), Select Grain (single grains), The Vatting (vatted malts) and Crafted Blend (blended Scotch from their own blending). The vast majority of the releases are single casks.

Compass Box Whisky Co

www.compassboxwhisky.com

John Glaser, founder and co-owner of the company, has a philosophy which is strongly influenced by meticulous selection of oak for the casks, clearly inspired by his time in the wine business. But he also has a lust for experimenting to test the limits, which was clearly shown when Spice Tree, matured in casks containing extra staves, was launched in 2005. Glaser and Compass Box are also advocating more transparency in the industry where the customer is given as much information as possible about the contents of the bottle. The company divides its ranges into a Signature Range and a Limited Range. Spice Tree (a blended malt), The Peat Monster

(a combination of peated islay whiskies and Highland malts), Oak Cross (American oak casks fitted with heads of French oak) and Hedonism (a vatted grain whisky) are included in the former. A new addition was made in autumn 2018 when The Story of the Spaniard was released. A blend of Highland and Speyside whiskies, it had been partially matured in Spanish red wine casks. At the same time, Asyla was removed from the range.

In the Limited range, whiskies are regularly replaced and at times only to resurface a couple of years later in new variations. Recent limited releases include Juveniles, a collaboration between Compass Box and Paris wine bar Juveniles Bistrot à Vins and Stranger & Stranger which technically isn´t a whisky. While 99% of the content come from three Speyside malt distilleries, the remaining one percent is an 18 month old grain whisky from Girvan. In 2019, the second edition of the peated No Name was released as well as The Circle. The latter was the result of a bartender programme initiated by Compass Box in 2018 with the aim of "connecting the creative processes of the blending room with those of the bar trade". In spring 2019Affinity, the world´s first Scotch and Calvados blend, was released as well as Tobias & the Angel, a blend of old (well over 20 years) Clynelish and Caol Ila. Finally, in autumn 2019, The Myths & Legends Limited Edition series was launched where two of the three releases were single malts.There are also two blended Scotch, Artist´s Blend and Glasgow Blend, with a 50% proportion of malt whisky. In autumn 2014, Compass Box made a long-term agreement with John Dewar & Sons where the Bacardi-owned company would supply Compass Box with stocks of whisky for future bottlings. In spring 2015 it was further announced that Bacardi had acquired a minority share of the independent bottler.

The Circle No. 1, 46%

Nose: Fresh and vibrant with notes of grapefruit, heather, candy, ginger and green apples.

Palate: Inviting notes of sweet barley, citrus, honey, vanilla ice cream and pears. Creamy texture and slightly peppery.

The Story of the Spaniard, 43%

Nose: A combination of leather, coffee beans, mustard seeds, dark chocolate, figs, sage, salsiccia and oranges.

Palate: Impressive and rich, dark berries, chocolate, muscovado sugar, sabayone and cinnamon.

North Star Spirits

www.northstarspirits.com

Founded in 2016, by Iain Croucher who was sales manager and brand ambassador for AD Rattray before deciding to go it alone. Apart from a number of single malt releases, Croucher also introduced the blended malt brand Vega in 2017. Two of the latest releases were a 28 year old (1990) and a 33 year old (1985). In addition there is a blended Scotch called Spica where a 29 year old was the latest release. A new range called The Millenial Range was recently introduced. Four un-named single malts (Speyside, Highland, Islay and Island), bottled at 50% were released. North Star is not focused on just whisky from Scotland. They have bottled a 13 year old Tennessee Whiskey and there will be a small batch of Glasgow gin in the not too distant future.

Master of Malt

www.masterofmalt.com

Master of Malt is one of the biggest and most innovative whisky retailers in the UK. The company also has its own range of single-cask bottlings, from big names like The Macallan, to world whiskies like the Paul John 6 year old, and even more unusual expressions, for example, the Croftengea 11 year old 2007 from the Loch Lomond distillery. All whiskies in the range are bottled at both natural cask strength and colour. Recent additions include Springbank 25 year old 1993, Littlemill 27 year old 1991 and Glenrothes 28 year old 1989. In addition, Master of Malt offers a Blend Your Own option, and stocks thousands of Drinks by the Dram 30ml sample-size bottles. Customers can entirely personalise contents of Drinks by the Dram tasting sets with these. The retailer also offers the Dram Club monthly whisky subscription service.

Springbank 1993 25 years old, 51%

Nose: Robust, fried mustard seeds, old wine cellar, ripe grapes, cured meat, fennel and roasted cashew nuts.

Palate: A peppery start leads to lemon sherbet, coffee, salted caramel, nutmeg, olive tapenade and a damp smokiness.

Atom Brands

Part of the Atom Group which includes online retailer Master of Malt and the UK Distributor Maverick Drinks, Atom Brands is home of their home grown brands as well as their independently bottled whiskies, rums gins and other spirits from around the world. The brands include Drinks by the Dram with 30ml samples, and the spirit-filled Advent Calendars, as well as Ableforth's Bathtub Gin, That Boutique-y Drinks Company, The Blended Whisky Company, The Handmade Cocktail Company, 1897 Quinine Gin, Darkness, Bitter Bastards, Origin, The 'Hot Enough' Vodka Company and Mr Lyan.

That Boutique-y Whisky Company
www.thatboutiqueywhiskycompany.com

Established in 2012, this independent bottler is best characterised by its graphic-novel-style labels, and uncompromising approach to flavour and quality. Having worked with over 160 distilleries to date, That Boutique-y Whisky Company's range is a global representation of the whisky category. With new Head of Whisky, Sam Simmons at the helm, the range has come to include some of the most renowned names in Scotch production, as well as whisky from smaller, independent distilleries across the globe, including whisky produced in South Africa, India and Switzerland, and covering blends, blended malts, single grains, bourbons, and more recently young ryes.

Bourbon Whiskey No. 1 24 year old, 48%

Nose: Intriguing notes of maple syrup, lilac and roses, marsipan, chocolate, roasted almonds, coffee beans and coriander.

Palate: Candied oranges, chocolate covered cashew nuts, demerara sugar, caraway, menthol and some bitter oak. Long finish – classic old bourbon!

The Character of Islay Whisky Company
www.characterofislay.com

The Character of Islay Whisky Company offers a range of distinctive whiskies with Islay at their heart. A modern, yet romanticised approach to the Islay whisky category, The Character of Islay

Company uses fabled characters to create an imaginary storyline that represent the expressions of Islay whisky, and the regional characteristics of Islay itself. The brand's first expression, Aerolite Lyndsay is a 10 year old single malt from an undisclosed distillery on Islay. The striking label gives a nod to bottles of yesteryear, tying in to the long history of distilling on the island.

Aerolite Lyndsay 10 year old, 46%

Nose: Lovely, floral peat with notes of sweet barley, apples, citrus and a hint of smoked chili.

Palate: Floral turns grassy and heathery at first, dry ashes, chili pepper, pears with chocolate cream and lemon curd.

Darkness!
www.darknesswhisky.com

The first whiskies from Darkness! were released in spring 2014 and the key words for these expressions are dark and heavily sherried. To create the character, single malts are filled into specially commissioned 50 litre first fill Sherry casks that they are finished for more than 3 months. Pedro Ximénez, Oloroso, Palo Cortado, Fino and Moscatel Sherry casks have all been used (specified on each bottling) as well as hybrid PX and Oloroso casks made up with staves from each.

Darkness 8 year old, 47,8%

Nose: Starts off autumnal with notes of dry leaves, then comes figs and dates, fudge and vanilla ice cream.

Palate: The dried fruit comes forward together with dark chocolate, nutmeg, cinnamon, a bit of pepper and a hint of smoke. Rounded and well balanced.

The Whisky Agency
www.whisky-agency.de

The man behind this company is Carsten Ehrlich, to many whisky aficionados known as one of the founders of the annual Whisky Fair in Limburg, Germany. His experience from sourcing casks for limited Whisky Fair bottlings led him to start as an independent bottler under the name The Whisky Agency, a business celebrating its 10th anniversary in 2018. There are several ranges including The Whisky Agency, The Perfect Dram and Specials with some unusual bottlings.

A Dewar Rattray Ltd
www.adrattray.com

The company was founded by Andrew Dewar Rattray in 1868. In 2004 the company was revived by Tim Morrison, previously of Morrison Bowmore Distillers and fourth generation descendent of Andrew Dewar, with a view to bottling single cask malts from different regions in Scotland. One of its best-sellers is a single malt named Stronachie which is actually sourced from Benrinnes. There are currently two expressions, a 10 year old and a 10 year old sherry finish. A peated, blended malt, Cask Islay, became available in 2011 and then again in 2013 but this time as a single malt. It was complemented in 2018 by Cask Orkney 18 year old and in 2019 by Cask Speyside 10 year old. The AD Rattray´s Cask Collection is a range of single cask whiskies bottled at cask strength and without colouring or chill-filtration. An exciting new experiment, The Octave Project, was launched in 2018. It involved taking a cask from a single distillery and divide the content into four smaller casks, so called octaves. The octaves had been made from staves taken from casks that had previously contained PX sherry, Oloroso sherry, Rioja or Rum. The original whisky was then left for a period of time in the new, smaller casks. In 2011, the company opened A Dewar Rattray´s Whisky Experience & Shop in Kirkoswald, South Ayrshire. Apart from having a large choice of whiskies for sale, there is a sample room, as well as a cask room. All the products in the shop, including personalised own label single cask bottlings, are also available on-line from www.adrattray.com.

Cask Speyside 10 year old, 46%
Nose: Lemon curd, apples and pears, vanilla fudge, honey and rose petals.
Palate: Rounded and balanced with more of the citrus notes as well as honey, pastry, cooked pears, caramel and a hint of pepper.

Cask Islay Sherry Edition, 59,9%
Nose: Sweet, peaty notes, smoked ham, chorizo, heather and sweet sobranie tobacco.
Palate: Profoundly smoky, malted barley, muscovado sugar, dried plums, black pepper, roasted hazelnuts, espresso and a long finish.

Douglas Laing & Co

www.douglaslaing.com

Established in 1948 by Douglas Laing, this firm was run for many years by his two sons, Fred and Stewart. In 2013, the brothers decided to go their separate ways. Douglas Laing & Co is now run by Fred Laing and his daughter, Cara. Douglas Laing has the following brands in their portfolio; Provenance (single casks bottled at 46%), Director´s Cut (old and rare single malts), Premier Barrel (single malts in ceramic decanters), Clan Denny (mainly old single grains), Double Barrel (two malts vatted together), and Old Particular, a range of single malts and grains. The latter has also been expanded with two brand extensions; XOP and XOP "The Black Series". Yet another new range was launched in spring 2019 - Elements Single Cask Collection..

Six years ago the company started a range that has become highly succesful. The first installment in th series that eventually was given the name Remarkable Region Malts, was Scallywag - a blended malt influenced by sherried whiskies from Speyside. More versions have followed with Scallywag 10 yer old, Scallywag Chocolate Edition and Scallywag Red-Nosed Reindeer being the latest. The range has been expanded over the years and now includes Timorous Beastie from the Highlands where a 10 year old in 2018 followed previous expressions. Rock Oyster (the name was changed in spring 2019 to Rock Island) is a blended malt combining whiskies from Islay, Arran, Orkney and Jura and can be found without age statement or as an 18 year old. The Epicurean represents the Lowlands while The Gauldrons is made from Campbeltown malts. The final regio-

nal whisky is Big Peat, a vatting of Islay malts. This was launched several years ago but was later included in the range. A limited 26 year old Big Peat was released end of 2018 together witbh a 10 year old. In the Remarkable Regional Malts is also includeed a limited 10 year old blended malt from 2018 - With A Twist. In July 2017, it was announced that Douglas Laing would also become distillers. Their chosen site in Glasgow is on the banks of the river Clyde just opposite the new Clydeside Distillery and the goal is to have the distillery operational by the end of 2019 or beginning of 2020.

Big Peat 26 year old, 51,5%
Nose: Barbecue smokiness, grape juice, tropical fruits and seaweed.
Palate: Bursting with tropical fruits, a mellow, delicious peatiness, liquorice, cinnamon, nutmeg and burnt toast with orange marmalade.

Timorous Beastie 10 years old, 46,8%
Nose: Dried nettles, herbs, lemon zest, malty crackers, fudge and almonds.
Palate: Fresh and vibrant with notes of citrus, ginger, goose berries, pepper, ice tea and lemon balm.

Malts of Scotland

www.malts-of-scotland.com

Thomas Ewers from Germany, bought casks from Scottish distilleries and decided in the spring of 2009 to start releasing them as single casks bottled at cask strength and with no colouring or chill filtration. The backbone of the assortment is the Basic Line with three blended malts; Classic (18yo), Sherry (15yo) and Peat (10yo). Apart from other ranges of Scotch single malts, Ewers has also added a range called Malts of Ireland. At the moment he has released more than 100 bottlings and apart from a large number of single casks, there are two special series, Amazing Casks and Angel´s Choice, both dedicated to very special and superior casks. Ewers was inducted as a Keeper of the Quaich in 2016.

Hunter Laing & Co

www.hunterlaing.com

This company was formed after the demerger between Fred and Stewart Laing in 2013 (see Douglas Laing). It is run by Stewart Laing and his two sons, Scott and Andrew. The relatively new company Edition Spirits, founded by Andrew has also been absorbed into Hunter Laing with the range of single malts called First Editions. From the demerger, the following ranges and brands ended up in the Hunter Laing portfolio; The Old Malt Cask (rare and old malts, bottled at 50%), The Old and Rare Selection (an exclusive range of old malts offered at cask strength) and The Sovereign (a range of old and rare grain whiskies). Another range with the name Hepburn´s Choice was launched in spring 2014. These single malts are younger than The Old Malt Cask expressions and bottled at 46%. In June 2019, Scarabus, a single malt from an undisclosed Islay distillery, was released as the first in a new range.

In January 2016, the company announced their intentions of building a distillery on Islay on the northeast coast near Bunnahabhain. Ardnahoe Distillery came on stream in November 2018 and was opened to the public in spring 2019. The distillery has a capacity of 1 million litres and Jim McEwan, who retired from Bruichladdich in 2015, joined the team as production director. McEwan was also responsible for selecting six rare Islay whiskies that were released in May under the label Feis Ile 2017 Kinship. A second range of Kinship bottlings was released in 2018 and in 2019 the range included single malts such as Caol Ila 40 year old, Ardbeg 26 year old and Bowmore 30 year old

Islay Journey, 46%

Nose: Green notes of vegetables and dry grass, citrus, heather, malted barley and ash.

Palate: Starts off on a sweet and fruity note followed by seaweed peat, pears, vanilla and roasted nuts.

Scarabus, 46%

Nose: Intense with notes of bacon, ashes, tobacco leaves, orange zest and herbs.

Palate: Sweet smoke and pepper followed by roasted almonds, burnt sugar, tobacco, clove and vanilla.

Wemyss Malts

www.wemyssmalts.com

Founded in 2005, the family-owned independent bottler opened up their own whisky distillery at Kingsbarns in Fife in 2014. The company is mainly known for its range of blended malts of which there are three core expressions – The Hive, Spice King and Peat Chimney. These are available at 46% un chill-filtered and also in limited edition batch strength, typically around 55-58%. In 2017, the family bottled a new part of their blended malt range called The Family Collection consisting of spirit sourced and fully matured by the family. There were two releases, Vanilla Burst and Treacle Chest and they were followed up in 2019 by Blooming Gorse, comprised of two Northern Highland malts matured in bourbon barrels and hogsheads and Flaming Feast which is a blend of one Eastern Highland and one Island single malt. Another blended malt is Nectar Grove, released in 2018 and finished in ex-Madeira casks.

Another side of the business involves single malts. There are two ranges; one of which consists of single casks bottled at 46% or the occasional cask strength. The names of the whiskies reflect what they taste like although for some time now, the distillery name is also printed on the label. All whiskies are un chill-filtered and without colouring and every year 60-70 bottlings are released.

In February 2019, Wemyss launched a brand new website which includes an online club for whisky fans to join called the Wemyss Malts Cask Club. The first two bottlings offered to members were a 35 year old Caol Ila (Smoky Nectar) and a 31 year old, sherried Bunnahabhain.

Blooming Gorse, 46%

Nose: Herbal with notes of pears, heather, vanilla, fudge, thyme and citrus.

Palate: Rich, roasted nuts, vanilla ice cream, apricots, ginger, lemon zest and a hint of pepper.

Peat Chimney batch strength 002, 57%

Nose: Smoky and ashy notes, cigars, smoked clams, roasted pine nuts and burnt toast.

Palate: Dry smoke with sea salt, cured meat, apples, honey, nuts and lemon zest.

Single Cask Nation

www.singlecasknation.com

In 2011, Jason Johnstone-Yellin and Joshua Hatton, two well-known whisky bloggers, started, in alliance with Seth Klaskin, a new career as independent bottlers. The initial idea with Single Cask Nation somewhat reminds you of Scotch Malt Whisky Society in the sense that you have to become a member of the nation in order to buy the bottlings. In 2017, the owners decided to develop an alternative way of selling their products and launched a special range of whiskies that could also be found at retailers in a number

of states in the USA. This new way of doing business proved succesful and from 2019 and onwards, their products will eventually be available in Europe and Canada as well. The 5th release of what they themselves call Single Cask Nation Retail Release in July 2019 included Port Dundas 42 year old, Cameronbridge 26 year old, Clynelish 23 year old and Imperial 23 year old. In order to focus on the new task the owners have stopped arranging (at least for the time being) their popular events, Whisky Jewbilee, in New York, Chicago and Seattle.

Meadowside Blending

www.meadowsideblending.com

The company may be a newcomer to the family of independent bottlers but the founder certainly isn´t. Donald Hart, a Keeper of the Quaich and co-founder of the well-known bottler Hart Brothers, runs the Glasgow company together with his son, Andrew. There are four sides to the business – blends sold under the name The Royal Thistle, single malts labelled The Maltman, single cask single grains under the label The Grainman and Excalibur with deluxe blended whiskies. A fifth range was recently introduced - Vital Spark focusing on single malts with "a maritime twist".

Elixir Distillers

www.elixirdistillers.com

The company is owned by Sukhinder Singh, known by most for his two very well-stocked shops in London, The Whisky Exchange. In the beginning of October every year, he is hosting The Whisky Show in London, one of the best whisky festivals in the world and for the last two years he has also been involved in the Old & Rare Show in Glagow. In 2005 he started as an independent bottler of malt whiskies operating under the brand name The Single Malts of Scotland. There are around 50 bottlings on offer at any time, either as single casks or as batches bottled at cask strength or at 46%. In 2009 a new range of Islay single malts under the name Port Askaig was introduced. The current core range consists of 100° Proof, 16, 19, 30 and 45 year old. Recent limited releases include Port Askaig 10th Anniversary, a 28 year old and a 25 year old which is an exclusive to the USA exclusive. Elements of Islay, a series of cask strength single malts in which all Islay distilleries are, or will be, represented was introduced a few years before Port Askaig. The list of the product range is cleverly constructed with periodical tables in mind in which each distillery has a two-letter acronym followed by a batch number. The most recent bottlings, released in spring 2019, are Cl_{12}, Pl_6, Ma_3 (peated Bunnahabhain) and Lp_{10}. In 2019, a new range of single casks, single malts, blends or single grain, was introduced. The first release of Whisky Trail consists of 8 bottles with ages ranging from 9 to 44 years. Finally, the company also has a range called Director´s Special which showcases exceptionally old and rare single malts. Sukhinder Singh also has plans to build a whisky distillery named Farkin on Islay on the outskirts of Port Ellen.

Port Askaig 10 year old, 55,85%

Nose: Delicious wood smoke paired with oranges, heather, bananas, honey and ginger.

Palate: Rich with a heathery smokiness and an oily, salty note, orange zest, spices and coffee.

Elements of Islay Cl_{12}, 57,5%

Nose: Briny with notes of almonds, vanilla, crème brûlée, dried grass and cocoa powder.

Palate: Dry and earthy with notes of Turkish delight, roasted almonds, oatmeal crackers and sweet notes of grilled lemons.

The Ultimate Whisky Company

www.ultimatewhisky.com

Founded in 1994 by Han van Wees and his son Maurice, this Dutch independent bottler has until now bottled close to 1,000 single malts. All whiskies are un chill-filtered, without colouring and bottled at either 46% or cask strength. The van Wees family also operate one of the finest spirits shops in Europe - Van Wees Whisky World in Amersfoort - with i.a. more than 1,000 different whiskies including more than 500 single malts. In spring 2018, Han van Wees was distinguished with the role of Master of the Quaich for his many years of promoting Scotch whisky.

The Vintage Malt Whisky Company

www.vintagemaltwhisky.com

Founded in 1992 by Brian Crook, who previously had twenty years experience in the malt whisky industry, the company today is run by his three children, Andrew, Caroline and Kim, supplying whisky to more than 35 countries. The company also owns and operates a sister company called The Highlands & Islands Scotch Whisky Co. In 2018, they acquired a former factory in Port Ellen on Islay where they hope to eventually distill a range of Islay based spirits although probably not malt whisky. The most famous brands in the range are two single Islay malts called Finlaggan and The Ileach. The latter comes in two versions, bottled at 40% and 58%. The Finlaggan range consists of Old Reserve, Eilean Mor, Port Finish, Sherry Finish and Cask Strength (56%). In 2019, a limited Finlaggan, matured for the full time in Rioja red wine casks was released. Other expressions in the company´s range are Islay Storm the blended malts Smokestack, Glenalmond and Black Cuillin and, not least, a wide range of single cask single malts under the name The Cooper´s Choice. They are bottled at 46% or at cask strength and are all non coloured and non chill-filtered.

Finlaggan Red Wine Cask Matured, 46%
Nose: Seductive notes of wine and smoke, cherry trifle, strawberries, leather and potpourri.
Palate: Smoke and wine starts on a bitter note but soon develops into marzipan, roasted almonds, red liquorice and milk chocolate.

Finlaggan Original Peaty, 40%
Nose: Buttery notes of cured ham, apples, Danish pastry, custard and smoky heather.
Palate: Dry, ashy and smoky at first, followed by yellow fruit, sponge cake and celeriac.

Svenska Eldvatten

www.eldvatten.se

Founded in 2011 by Tommy Andersen and Peter Sjögren. Since the start, more than 100 single casks, bottled at cask strength, have been released. In their range of spirits they have aged tequila and rum and they have also launched their own rum, WeiRon, as well as gin and aquavit. A new, limited range called Silent Swede was launched in 2018. The seven expressions all came from the closed Swedish distillery Grythyttan. Svenska Eldvatten are also importers to Sweden of whisky from Murray McDavid, AD Rattray, North Star Spirits, Sansibar and Hidden Spirits.

Wm Cadenhead & Co

www.cadenhead.scot

This company was established in 1842 and is owned by J & A Mitchell (who also owns Springbank) since 1972. The single malts

from Cadenheads are neither chill filtered nor coloured. The current range consists of Authentic Collection (single cask cask strength whiskies, exclusively sold in their own shops), World Whiskies (single malts from non Scottish distillers as well as from Scottish grain distillers) and Small Batch, a range which can be divided into three separate ranges; Gold Label (single casks bottled at cask strength), Small Batch Cask Strength (2-4 casks of whisky from the same vintage, bottled at cask strength) and Small Batch 46% (same as the previous but diluted to 46%). A fourth range is William Cadenhead Range, which consists of blended whisky as well as single malts from undisclosed distilleries. There are nine dedicated Cadenhead´s Whisky Shops in Europe (see the Shop directory on pages 282-287) and also an online retail site - www.cadenhead.shop.

Benrinnes 1997 21 year old, 53,7%
Nose: Fresh and floral with notes of lilac, nettles and heather, bees wax, furniture polish and green apples.
Palate: Intense and peppery yet perfectly balanced, tropical fruits (mango and banana), herbs and eucalyptus.

Speyburn 2008 10 year old, 46%
Nose: Fresh, malty and slightly buttery with notes of grapes, candy, red berries and menthol.
Palate: Rich and peppery with notes of honey, vanilla, liquorice, shortbread, tobacco and with a hint of smoke.

Adelphi Distillery

www.adelphidistillery.com

Adelphi Distillery is named after a distillery which closed in 1902. The company is owned by Keith Falconer and Donald Houston, who recruited Alex Bruce from the wine trade to act as Managing Director. The company offers a range of single malts every year where the whiskies are always bottled at cask strength, uncoloured and non chill-filtered. There are also two recurrent brands, Fascadale and Liddesdale, where the single malt differs from batch to batch. In 2015, the first two bottlings of a new brand saw the light of day. Together with Fusion Whisky, Adelphi launched The Glover – a unique vatting of single malt from the closed Japanese distillery Hanyu and two Scottish single malts, Longmorn and Glen Garioch. A 14 and a 22 year old (and later an 18 year old) were released. This was followed by The E&K where Amrut single malt from India was blended with Scotch malt whisky and, in spring 2018, yet another two combinations appeared – The Brisbane (Starward single malt from Australia and Glen Garioch and Glen Grant from Scotland) and The Winter Queen (malt whisky from Zuidam distillery in the Netherlands, blended with Longmorn and Glenrothes).

Since 2014, Adelphi is also operating its own distillery in Glenbeg on the Ardnamurchan peninsula. Since the opening, the owners have regularly released malt spirit (less than 3 years old) with the latest being Spirit 2019 AD. There are no plans to release the first Ardnamurchan single malt whisky until 2021.

Ardnamurchan 2018 AD, 55.3%
Nose: Vibrant with notes of citrus, barley, green grass, nutmeg, furniture polish, saw dust and a whiff of smoke.
Palate: Lively with notes of granola, toffee, roasted marshmallows, a hint of pepper and some barbecue peat.

Blair Athol, 1997, 57.2%
Nose: Seductive and full of dried fruits, walnuts, maple syrup, tobacco leaves, chocolate, honey and a touch of lime.
Palate: Very powerful and dry, tobacco, espresso, tiramisu, dried plums and dark chocolate.

Deerstalker Whisky Co

www.deerstalkerwhisky.com

The Deerstalker brand, which dates from 1880 was originally owned by J.G. Thomson & Co of Leith and subsequently Tennent Caledonian Breweries. It was purchased by Glasgow based Aberko Ltd in 1994 and is managed by former Tennent's Export Director Paul Aston. The Deerstalker range covers single malts as well as blended malt whiskies. Currently there is only one single malts, a 12 year old. A Deerstalker Blended Malt (Highland Edition) was launched in 2014 and in autumn 2016 the Peated Edition of Deerstalker was released.

Deerstalker Blended Malt Peated Edition, 43%
Nose: Starts off with a good whiff of smoke followed by honey, oranges, mushrooms and roasted turnips.
Palate: Smoky with a touch of honey sweetness, vanilla fudge, citrus, dark chocolate, salted caramel, eucalyptus and a hint of pepper.

Morrison & MacKay Whisky

www.mandmwhisky.co.uk

A relative newcomer as an independent bottler, there is nonetheless plenty of experience in the company. The Morrison part of the business name is represented by Brian Morrison (as well as his son Jamie) who´s father was the legendary Stanley P Morrison, founder of Morrison Bowmore Distilleries and at one time owner of Bowmore, Auchentoshan and Glen Garioch. After leaving Morrison Bowmore, Brian started the Scottish Liqueur Centre and the family has recently also opened a distillery in Aberargie, just south of Perth. Meanwhile the liqueur business has been expanded to also include malt whisky under the name Carn Mor. Currently there are three ranges; Carn Mor Strictly Limited, usually bottled at 46%, Celebration of the Cask which are single casks bottled at cask strength and Celebration of the Cask Black Gold with heavily sherried whiskies in focus. A fourth range of blended malts is called Old Perth.

Edinburgh Whisky Ltd.

www.edinburghwhisky.com

Two friends, Gordon Watt, a former sales director at Moët Hennessy and Gregor Mathieson with a background in the wine business and partner in the Michelin awarded Andrew Fairlie restaurant at Gleneagles, founded the company in 2013. They were later joined by Iain Hamilton. Single malt single casks are bottled under the name The Library Collection while small batch blended malts

are sold under the name New Town. Another range to be introduced soon is Old Town blended malts.

Sansibar Whisky

www.sansibar-whisky.com

Started in 2012, this was the brainchild of the current majority owner and CEO Jens Drewitz and Carsten Ehrlich, the organizer of the famous Whisky Fair Limburg. Their idea was to create a range of high quality single malts from Scotland and to market them in connection with the well known Sansibar restaurant on the island of Sylt in northern Germany. Around 60 bottlings are produced per year and the range also includes rum.

Dramfool

www.dramfool.com

Bruce Farquhar, a whisky fan and collector for 20 years, decided in 2015 to start as an independent bottler. He sources his whisky from private individuals as well as from brokers and included in his latest releases are an 11 year old Girvan single grain and, for Feis Ile 2019, a heavily sherried, 14 year old Port Charlotte.

Angel´s Nectar

www.angelsnectar.co.uk

For ten years Robert Ransom worked as the sales and marketing director at Glenfarclas. He left in 2014 and founded Highfern Ltd. The main product is the blended malt Angels´s Nectar which is available in two versions - Original (Speyside and Highland, bottled at 40%) and Rich Peat Edition (Highland, bottled at 46%). Highfern is also the UK importer for Smögen single malt and gin and Langatun Swiss single malt.

The Single Cask Ltd

www.thesinglecask.co.uk

Founded by Ben Curtis who was distributor for a number of Scottish distilleries in south-east Asia before he started as an independent bottler in Singapore in 2010. Since then he has moved back to the UK but the company still runs an impressive whisky bar in Singapore. The bottling side of the business has grown over the years and the brand is now sold in the UK, Europe and Asia. The company also acts as a broker selling casks with both newmake and maturing whisky. A nice feature on their website are the drone videos of a number of Scottish distilleries.

Selected Malts

www.selectedmalts.se

A new bottler from Sweden, founded by Patrik Barkevall and Mikael Westerberg in 2017. Until now, they have specialised in fairly young single malts but with a maturation story that stands out from the ordinary. In September 2019, they releasecd their own blended malt, Zippin, which was made up by malts from Tullibardine, Glen Ord, Macduff and Ardmore (peated). They also recently acquired the distribution rights for GlenAllachie in Sweden.

The Alistair Walker Whisky Co.

Alistair Walker, from the Walker family who used to own BenRiach, GlenDronach and Glenglassaugh, has spent almost twenty years in the whisky business. When the family sold the distilleries in 2016, he started thinking about what to do next and decided to start up as an independent bottler. The brand is called Infrequent Flyers and the first releases (nine single casks) appeared in August 2019 with a second batch being launched in October. Alistair will be focusing on single casks from Scottish distilleries.

Whisky
shops

AUSTRALIA

The Odd Whisky Coy
PO Box 471
Glenside, SA, 5065
Phone: +61 (0)417 85 22 96
www.theoddwhiskycoy.com.au
This on-line whisky specialist has an impressive range. They are agents for famous brands such as Springbank, Benromach and Berry Brothers and arrange recurrent seminars on the subject.

World of Whisky
Shop G12, Cosmopolitan Centre
2-22 Knox Street
Double Bay NSW 2028
Phone: +61 (0)2 9363 4212
www.worldofwhisky.com.au
A whisky specialist which offers a range of 400 different expressions, most of them single malts. The shop is also organising and hosting regular tastings.

AUSTRIA

Potstill
Laudongasse 18
1080 Wien
Phone: +43 (0)664 118 85 41
www.potstill.org
Austria's premier whisky shop with over 1100 kinds of which c 900 are malts, including some real rarities. Arranges tastings and seminars and ships to several European countries. On-line ordering.

Cadenhead Austria
Döblinger Hauptstraße 32
1190 Wien
Phone: +43 (0)677 622 476 40
www.cadenhead-vienna.at
The former shop in Salzburg is closed and a new shop opened up in Vienna in August 2018. Focusing on the Cadenhead range but with a wide range of other whiskies and spirits as well.

Pinkernells Whisky Market
Alter Markt 1
5020 Salzburg
Phone: +43 (0)662 84 53 05
www.pinkernells.at
More than 400 whiskies are on offer and they are also importers of Maltbarn, The Whisky Chamber and Jack Wiebers. Regular tastings.

BELGIUM

Whiskycorner
Kraaistraat 16
3530 Houthalen
Phone: +32 (0)89 386233
www.whiskycorner.be
A very large selection of single malts, no less than 2000 different! Also other whiskies, calvados and grappas. The site is in both French and English.

Jurgen´s Whiskyhuis
Gaverland 70
9620 Zottegem
Phone: +32 (0)9 336 51 06
www.whiskyhuis.be
A huge assortment of more than 2,000 different single malts. Also a good range of grain whiskies and bourbons.

Huis Crombé
Doenaertstraat 20
8510 Marke
Phone: +32 (0)56 21 19 87
www.crombewines.com
A wine retailer which also covers all kinds of spirits. A large assortment of Scotch is supplemented with whiskies from Japan, the USA and Ireland to mention a few.

Anverness Whisky & Spirits
Grote Steenweg 74
2600 Berchem – Antwerpen
Phone: +32 (0)3 218 55 90
www.anverness.be
Peter de Decker has established himself as one of the best Belgian whisky retailers where, apart from an impressive range of whiskies, recurrent tastings and whisky dinners play an important role.

We Are Whisky
Avenue Rodolphe Gossia 33
1350 Orp-Jauche
Phone: +32 (0)471 134556
www.wearewhisky.com
On-line retailer with a range of more than 800 different whiskies. They also arrange 3-4 tastings every month.

Dram 242
Rijgerstraat 60
9310Moorsel
Phone: +32 (0)477 26 09 93
www.dram242.be
A wide range of whiskies. Apart from the core official bottlings, they have focused on rare, old expressions as well as whiskies from small, independent bottlers.

CANADA

Kensington Wine Market
1257 Kensington Road NW
Calgary, Alberta T2N 3P8
Phone: +1 403 283 8000
www.kensingtonwinemarket.com
Taken over four years ago by long time employee Andrew Ferguson, the shop has a very large range of whiskies (more than 1500) as well as other spirits and wines. More than 80 tastings in the shop every year. Also the home of the Scotch Malt Whisky Society in Canada.

World of Whisky
Unit 240, 333 5 Avenue SW
Calgary
Alberta T2P 3B6
Phone: +1 587 956 8511
www.coopwinespiritsbeer.com/stores/world-of-whisky/
Specialising in whisky from all corners of the world. Currently there are over 1100 different whiskies in the range including some extremely rare ones from Scotland.

DENMARK

Juul´s Vin & Spiritus
Værnedamsvej 15
1819 Frederiksberg
Phone: +45 33 31 13 29
www.juuls.dk
A very large range of wines, fortified wines and spirits with more than 1100 different whiskies (800 single malts).

Cadenhead´s WhiskyShop Denmark
Kongensgade 69 F
5000 Odense C
Phone: +45 66 13 95 05
www.cadenheads.dk
Whisky specialist with a very good range, not least from Cadenhead's. Nice range of champagne, cognac and rum. Arranges whisky and beer tastings. On-line ordering.

Whisky.dk
Vejstruprødvej 15
6093 Sjølund
Phone: +45 5210 6093
www.whisky.dk
Henrik Olsen and Ulrik Bertelsen are well-known in Denmark for their whisky shows but they also run an on-line spirits shop with an emphasis on whisky but also including an impressive stock of rums.

ENGLAND

The Whisky Exchange
2 Bedford Street, Covent Garden
London WC2E 9HH
Phone: +44 (0)20 7100 0088
90-92 Great Portland Street, Fitzrovia
London W1W 7NT
Phone: +44 (0)20 7100 9888
www.thewhiskyexchange.com
An excellent whisky shop owned by

Sukhinder Singh. Started off as a mail order business, run from a showroom in Hanwell, but later opened up at Vinopolis in downtown London. Moved to a new and bigger location in Covent Garden a couple of years ago and recently opened up a second shop in Fitzrovia. The assortment is huge with well over 1000 single malts to choose from. Some rarities which can hardly be found anywhere else are offered thanks to Singh's great interest for antique whisky. There are also other types of whisky and cognac, calvados, rum etc. On-line ordering and ships all over the world.

The Whisky Shop
(See also Scotland, The Whisky Shop)
11 Coppergate Walk
York YO1 9NT
Phone: +44 (0)1904 640300

510 Brompton Walk
Lakeside Shopping Centre
Thurrock Grays, Essex RM20 2ZL
Phone: +44 (0)1708 866255

7 Turl Street
Oxford OX1 3DQ
Phone: +44 (0)1865 202279

3 Swan Lane
Norwich NR2 1HZ
Phone: +44 (0)1603 618284

70 Piccadilly
London W1J 8HP
Phone: +44 (0)207 499 6649

Unit 7 Queens Head Passage
Paternoster
London EC4M 7DZ
Phone: +44 (0)207 329 5117

3 Exchange St
Manchester M2 7EE
Phone: +44 (0)161 832 6110

25 Chapel Street
Guildford GU1 3UL
Phone: +44 (0)1483 450900

Unit 9 Great Western Arcade
Birmingham B2 5HU
Phone: +44 (0)121 233 4416

64 East Street
Brighton BN1 1HQ
Phone: +44 (0)1273 327 962

3 Cheapside
Nottingham NG1 2HU
Phone: +44 (0)115 958 7080

9-10 High Street
Bath BA1 5AQ
Phone: +44 (0)1225 423 535

Unit 1/9 Red Mall,
Intu Metro Centre
Gateshead NE11 9YP
Phone: +44 (0)191 460 3777

Unit 201 Trentham Gardens
Stoke on Trent ST4 8AX
Phone: +44 (0)1782 644 483
www.whiskyshop.com
The largest specialist retailer of whiskies in the UK with 20 outlets. A large product range with over 700 kinds, including 400 malt whiskies and 140 miniature bottles, as well as accessories and books. They also run The W Club, the leading whisky club

in the UK where the excellent Whiskeria magazine is one of the member´s benefits. Shipping all over the world.

Royal Mile Whiskies
3 Bloomsbury Street
London WC1B 3QE
Phone: +44 (0)20 7436 4763
www.royalmilewhiskies.com
The London branch of Royal Mile Whiskies. See also Scotland, Royal Mile Whiskies.

Berry Bros. & Rudd
63 Pall Mall
London SW1Y 5HZ
Phone: +44 (0)800 280 2440
www.bbr.com/whisky
A legendary company that recently opened a new shop in Pall Mall. One of the world's most reputable wine shops but with an exclusive selection of malt whiskies, some of them bottled by Berry Bros. themselves.

The Wright Wine & Whisky Company
The Old Smithy, Raikes Road, Skipton,
North Yorkshire BD23 1NP
Phone: +44 (0)1756 700886
www.wineandwhisky.co.uk
An eclectic selection of near to 1000 different whiskies. 'Tasting Cupboard' of nearly 100 opened bottles for sampling with regular hosted tasting evenings. Great 'Collector to Collector' selection of old whiskies plus a fantastic choice of 1200+ wines, premium spirits and liqueurs.

Master of Malt
Unit 1, Ton Business Park, 2-8 Morley Rd.
Tonbridge, Kent, TN9 1RA
Phone: 0800 5200 474
www.masterofmalt.com
Online retailer and independent bottler with a very impressive range of more than 2,500 whiskies, including over 2,000 Scotch whiskies and over 1,500 single malts. In addition to whisky there is an enormous selection of gins, rums, cognacs, armagnacs, tequilas and more. The website contains a wealth of information and news about the distilleries and innovative per-sonalised gift ideas. Drinks by the Dram 30ml samples of more than 3,300 different whiskies are also available to try before you buy a full bottle as well as a Build Your Own Tasting Set option and Dram Club monthly subscription services.

Whiskys.co.uk
The Square, Stamford Bridge
York YO4 11AG
Phone: +44 (0)1759 371356
www.whiskys.co.uk
Good assortment with more than 600 different whiskies. Also a nice range of armagnac, rum, calvados etc. The owners also have another website, www.whiskymerchants.co.uk with a huge amount of information on just about every whisky distillery in the world.

The Wee Dram
5 Portland Square, Bakewell
Derbyshire DE45 1HA
Phone: +44 (0)1629 812235
www.weedram.co.uk

Large range of Scotch single malts with whiskies from other parts of the world and a good range of whisky books. Run 'The Wee Drammers Whisky Club' with tastings and seminars. End of October they arrange the yearly Wee Dram Fest whisky festival.

Hard To Find Whisky
1 Spencer Street
Birmingham B18 6DD
Phone: +44 (0)121 448 84 84
www.htfw.com
As the name says, this family owned shop specialises in rare, collectable and new releases of single malt whisky. The range is astounding - more than 3,000 different bottlings including no less than 441 dif-ferent Macallan. World wide shipping.

Nickolls & Perks
37 High Street, Stourbridge
West Midlands DY8 1TA
Phone: +44 (0)1384 394518
www.nickollsandperks.com
Mostly known as wine merchants but also has a huge range of whiskies with 1,900 different kinds including 1,300 single malts. Since 2011, they also organize the acclaimed Midlands Whisky Festival, see www.whiskyfest.co.uk

Gauntleys of Nottingham
4 High Street
Nottingham NG1 2ET
Phone: +44 (0)115 9110555
www.gauntleys.com
A fine wine merchant established in 1880. The range of wines are among the best in the UK. All kinds of spirits, not least whisky, are taking up more and more space and several rare malts can be found.

Hedonism Wines
3-7 Davies St.
London W1K 3LD
Phone: +44 (020) 729 078 70
www.hedonism.co.uk
Located in the heart of London, this is a temple for wine lovers but also with an im-pressive range of whiskies. They have over 1,500 different bottlings from Scotland and the rest of the world.

The Lincoln Whisky Shop
87 Bailgate
Lincoln LN1 3AR
Phone: +44 (0)1522 537834
www.lincolnwhiskyshop.co.uk
Mainly specialising in whisky with more than 400 different whiskies but also 500 spirits and liqueurs. Mailorder worldwide.

Milroys of Soho
3 Greek Street
London W1D 4NX
Phone: +44 (0)207 734 2277
shop.milroys.co.uk
A classic whisky shop in Soho with a very good range with over 700 malts and a wide selection of whiskies from around the world. Also a whisky bar within the shop.

Arkwrights
114 The Dormers
Highworth
Wiltshire SN6 7PE
Phone: +44 (0)1793 765071

www.whiskyandwines.com
A good range of whiskies (over 700 in stock) as well as wine and other spirits. Regular tastings in the shop. On-line ordering with shipping all over the world.

Edencroft Fine Wines
8-10 Hospital Street, Nantwich
Cheshire, CW5 5RJ
Phone: +44 (0)1270 629975
www.edencroft.co.uk
Family owned wine and spirits shop since 1994. Around 250 whiskies and also a nice range of gin, cognac and other spirits including cigars. Worldwide shipping.

Cadenhead´s Whisky Shop
26 Chiltern Street
London W1U 7QF
Phone: +44 (0)20 7935 6999
www.whiskytastingroom.com
One in a chain of shops owned by independent bottlers Cadenhead. Sells Cadenhead's product range and c. 200 other whiskies. Regular tastings.

Constantine Stores
30 Fore Street
Constantine, Falmouth
Cornwall TR11 5AB
Phone: +44 (0)1326 340226
www.drinkfinder.co.uk
A full-range wine and spirits dealer with a good selection of whiskies from the whole world (around 800 different, of which 600 are single malts).Worldwide shipping.

House of Malt
48 Warwick Road
Carlisle CA1 1DN
Phone: +44 (0)1228 658 422
www.houseofmalt.co.uk
A wide selection of whiskies from Scotland and the world as well as other spirits and craft ales. Regular tasting evenings and events.

The Vintage House
42 Old Compton Street
London W1D 4LR
Phone: +44 (0)20 7437 5112
www.sohowhisky.com
A huge range of 1400 kinds of malt whisky, many of them rare. Supplementing this is also a selection of fine wines.

Whisky On-line
Units 1-3 Concorde House, Charnley Road, Blackpool, Lancashire FY1 4PE
Phone: +44 (0)1253 620376
www.whisky-online.com
A good selection of whisky and also cognac, rum, port etc. On-line ordering with shipping all over the world.

FRANCE
La Maison du Whisky
20 rue d´Anjou
75008 Paris
Phone: +33 (0)1 42 65 03 16

6 carrefour de l´Odéon
75006 Paris
Phone: +33 (0)1 46 34 70 20

(2 shops outside France)
47 rue Jean Chatel
97400 Saint-Denis, La Réunion

Phone: +33 (0)2 62 21 31 19
The Pier at Robertson Quay
80 Mohamed Sultan Road, #01-10
Singapore 239013
Phone: +65 6733 0059
www.whisky.fr
France's largest whisky specialist with over 1200 whiskies in stock. Also a number of own-bottled single malts. La Maison du Whisky acts as a EU distributor for many whisky producers around the world. Also run a specialist rum shop and a whisky bar in Paris.

The Whisky Shop
7 Place de la Madeleine
75008 Paris
Phone: +33 (0)1 45 22 29 77
www.whiskyshop.fr
The large chain of whisky shops in the UK has now opened up a store in Paris as well.

GERMANY
Celtic Whisk(e)y & Versand
Otto Steudel
Bulmannstrasse 26
90459 Nürnberg
Phone: +49 (0)911 45097430
www.celtic-whisky.de
A very impressive single malt range with well over 1000 different single malts and a good selection from other parts of the world.

SCOMA
Am Bullhamm 17
26441 Jever
Phone: +49 (0)4461 912237
www.scoma.de
Very large range of c 750 Scottish malts and many from other countries. Holds regular seminars and tastings. The excellent, monthly whisky newsletter SCOMA News is produced and can be downloaded as a pdf-file from the website.

The Whisky Store
Am Grundwassersee 4
82402 Seeshaupt
Phone: +49 (0)8801 30 20 000
www.whisky.de
A very large range comprising c 700 kinds of whisky of which 550 are malts. Also sells whisky liqueurs, books and accessories. The website is a goldmine of information. On-line ordering.

Cadenhead´s Whisky Market
Luxemburger Strasse 257
50939 Köln
Phone: +49 (0)221-2831834
www.cadenheads.de
Good range of malt whiskies (c 350 different kinds) with emphasis on Cadenhead's own bottlings. Other products include wine, cognac and rum etc. Arranges recurring tastings and also has an on-line shop.

Pinkernells Whisky Market
Boxhagener Straße 36
10245 Berlin
Phone: +49 (0)30-22 600 610
www.pinkernells.de
An extensive range of whiskies (more than 700) and they arrange 4-5 tastings

monthly. Also work as whisky consultants doing corporate events all over Germany.

Home of Malts
Hosegstieg 11
22880 Wedel
Phone: +49 (0)4103 965 9695
www.homeofmalts.com
Large assortment with over 800 different single malts as well as whiskies from many other countries. Also a nice selection of cognac, rum etc. On-line ordering.

Reifferscheid
Mainzer Strasse 186
53179 Bonn / Mehlem
Phone: +49 (0)228 9 53 80 70
www.whisky-bonn.de
A well-stocked shop with a large range of whiskies, wine, spirit, cigars and a delicatessen. Regular tastings.

Whisky-Doris
Germanenstrasse 38
14612 Falkensee
Phone: +49 (0)3322-219784
www.whisky-doris.de
Large range of over 300 whiskies and also sells own special bottlings. Orders via email. Shipping also outside Germany.

Finlays Whisky Shop
Hofheimer Str. 30
65719 Hofheim-Lorsbach
Phone: +49 (0)6192 30 90 335
www.finlayswhiskyshop.de
Whisky specialists with a large range of over 1,600 whiskies. Finlays also work as the importer to Germany of Douglas Laing, James MacArthur and Wilson & Morgan.

Weinquelle Lühmann
Lübeckerstrasse 145
22087 Hamburg
Phone: +49 (0)40-300 672 950
www.weinquelle.com
An impressive selection of both wines and spirits with over 1000 different whiskies of which 850 are malt whiskies. Also an impressive range of rums.

The Whisky-Corner
Reichertsfeld 2
92278 Illschwang
Phone: +49 (0)9666-951213
www.whisky-corner.de
A small shop but large on mail order. A very large assortment of over 2000 whiskies. Also sells blended and American whiskies. The website is very informative with features on, among others, whisky-making, tasting and independent bottlers.

World Wide Spirits
Hauptstrasse 12
84576 Teising
Phone: +49 (0)8633 50 87 93
www.worldwidespirits.de
A nice range of more than 1,000 whiskies with some rarities from the twenties. Also large selection of other spirits.

WhiskyKoch
Weinbergstrasse 2
64285 Darmstadt
Phone: +49 (0)6151 99 27 105
www.whiskykoch.de
A combination of a whisky shop and

restaurant. The shop has a nice selection of single malts as well as other Scottish products and the restaurant has specialised in whisky dinners and tastings.

Kierzek
Weitlingstrasse 17
10317 Berlin
Phone: +49 (0)30 525 11 08
www.kierzek-berlin.de
Over 400 different whiskies in stock. In the product range 50 kinds of rum and 450 wines from all over the world are found among other products. Mail order is available.

House of Whisky
Ackerbeeke 6
31683 Obernkirchen
Phone: +49 (0)5724-399420
www.houseofwhisky.de
Aside from over 1,200 different malts also sells a large range of other spirits (including over 100 kinds of rum). On-line ordering.

HUNGARY
Whisky Shop Budapest
Veres Pálné utca 7.
1053 Budapest
Phone: +36 1 267-1588
www.whiskynet.hu
www.whiskyshop.hu
Largest selection of whisky in Hungary. More than 900 different whiskies from all over the world. Even Hungarian whisky and a large selection of other fine spirits are available. Most of them can be tasted in the GoodSpirit Whisky & Cocktail Bar which operates in the same venue.

IRELAND
Celtic Whiskey Shop
27-28 Dawson Street
Dublin 2
Phone: +353 (0)1 675 9744
www.celticwhiskeyshop.com
More than 500 kinds of Irish whiskeys but also a good selection of Scotch, wines and other spirits. World wide shipping.

ITALY
Whisky Shop
by Milano Whisky Festival
Via Cavaleri 6, Milano
Phone: +39 (0)2 48753039
www.whiskyshop.it
The team behind the excellent Milano Whisky Festival also have an on-line whiskyshop with almost 500 different single malts including several special festival bottlings.

Whisky Antique S.R.L.
Via Giardini Sud
41043 Formigine (MO)
Phone: +39 (0)59 574278
www.whiskyantique.com
Long-time whisky enthusiast and collector Massimo Righi owns this shop specialising in rare and collectable spirits – not only whisky but also cognac, rum, armagnac

etc. He also acts as an independent bottler with the brand Silver Seal. They are the Italian importer for brands like Jack Wiebers, The Whisky Agency and Perfect Dram.

JAPAN
Liquor Mountain Co.,Ltd.
4F Kyoto Kowa Bldg.
82 Tachiurinishi-Machi,
Takakura-Nishiiru,
Shijyo-Dori, Shimogyo-Ku,
Kyoto, 600-8007
Phone: +81 (0)75 213 8880
www.likaman.co.jp
The company has more than 150 shops specialising in spirits, beer and food. Around 20 of them are designated whisky shops under the name Whisky Kingdom (although they have a full range of other spirits) with a range of 500 different whiskies. The three foremost shops are;

Rakzan Sanjyo Onmae
1-8, HigashiGekko-cho, Nishinokyo,
Nakagyo-ku, Kyoto-shi
Kyoto
Phone: +81 (0)75-842-5123

Nagakute
2-105, Ichigahora, Nagakute-shi
Aichi
Phone: +81 (0)561-64-3081

Kabukicho 1chome
1-2-16, Kabuki-cho, Shinjuku-ku
Tokyo
Phone: +81 (0)3-5287-2080

THE NETHERLANDS
Whiskyslijterij De Koning
Hinthamereinde 41
5211 PM 's Hertogenbosch
Phone: +31 (0)73-6143547
www.whiskykoning.nl
An enormous assortment with more than 1400 kinds of whisky including c 800 single malts. Arranges recurring tastings. On-line ordering. Shipping all over the world.

Van Wees - Whiskyworld.nl
Leusderweg 260
3817 KH Amersfoort
Phone: +31 (0)33-461 53 19
www.whiskyworld.nl
A very large range of 1000 whiskies including over 500 single malts. Also have their own range of bottlings (The Ultimate Whisky Company). On-line ordering.

Wijnhandel van Zuylen
Loosduinse Hoofdplein 201
2553 CP Loosduinen (Den Haag)
Phone: +31 (0)70-397 1400
www.whiskyvanzuylen.nl
Excellent range of whiskies (circa 1100) and wines. Email orders with shipping to some ten European countries.

Wijnwinkel-Slijterij
Ton Overmars
Hoofddorpplein 11
1059 CV Amsterdam
Phone: +31 (0)20-615 71 42
www.tonovermars.nl

A very large assortment of wines, spirits and beer which includes more than 400 single malts. Arranges recurring tastings. Online ordering.

Wijn & Whisky Schuur
Blankendalwei 4
8629 EH Scharnegoutem
Phone: +31 (0)515-520706
www.wijnwhiskyschuur.nl
Large assortment with 1000 different whiskies and a good range of other spirits as well. Arranges recurring tastings.

Versailles Dranken
Lange Hezelstraat 83
6511 Cl Nijmegen
Phone: +31 (0)24-3232008
www.versaillesdranken.nl
A very impressive range with more than 1500 different whiskies, most of them from Scotland but also a surprisingly good selection (more than 60) of Bourbon. Arranges recurring tastings.

Wine and Whisky Specialist van der Boog
Prinses Irenelaan 359-361
2285 GA Rijswijk
Phone: +31 70 - 394 00 85
www.passionforwhisky.com
A very good range of almost 700 malt whiskies (as well as a wide range of other spirits). World wide shipping.

NEW ZEALAND
Whisky Galore
834 Colombo Street
Christchurch 8013
Phone: +64 (0) 800 944 759
www.whiskygalore.co.nz
The best whisky shop in New Zealand with 550 different whiskies, approximately 350 which are single malts. There is also online mail-order with shipping all over the world except USA and Canada.

POLAND
George Ballantine´s
Krucza str 47 A, Warsaw
Phone: +48 22 625 48 32

Pulawska str 22, Warsaw
Phone: +48 22 542 86 22

Marynarska str 15, Warsaw
Phone: +48 22 395 51 60

Zygmunta Vogla str 62, Warsaw
Phone: +48 22 395 51 64
www.sklep-ballantines.pl
A huge range of single malts and apart from whisky there is a full range of spirits and wines from all over the world. Recurrent tastings and organiser of Whisky Live Warsaw.

Dom Whisky
Wejherowska 67, Reda
Phone: +48 691 760 000, shop
Phone: +48 691 930 000, mailorder
www.sklep-domwhisky.pl
On-line retailer who recently opened a shop in Reda. A very large range of whiskies and other spirits. Organiser of a whisky festival in Jastrzębia Góra.

RUSSIA

Whisky World Shop
9, Tverskoy Boulevard
123104 Moscow
Phone: +7 495 787 9150
www.whiskyworld.ru
Huge assortment with more than 1,000 different single malts. The range is supplemented with a nice range of cognac, armagnac, calvados, grappa and wines.

SCOTLAND

Gordon & MacPhail
58 - 60 South Street, Elgin
Moray IV30 1JY
Phone: +44 (0)1343 545110
www.gordonandmacphail.com
This legendary shop opened already in 1895 in Elgin. The owners are perhaps the most well-known among independent bottlers. The shop stocks more than 800 bottlings of whisky and more than 600 wines and there is also a delicatessen counter with high-quality products. Tastings are arranged in the shop and there are shipping services within the UK and overseas. The shop attracts visitors from all over the world.

Royal Mile Whiskies (2 shops)
379 High Street, The Royal Mile
Edinburgh EH1 1PW
Phone: +44 (0)131 2253383

3 Bloomsbury Street
London WC1B 3QE
Phone: +44 (0)20 7436 4763
www.royalmilewhiskies.com
Royal Mile Whiskies is one of the most well-known whisky retailers in the UK. It was established in Edinburgh in 1991. There is also a shop in London since 2002 and a cigar shop close to the Edinburgh shop. The whisky range is outstanding with many difficult to find elsewhere. They have a comprehensive site regarding information on regions, distilleries, production, tasting etc. Royal Mile Whiskies also arranges 'Whisky Fringe' in Edinburgh, a two-day whisky festival which takes place annually in mid August. On-line ordering with worldwide shipping.

The Whisky Shop
(See also England, The Whisky Shop)
Unit L2-02 Buchanan Galleries
220 Buchanan Street
Glasgow G1 2GF
Phone: +44 (0)141 331 0022
17 Bridge Street
Inverness IV1 1HD
Phone: +44 (0)1463 710525
93 High Street
Fort William PH33 6DG
Phone: +44 (0)1397 706164
52 George Street
Oban PA34 5SD
Phone: +44 (0)1631 570896

Unit 23 Waverley Mall
Waverley Bridge
Edinburgh EH1 1BQ
Phone: +44 (0)131 558 7563
28 Victoria Street

Edinburgh EH1 2JW
Phone: +44 (0)131 225 4666
www.whiskyshop.com
The first shop opened in 1992 in Edinburgh and this is now the United Kingdom's largest specialist retailer of whiskies with 20 outlets (plus one in Paris). A large product range with over 700 kinds, including 400 malt whiskies and 140 miniature bottles, as well as accessories and books. The own range 'Glenkeir Treasures' is a special assortment of selected malt whiskies. The also run The W Club, the leading whisky club in the UK where the excellent Whiskeria magazine is one of the member´s benefits. On-line ordering.

Loch Fyne Whiskies
Main Street, Inveraray
Argyll PA32 8UD
Phone: +44 (0)800 107 1936

36 Cockburn St
Edinburgh EH1 1PB
Phone: +44 (0)131 226 2134
www.lochfynewhiskies.com
A legendary shop and with a second shop in Edinburgh since June 2018. The range of malt whiskies is large and they have their own house blend, the prize-awarded Loch Fyne, as well as their 'The Loch Fyne Whisky Liqueur'. There is also a range of house malts called 'The Inverarity'. On-line ordering with worldwide shipping.

Whiskies of Scotland
36 Gordon Street
Huntly
Aberdeenshire AB54 8EQ
Phone: +44 (0) 845 606 6145
www.thespiritsembassy.com
Owned by independent bottler Duncan Taylor. In the assortment is of course the whole Duncan Taylor range but also a selection of their own single malt bottlings called Whiskies of Scotland. A total of almost 700 different expressions. On-line shop with shipping worldwide.

The Whisky Shop Dufftown
1 Fife Street, Dufftown
Moray AB55 4AL
Phone: +44 (0)1340 821097
www.whiskyshopdufftown.com
Whisky specialist in Dufftown in the heart of Speyside, wellknown to many of the Speyside festival visitors. More than 500 single malts as well as other whiskies. Arranges tastings as well as special events during the Festivals. On-line ordering.

Cadenhead's Whisky Shop
30-32 Union Street
Campbeltown PA28 6JA
Phone: +44 (0)1586 551710
www.cadenhead.shop
Part of the chain of shops owned by independent bottlers Cadenhead. Sells Cadenhead's products and other whiskies with a good range of Springbank. On-line ordering.

Cadenhead´s Whisky Shop
172 Canongate, Royal Mile
Edinburgh EH8 8DF
Phone: +44 (0)131 556 5864

www.cadenhead.shop
The oldest shop in the chain owned by Cadenhead. Sells Cadenhead's product range and a good selection of other whiskies and spirits. Recurrent tastings. On-line ordering.

The Good Spirits Co.
23 Bath Street,
Glasgow G2 1HW
Phone: +44 (0)141 258 8427
www.thegoodspiritsco.com
A specialist spirits store selling whisky, bourbon, rum, vodka, tequila, gin, cognac and armagnac, liqueurs and other spirits. They also stock quality champagne, fortified wines and cigars. There are more than 400 single malts in the range as well as over 100 whiskies from the rest of the world.

The Carnegie Whisky Cellars
The Carnegie Courthouse, Castle Street
IV25 3SD Dornoch
Phone: +44 (0)1862 811791
www.thecarnegiecourthouse.co.uk/whisky-cellars/
Opened by Michael Hanratty in 2016, this shop has already become a destination for whisky enthusiasts from the UK and abroad. The interior of the shop is ravishing and the extensive range includes all the latest releases as well as rare and collectable bottles. UK and international shipping.

Abbey Whisky
Dunfermline KY11 3BZ
Phone: +44 (0)800 051 7737
www.abbeywhisky.com
Family run online whisky shop specialising in exclusive, rare and old whiskies from Scotland and the world. Apart from a wide range of official and independent bottlings, Abbey Whisky also selects their own casks and bottle them under the name 'The Rare Casks' and 'The Secret Casks'.

The Scotch Whisky Experience
354 Castlehill, Royal Mile
Edinburgh EH1 2NE
Phone: +44 (0)131 220 0441
www.scotchwhiskyexperience.co.uk
The Scotch Whisky Experience is a must for whisky devotees visiting Edinburgh. An interactive visitor centre dedicated to the history of Scotch whisky. This five-star visitor attraction has an excellent whisky shop with almost 300 different whiskies in stock. Recently, after extensive refurbishment, a brand new and interactive shop was opened.

A.D. Rattray´s Whisky Experience & Whisky Shop
32 Main Road
Kirkoswald
Ayrshire KA19 8HY
Phone: +44 (0) 1655 760308
www.adrattray.com
A combination of whisky shop, sample room and educational center owned by the independent bottler A D Rattray. Tasting menus with different themes are available.

Whiski Shop
4 North Bank Street
Edinburgh EH1 2LP
Phone: +44 (0)131 225 7224
www.whiskishop.com
www.whiskirooms.co.uk
A new concept located near Edinburgh
Castle, combining a shop, a tasting
room and a bistro. Also regular whisky
tastings. Online mail order with worldwide
delivery.

Robbie's Drams
3 Sandgate, Ayr
South Ayrshire KA7 1BG
Phone: +44 (0)1292 262 135
www.robbieswhiskymerchants.com
An extensice range of whiskies available
both in store and from their on-line shop.
Specialists in single cask bottlings, closed
distillery bottlings, rare malts, limited
edition whisky and a nice range of their
own bottlings. Worldwide shipping.

The Whisky Barrel
PO Box 23803, Edinburgh, EH6 7WW
Phone: +44 (0)845 2248 156
www.thewhiskybarrel.com
Online specialist whisky shop based in
Edinburgh. They stock over 1,000 single
malt and blended whiskies including
Scotch, Japanese, Irish, Indian, Swedish
and their own casks. Worldwide shipping.

The Scotch Malt Whisky Society
www.smws.com
A legendary society with more than 20 000
members worldwide, specialised in own
bottlings of single cask Scotch whisky,
releasing between 150 and 200 bottlings
every year. Recently, the Society has also
started bottling whisky from other parts of
the world as well as gin, rum, armagnac
and other spirits.

Drinkmonger
100 Atholl Road
Pitlochry PH16 5BL
Phone: +44 (0)1796 470133

11 Bruntsfield Place
Edinburgh EH10 4HN
Phone: +44 (0)131 229 2205
www.drinkmonger.com
Owned by Royal Mile Whiskies, the idea
is to have a 50:50 split between wine and
specialist spirits with the addition of a
cigar assortment. The whisky range is a
good cross-section with some rarities and a
focus on local distilleries.

Luvian's
93 Bonnygate, Cupar
Fife KY15 4LG
Phone: +44 (0)1334 654 820
66 Market Street, St Andrews
Fife KY16 9NU
Phone: +44 (0)1334 477752
www.luvians.com
Wine and whisky merchant with a very
nice selection of more than 600 malt
whiskies.

The Stillroom by Deseo
Gleneagles Hotel
Auchterarder, Perthshire PH3 1NF
Phone: +44 (0) 1764 694 188
www.gleneagles.com

Located in the famous hotel, George Bry-
ers has selected a nice range of both rare
and collectible whiskies as well as single
malts from a large number of Scottish
distilleries.

Robertsons of Pitlochry
44-46 Atholl Road
Pitlochry PH16 5BX
Phone: +44 (0) 1796 472011
www.robertsonsofpitlochry.co.uk
With new owner since 2013, the shop has
grown to become one of Scotland's best.
An extensive range of both whisky and gin
is complemented by single malts bottled
under their own label. There's also an
excellent tasting room (The Bothy).

Robert Graham Ltd (3 shops)
194 Rose Street
Edinburgh EH2 4AZ
Phone: +44 (0)131 226 1874

111 West George Street
Glasgow G2 1QX
Phone: +44 (0)141 248 7283

254 Canongate
Royal Mile
Edinburgh EH8 8AA
Phone: +44 (0)131 556 2791
www.robertgraham1874.com
Established in 1874 this company
specialises in Scotch whisky and cigars.
A nice assortment of malt whiskies is
complemented by an impressive range of
cigars. They also bottle whiskies under
their own label.

The Speyside Whisky Shop
110A High Street
Aberlour AB38 9NX
Phone: +44 (0) 1340 871260
www.thespeysidewhisky.com
Opened in 2018, the shop is situated in the
very heart of Speyside, in Aberlour. The
owners specialise in highly collectable
single malts from a variety of distilleries.
Also a wide selection of craft gins.

The Jar
33 Ayr St
Troon KA10 6EB
Phone: +44 (0) 1292 319877
www.thejartroon.com
An extensive range of single malts (over
300) and Scottish gins. Specialises in rare
and collectable releases.

SOUTH AFRICA
WhiskyBrother
Hyde Park Corner
(middle level inside shopping mall)
Johannesburg
Phone: +27 (0)11 325 6261
www.whiskybrother.com
A shop specialising in all things whisky -
apart from 400 different bottlings they also
sell glasses, books etc. Also sell whiskies
bottled exclusively for the shop. Regular
tastings and online shop. Recently opened
their own whisky bar in Johannesburg with
more than 1,000 different whiskies to try.
The owner, Marc Pendlebury, is also the
organiser of The Only Whisky Show.

SWITZERLAND
P. Ullrich AG
Schneidergasse 27
4051 Basel
Phone: +41 (0)61 338 90 91
Another two shops in Basel:
Laufenstrasse 16 & Unt. Rebgasse 18
and one in Talacker 30 in Zürich
www.ullrich.ch
A very large range of wines, spirits, beers,
accessories and books. Over 800 kinds of
whisky with almost 600 single malt. On-
line ordering. Recently, they also founded
a whisky club with regular tastings (www.
whiskysinn.ch).

Eddie's Whiskies
Bahnhofstrasse/Dorfgasse 27
8810 Horgen
Phone: +41 (0)43 244 63 00
www.eddies.ch
A whisky specialist with more than 700
different whiskies in stock with emphasis
on single malts (more than 500 different).
Also arranges tastings.

Angels Share Shop
Unterdorfstrasse 15
5036 Oberentfelden
Phone: +41 (0)62 724 83 74
www.angelsshare.ch
A combined restaurant and whisky shop.
More than 600 different kinds of whisky
as well as a good range of cigars. Scores
extra points for short information and
photos of all distilleries. On-line ordering.

UKRAINE
WINETIME
Mykoly Bazhana 1E
Kyiv 02068
Phone: +38 (0)44 338 08 88
www.winetime.ua
WINETIME is the largest specialized
chain of wine and spirits shops in Ukraine.
The company runs 21 stores in 14 regions
of Ukraine.An impressive selection of
spirits with over 1000 whiskies of which
600 are malt whiskies. On-line ordering.
Also regular whisky tastings.

USA
Binny's Beverage Depot
5100 W. Dempster (Head Office)
Skokie, IL 60077
Phone:
Internet orders, 888-942-9463 (toll free)
www.binnys.com
A chain of no less than 40 stores in the
Chicago area, covering everything within
wine and spirits. Some of the stores also
have a gourmet grocery, cheese shop and,
for cigar lovers, a walk-in humidor. Also
lots of regular events in the stores. The
range is impressive with more than 2200
whisk(e)y including 750 single malts, 400
bourbons and more. Among other products
more than 500 kinds of tequila and mezcal,
450 vodkas, 400 rums and 250 gins.

Statistics

The information on the following pages is based
on figures from Scotch Whisky Association (SWA), Drinks International
and directly from the producers.

The Top 30 Whiskies of the World

Sales figures for 2018 (units in million 9-litre cases)

Officer's Choice (Allied Blenders & Distillers), Indian whisky — 34,0
McDowell's No. 1 (Diageo/United Spirits), Indian whisky — 29,0
Imperial Blue (Pernod Ricard), Indian whisky — 22,7
Royal Stag (Pernod Ricard), Indian whisky — 21,6
Johnnie Walker (Diageo), Scotch whisky — 18,9
Jack Daniel's (Brown-Forman), Tennessee whiskey — 13,3
Original Choice (John Distilleries), Indian whisky — 11,5
Jim Beam (Beam Suntory), Bourbon — 9,7
Hayward's Fine (Diageo/United Spirits), Indian whisky — 9,4
Jameson (Pernod Ricard), Irish whiskey — 7,5
Ballantine's (Pernod Ricard), Scotch whisky — 7,4
Crown Royal (Diageo), Canadian whisky — 7,3
Blenders Pride (Pernod Ricard), Indian whisky — 7,3
Director's Special Black (Diageo/United Spirits), Indian whisky — 6,2
Old Tavern (Diageo/United Spirits), Indian whisky — 6,2
Bagpiper (Diageo/United Spirits), Indian whisky — 5,6
Royal Challenge (Diageo/United Spirits), Indian whisky — 5,6
Bangalore Malt (John Distilleries), Indian whisky — 5,1
Kakubin (Suntory), Japanese whisky — 5,0
Grant's (Wm Grand & Sons) Scotch whisky — 4,6
Chivas Regal (Pernod Ricard), Scotch whisky — 4,5
Director's Special (Diageo/United Spirits), Indian whisky — 4,2
William Lawson's (Bacardi), Scotch whisky — 3,3
Black Nikka Clear (Asahi Breweries), Japanese whisky — 3,2
J&B (Diageo), Scotch whisky — 3,1
William Peel (Belvédère), Scotch whisky — 3,0
Famous Grouse (Edrington), Scotch whisky — 3,0
Dewar's (Bacardi) Scotch whisky — 2,8
Black & White (Diageo), Scotch whisky — 2,7
Seagram's 7 Crown (Diageo), Scotch whisky — 2,7

Source: Drinks International, The Millionaires Club 2019

Global Exports of Scotch by Region

Volume (litres of pure alcohol)			chg	Value (£ Sterling)			chg
Region	2018	2017	%	Region	2018	2017	%
Africa	17,914,309	20,645,837	-13	Africa	173,491,072	189,172,758	-8
Asia	78,313,083	68,520,034	+14	Asia	1,004,589,064	866,853,336	+16
Australasia	9,375,830	9,282,299	+1	Australasia	122,942,724	113,526,648	+8
C&S America	37,455,714	33,651,026	+11	C&S America	368,440,815	336,371,876	+10
Eastern Europe	6,665,125	4,654,954	+43	Eastern Europe	30,829,161	23,704,619	+30
Europe (other)	5,962,997	6,397,269	-7	Europe (other)	84,680,517	94,934,930	-11
European Union	129,941,905	133,663,520	-3	European Union	1,407,472,347	1,382,134,101	+2
Middle East	15,010,949	14,655,461	+2	Middle East	259,288,893	245,071,851	+6
North America	58,615,130	54,114,317	+8	North America	1,260,321,771	1,116,054,150	+7
Total	**359,245,042**	**345,584,717**	**+4**	**Total**	**4,712,056,364**	**4,367,824,269**	**+8**

Source: Scotch Whisky Association

Export of Bottled Blended Scotch 2018

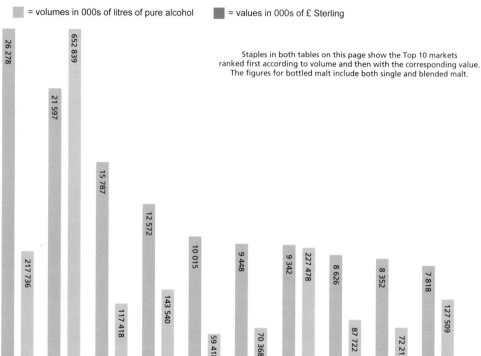

▉ = volumes in 000s of litres of pure alcohol   ▉ = values in 000s of £ Sterling

Staples in both tables on this page show the Top 10 markets
ranked first according to volume and then with the corresponding value.
The figures for bottled malt include both single and blended malt.

France	USA	Mexico	Spain	Brazil	South Africa	Singapore	Latvia	Germany	UAE
26 278 / 217 736	21 597 / 652 839	15 787 / 117 418	12 572 / 143 540	10 015 / 59 418	9 448 / 70 368	9 342 / 227 478	8 626 / 87 722	8 352 / 72 213	7 818 / 127 509

Source: Scotch Whisky Association

Export of Bottled Malt Scotch 2018

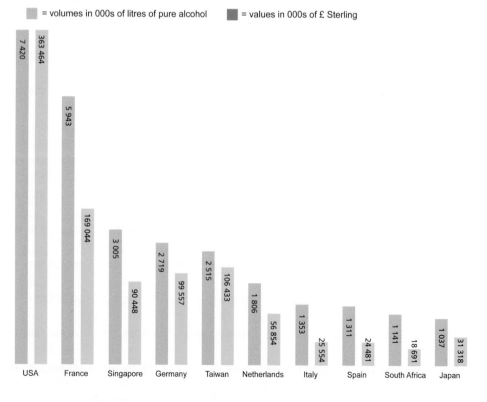

▉ = volumes in 000s of litres of pure alcohol   ▉ = values in 000s of £ Sterling

USA	France	Singapore	Germany	Taiwan	Netherlands	Italy	Spain	South Africa	Japan
7 420 / 363 464	5 943 / 169 044	3 005 / 90 448	2 719 / 99 557	2 515 / 106 433	1 806 / 56 854	1 353 / 25 554	1 311 / 24 481	1 141 / 18 691	1 037 / 31 318

Source: Scotch Whisky Association

Distillery Capacity

Litres of pure alcohol - Scottish, active distilleries only

Distillery	Litres	Distillery	Litres	Distillery	Litres
Glenlivet	21 000 000	Ben Nevis	2 000 000	Benromach	700 000
Macallan	15 000 000	The Borders	2 000 000	Kingsbarns	600 000
Glenfiddich	13 700 000	Bowmore	2 000 000	Speyside	600 000
Roseisle	12 500 000	Inchdairnie	2 000 000	Annandale	500 000
Ailsa Bay	12 000 000	Knockdhu	2 000 000	Ardnamurchan	500 000
Glen Ord	11 000 000	Balblair	1 800 000	The Clydeside	500 000
Teaninich	10 200 000	Pulteney	1 800 000	Royal Lochnagar	500 000
Dalmunach	10 000 000	Bruichladdich	1 500 000	Torabhaig	500 000
Balvenie	7 000 000	Bladnoch	1 500 000	Kilchoman	460 000
Caol Ila	6 500 000	Glendronach	1 400 000	Brew Dog	450 000
Glen Grant	6 200 000	Glen Spey	1 400 000	Harris	400 000
Glenmorangie	6 200 000	Knockando	1 400 000	Glenturret	340 000
Dufftown	6 000 000	Glen Garioch	1 370 000	Glasgow	270 000
Glen Keith	6 000 000	Glencadam	1 300 000	Edradour	260 000
Mannochmore	6 000 000	Scapa	1 300 000	Lindores Abbey	260 000
Auchroisk	5 900 000	Arran	1 200 000	Holyrood	250 000
Miltonduff	5 800 000	Glenglassaugh	1 100 000	Arbikie	200 000
Glen Moray	5 700 000	Glengoyne	1 100 000	Isle of Raasay	200 000
Glenrothes	5 600 000	Ardnahoe	1 000 000	Glen Wyvis	140 000
Linkwood	5 600 000	Ardross	1 000 000	Wolfburn	135 000
Ardmore	5 550 000	Tobermory	1 000 000	Ballindalloch	100 000
Dailuaine	5 200 000	Oban	870 000	Eden Mill	100 000
Glendullan	5 000 000	Glen Scotia	800 000	Ncn'ean	100 000
Loch Lomond	5 000 000	Aberargie	750 000	Daftmill	65 000
Tomatin	5 000 000	Glengyle	750 000	Dornoch	30 000
Clynelish	4 800 000	Lagg	750 000	Strathearn	30 000
Kininvie	4 800 000	Springbank	750 000	Abhainn Dearg	20 000
Tormore	4 800 000				
Longmorn	4 500 000				
Speyburn	4 500 000				
Dalmore	4 300 000				
Glenburgie	4 250 000				
Allt-a-Bhainne	4 200 000				
Braeval	4 200 000				
Glentauchers	4 200 000				
Craigellachie	4 100 000				
Royal Brackla	4 100 000				
Glenallachie	4 000 000				
Tamdhu	4 000 000				
Tamnavulin	4 000 000				
Aberlour	3 800 000				
Mortlach	3 800 000				
Glenlossie	3 700 000				
Benrinnes	3 500 000				
Glenfarclas	3 500 000				
Aberfeldy	3 400 000				
Cardhu	3 400 000				
Macduff	3 400 000				
Laphroaig	3 300 000				
Talisker	3 300 000				
Tomintoul	3 300 000				
Aultmore	3 200 000				
Bunnahabhain	3 200 000				
Fettercairn	3 200 000				
Inchgower	3 200 000				
Deanston	3 000 000				
Tullibardine	3 000 000				
Balmenach	2 800 000				
Benriach	2 800 000				
Blair Athol	2 800 000				
Glen Elgin	2 700 000				
Strathmill	2 600 000				
Lagavulin	2 530 000				
Glenkinchie	2 500 000				
Highland Park	2 500 000				
Strathisla	2 450 000				
Ardbeg	2 400 000				
Jura	2 400 000				
Cragganmore	2 200 000				
Dalwhinnie	2 200 000				
Auchentoshan	2 000 000				

Summary of Malt Distillery Capacity by Owner

Owner (number of distilleries)	Litres of alcohol	% of Industry
Diageo (28)	121 300 000	30,0
Pernod Ricard (13)	76 500 000	18,9
William Grant (4)	37 500 000	9,3
Edrington Group (3)	23 100 000	5,7
Bacardi (John Dewar & Sons) (5)	18 200 000	4,5
Beam Suntory (5)	14 220 000	3,5
Emperador Inc (Whyte & Mackay) (4)	13 900 000	3,4
Pacific Spirits (Inver House) (5)	12 900 000	3,2
Moët Hennessy (Glenmorangie) (2)	8 600 000	2,1
Distell (Burn Stewart) (3)	7 200 000	1,8
Campari (Glen Grant) (1)	6 200 000	1,5
Loch Lomond Group (2)	5 800 000	1,4
La Martiniquaise (Glen Moray) (1)	5 700 000	1,4
Benriach Distillery Co (3)	5 300 000	1,3
Ian Macleod Distillers (2)	5 100 000	1,3
Tomatin Distillery Co (1)	5 000 000	1,2
Angus Dundee (2)	4 600 000	1,1
The Glenallachie Consortium (1)	4 000 000	1,0
J & G Grant (Glenfarclas) (1)	3 500 000	0,9
Picard (Tullibardine) (1)	3 000 000	0,8
John Fergus & Co. (Inchdairnie) (1)	2 000 000	0,5
Nikka (Ben Nevis Distillery) (1)	2 000 000	0,5
The Three Stills Co. (The Borders) (1)	2 000 000	0,5
Isle of Arran Distillers (2)	1 950 000	< 0,5
Rémy Cointreau (Bruichladdich) (1)	1 500 000	< 0,5
J & A Mitchell (2)	1 500 000	< 0,5
David Prior (Bladnoch) (1)	1 500 000	< 0,5
Hunter Laing (Ardnahoe) (1)	1 000 000	< 0,5
Greenwood Distillers (Ardross) (1)	1 000 000	< 0,5
The Perth Distilling Co. (Aberargie) (1)	750 000	< 0,5
Gordon & MacPhail (Benromach) (1)	700 000	< 0,5
Wemyss Malts (Kingsbarns) (1)	600 000	< 0,5
Harvey's of Edinburgh (Speyside) (1)	600 000	< 0,5
Adelphi Distillery (Ardnamurchan) (1)	500 000	< 0,5
Annandale Distillery Co. (1)	500 000	< 0,5
Morrison Glasgow Distillers (Clydeside) (1)	500 000	< 0,5
Mossburn Distillers (Torabhaig) (1)	500 000	< 0,5
Kilchoman Distillery Co. (1)	460 000	< 0,5
BrewDog plc (1)	450 000	< 0,5
Others (17)	2 900 000	0,7
Total (125)	**404 530 000**	

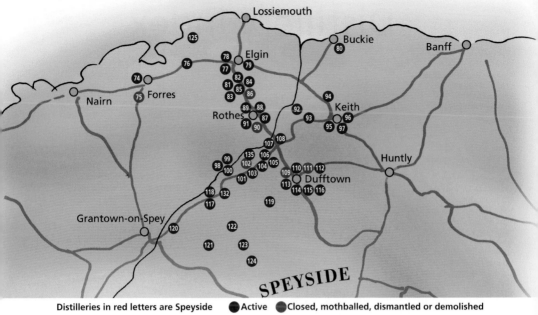

Distilleries in red letters are Speyside ● Active ● Closed, mothballed, dismantled or demolished

c = Closed, m = Mothballed, dm = Dismantled, d = Demolished

148 Aberargie	45 Deanston	51 Kinclaith (d)	2 Highland Park	52 Glen Flagler (d)	102 Imperial (d)
39 Aberfeldy	144 Dornoch	43 Kingsbarns	3 Scapa	53 Rosebank (c)	103 Dailuaine
106 Aberlour	110 Dufftown	114 Kininvie	4 Pulteney	54 St Magdalene (dm)	104 Benrinnes
127 Abhainn Dearg	136 Eden Mill	100 Knockando	5 Brora (c)	55 Glenkinchie	105 Glenallachie
126 Ailsa Bay	38 Edradour	21 Knockdhu	6 Clynelish	56 Ladyburn (dm)	106 Aberlour
119 Allt-a-Bhainne	32 Fettercairn	56 Ladyburn (dm)	7 Balblair	57 Bladnoch	107 Macallan
128 Annandale	141 Glasgow	63 Lagavulin	8 Glenmorangie	58 Arran	108 Craigellachie
134 Arbikie	13 Glen Albyn (d)	150 Lagg	9 Ben Wyvis (c)	59 Springbank	109 Convalmore (dm)
62 Ardbeg	105 Glenallachie	64 Laphroaig	10 Teaninich	60 Glengyle	110 Dufftown
25 Ardmore	76 Glenburgie	137 Lindores Abbey	11 Dalmore	61 Glen Scotia	111 Pittyvaich (d)
147 Ardnahoe	34 Glencadam	79 Linkwood	12 Glen Ord	62 Ardbeg	112 Glenfiddich
131 Ardnamurchan	23 Glendronach	48 Littlemill (d)	13 Glen Albyn (d)	63 Lagavulin	113 Balvenie
151 Ardross	116 Glendullan	46 Loch Lomond	14 Glen Mhor (d)	64 Laphroaig	114 Kininvie
58 Arran	85 Glen Elgin	36 Lochside (d)	15 Millburn (dm)	65 Port Ellen (dm)	115 Mortlach
49 Auchentoshan	35 Glenesk (dm)	143 Lone Wolf	16 Royal Brackla	66 Bowmore	116 Glendullan
92 Auchroisk	101 Glenfarclas	84 Longmorn	17 Tomatin	67 Bruichladdich	117 Tormore
94 Aultmore	112 Glenfiddich	107 Macallan	18 Glenglassaugh	68 Kilchoman	118 Cragganmore
7 Balblair	52 Glen Flagler (d)	20 Macduff	19 Banff (d)	69 Caol Ila	119 Allt-a-Bhainne
132 Ballindalloch	24 Glen Garioch	81 Mannochmore	20 Macduff	70 Bunnahabhain	120 Balmenach
120 Balmenach	18 Glenglassaugh	15 Millburn (dm)	21 Knockdhu	71 Jura	121 Tomintoul
113 Balvenie	50 Glengoyne	77 Miltonduff	22 Glenugie (dm)	72 Tobermory	122 Glenlivet
19 Banff (d)	87 Glen Grant	115 Mortlach	23 Glendronach	73 Talisker	123 Tamnavulin
30 Ben Nevis	60 Glengyle	145 Ncn´ean	24 Glen Garioch	74 Benromach	124 Braeval
82 Benriach	96 Glen Keith	33 North Port (d)	25 Ardmore	75 Dallas Dhu (c)	125 Roseisle
104 Benrinnes	55 Glenkinchie	40 Oban	26 Speyside	76 Glenburgie	126 Ailsa Bay
74 Benromach	122 Glenlivet	111 Pittyvaich (d)	27 Royal Lochnagar	77 Miltonduff	127 Abhainn Dearg
9 Ben Wyvis (c)	31 Glenlochy (d)	65 Port Ellen (dm)	28 Glenury Royal (d)	78 Glen Moray	128 Annandale
57 Bladnoch	83 Glenlossie	4 Pulteney	29 Dalwhinnie	79 Linkwood	129 Wolfburn
37 Blair Athol	14 Glen Mhor (d)	53 Rosebank (c)	30 Ben Nevis	80 Inchgower	130 Strathearn
138 Borders	8 Glenmorangie	125 Roseisle	31 Glenlochy (d)	81 Mannochmore	131 Ardnamurchan
66 Bowmore	78 Glen Moray	16 Royal Brackla	32 Fettercairn	82 Benriach	132 Ballindalloch
124 Braeval	12 Glen Ord	27 Royal Lochnagar	33 North Port (d)	83 Glenlossie	133 Inchdairnie
5 Brora (c)	89 Glenrothes	54 St Magdalene (dm)	34 Glencadam	84 Longmorn	134 Arbikie
67 Bruichladdich	61 Glen Scotia	3 Scapa	35 Glenesk (dm)	85 Glen Elgin	135 Dalmunach
70 Bunnahabhain	91 Glenspey	88 Speyburn	36 Lochside (d)	86 Coleburn (dm)	136 Eden Mill
69 Caol Ila	93 Glentauchers	26 Speyside	37 Blair Athol	87 Glen Grant	137 Lindores Abbey
90 Caperdonich (c)	41 Glenturret	59 Springbank	38 Edradour	88 Speyburn	138 Borders
99 Cardhu	22 Glenugie (dm)	130 Strathearn	39 Aberfeldy	89 Glenrothes	139 Torabhaig
142 Clydeside	28 Glenury Royal (d)	97 Strathisla	40 Oban	90 Caperdonich (c)	140 Harris
6 Clynelish	149 Glen Wyvis	95 Strathmill	41 Glenturret	91 Glenspey	141 Glasgow
86 Coleburn (dm)	140 Harris	73 Talisker	42 Daftmill	92 Auchroisk	142 Clydeside
109 Convalmore (dm)	2 Highland Park	98 Tamdhu	43 Kingsbarns	93 Glentauchers	143 Lone Wolf
118 Cragganmore	152 Holyrood	123 Tamnavulin	44 Tullibardine	94 Aultmore	144 Dornoch
108 Craigellachie	133 Inchdairnie	10 Teaninich	45 Deanston	95 Strathmill	145 Ncn´ean
42 Daftmill	102 Imperial (d)	72 Tobermory	46 Loch Lomond	96 Glen Keith	146 Isle of Raasay
103 Dailuaine	80 Inchgower	17 Tomatin	47 Inverleven (d)	97 Strathisla	147 Ardnahoe
75 Dallas Dhu (c)	47 Inverleven (d)	121 Tomintoul	48 Littlemill (d)	98 Tamdhu	148 Aberargie
11 Dalmore	146 Isle of Raasay	139 Torabhaig	49 Auchentoshan	99 Cardhu	149 Glen Wyvis
135 Dalmunach	71 Jura	117 Tormore	50 Glengoyne	100 Knockando	150 Lagg
29 Dalwhinnie	68 Kilchoman	44 Tullibardine	51 Kinclaith (d)	101 Glenfarclas	151 Ardross
		129 Wolfburn			152 Holyrood

SPEYSIDE

Distillery Index

Distillery Index

Distillery Index